W9-DHX-259

Consumer Reports Books
Guide to Appliances

Consumer Reports Books Guide to Appliances

By the Editors of Consumer Reports Books with Monte Florman

Consumers Union
Mount Vernon, New York

First printing, February 1986
Library of Congress Catalog Card Number: 86-70017
International Standard Book Number: 0-89043-035-7
Manufactured in the United States of America

Consumer Reports Books Guide to Appliances is a Consumer Reports
Book published by Consumers Union, the nonprofit organization that
publishes **Consumer Reports**, the monthly magazine of test reports,
product Ratings, and buying guidance. Established in 1936, Consumers
Union is chartered under the Not-For-Profit Corporation Law of the State
of New York.

The purposes of Consumers Union, as stated in its charter, are to
provide consumers with information and counsel on consumer goods
and services, to give information on all matters relating to the
expenditure of the family income, and to initiate and to cooperate with
individual and group efforts seeking to create and maintain decent living
standards.

Consumers Union derives its income solely from the sale of **Consumer
Reports** and other publications. In addition, expenses of occasional
public service efforts may be met, in part, by nonrestrictive,
noncommercial contributions, grants, and fees. Consumers Union
accepts no advertising or product samples and is not beholden in any
way to any commercial interest. Its Ratings and reports are solely for the
use of the readers of its publications. Neither the Ratings nor the reports
nor any Consumers Union publication, including this book, may be used
in advertising or for any commercial purpose. Consumers Union will take
steps open to it to prevent such uses of its material, its name, or the
name of **Consumer Reports**.

CONTENTS

How to get the most from this book

Appliances are an essential component of day-to-day household routine. Unfortunately, using an appliance is sometimes a lot easier than buying the right brand and model. The variety in stores, catalogs, and catalog showrooms is usually bewildering. There are any number of brand names, sizes, shapes, and models, with and without a multitude of claimed characteristics and attributes. Advertisers' claims can be dazzling and confusing. How can the consumer determine which of the competing brands offers the best value for the money, which offers the highest quality regardless of price, and which claims are accurate? Should a well-known brand name be your guide? How can you tell the difference between a genuine advance in technology and a "miracle" of advertising? This guide will provide invaluable information that will help you, the consumer, make intelligent choices.

The editors of Consumer Reports Books, with Monte Florman, Consumers Union's emeritus Technical Director, have developed this unique book that gives you in-depth information and advice about household appliances for nearly every room in your home—facts you need **before** you buy. The material in this guide is based on tests performed by experienced engineers, technicians, and home economists in Consumers Union's laboratories. The results of these tests were originally published in **Consumer Reports,** but have been reviewed and revised for inclusion in this appliance guide.

Most of the product entries in this book have two parts. The first provides basic information about the appliance, including what it will do (and won't do), the features of various brands and models tested, how to maintain the appliance, along with Consumers Union's recommendations on convenience, performance, safety, and durability.

The Ratings charts

The second part of most entries is a chart of detailed test findings and other information for particular brands and models, along with the date that the test results originally appeared in **Consumer Reports.** It is advisable to read the Ratings charts thoroughly. You may be tempted to choose the model that appears at the top of the Ratings order. Instead, begin with the introductory section that precedes each set of Ratings, and with the specifications and feature notes near the bottom of the Ratings chart, where you will find the characteristics, qualities, or deficiencies shared by the products in the test group. You may find out, in a sentence that starts with "Except as noted," what qualities most of them have in common. You will also discover the basis on which the Ratings order was decided.

Models are check rated ($\checkmark$) when the samples tested prove to be of high quality and appreciably superior to other products tested. Best Buy Ratings are given to products that not only rate high but are also relatively low priced, therefore giving better value per dollar than other products tested. (See next page for explanation of prices listed.)

Sometimes the products are listed alphabetically. In other cases, the first sentence of the chart states, "Listed in order of estimated overall quality." That means CU's engineers judged the brand (or product type) listed first to be the best; the one listed next, second best, and so on, in descending order. "Listed, except as noted, in order of estimated overall quality," means that somewhere in the Ratings one or more groups of products are about equal to each other and are therefore listed in a special fashion—for example, "Models judged approximately equal in overall quality are bracketed and listed alphabetically."

Models and model changes

Manufacturers commonly change models, once a year or more frequently to match their competition, to try to stimulate sales by offering something "new," and to incorporate product improvements in technology and overall design. But quite often dealers will carry over older models when a manufacturer changes models. Where the dealer carryover is heavy, older models may remain available for months, even years, after the manufacturer has discontinued them. On the other hand, there is a good chance that the particular brand and model of appliance that you select from one of the Ratings charts in this guide will either be out of stock when you try to buy it, or it will have been superseded by a later version. If you substitute a more recent equivalent of an older model, the probability is that its performance will be comparable to the model in the chart.

In our experience, appliances that have ranked high in **Consumer Reports'** Ratings have tended to remain high ranked when a subsequent version of the product has been tested. As a consequence of market pressures and consumers' increasing knowledge and sophistication about product selection, there has been a tendency toward quality "compression" in appliances as well as other products, with high-quality products remaining high in quality through successive model changes, and with lower-quality items tending to improve gradually. Thus quality differences between top- and bottom-rated brands are less dramatic, and the features and convenience differences you can see for yourself in the store become more important in your buying decision.

Prices

Prices given in the Ratings charts, unless otherwise indicated, are list prices, generally dating from the time of the original report. Retail prices at mail-order company stores are often higher than the catalog prices.

Of course, the term "list price" is difficult to define. Not only are discounts from "list" frequently available, but the list price itself is sometimes a fiction designed to enable the retailer to appear to be offering a bargain. Indeed, meaningful list prices have all but disappeared from some product categories—including appliances. Sometimes, in order to obtain a more meaningful price profile, surveys are conducted by Consumers Union shoppers across the country. For most appliances, however, list prices are a rough guide that may be useful in comparing prices of competing models.

To keep yourself as up to date as possible on the rapidly changing appliance marketplace, you should refer to current issues of **Consumer Reports** for additional information and new Ratings on appliances.

Air cleaners

Cooking, cleaning, pets, and especially cigarette smoking contribute to indoor air pollution. Heating systems also blow dust about the house. In the past, when the air became unpleasant, you opened a window until the air cleared—behavior now likely to earn you a snarl from whoever pays the heating bill.

But many years ago, opening the window was often unnecessary because the average house was so leaky that the air within changed a couple of times an hour. Today, that house, if it's been thoroughly weatherized, receives fresh air at half the old rate. That is still adequate ventilation, but smoke and smells linger much longer.

While smoke and smells may be unpleasant, at least they're easily perceived. Other pollutants are more insidious—gases such as combustion byproducts from a gas range, formaldehyde from particleboard, radon from some stone foundations or wells. These pollutants can become a concern in a new house built to be airtight; in such a house, the air may be exchanged only once every two or three hours.

In the late 1970s, a new product came along that promised to clear the air of tobacco smoke, odors, dust, pollen, and other pollutants—the tabletop air cleaner. **Pollenex** and **Ecologizer**, the first popular brands, were an immediate success. They were soon joined by competing models, notably the **"Clean Air Machines"** from Norelco.

Those early air cleaners worked by using a small fan to pull air through a filter.

The small fan/filter models paved the way for air cleaners using more expensive technologies. Some use a long-established method of removing particles from air, electrostatic precipitation. Others use a once-controversial device called the negative-ion generator. (The box on page 6 explains those technologies.) Nowadays, the range of designs and prices among air cleaners is wide.

Purifying?

The air cleaners tested by Consumers Union's engineers are of limited usefulness. They cannot dispose of many serious pollutants.

Any claim that an air cleaner can remove formaldehyde or other pollutant gases should be viewed with suspicion. In the tests, none of the air cleaners removed more than a trivial amount of formaldehyde gas. Studies have shown that reducing the level of gaseous contaminants in the air requires a much more elaborate filtration system than the ones these machines have.

Furthermore, air cleaners apparently aren't an effective way to deal with the health risk associated with radon gas and its by-products, a problem in some parts of the country.

If your windows can be opened and you can tolerate the chill, that may still be the best air cleaner of all. Contrary to popular opinion, opening a window in the dead of winter is not like shoveling dollars out the window. With the temperature 20° F outside and 70°

inside, opening the window a couple of inches—ventilation likely to be equivalent to that provided by a small exhaust fan—costs no more than a few cents an hour in lost heat.

Are air cleaners, then, completely useless? They certainly don't "purify" the air. But if all you want is to remove smoke or dust, some can do a creditable job.

Smoke removal

The smell of tobacco smoke is one reason people buy an air cleaner. Getting rid of the smell, however, is more difficult than getting rid of the smoke. Smoke is particulate—resins and small particles are suspended in the air. Although the particles are as small as one one-hundredth of a micron (less than a millionth of an inch) in diameter, they can be trapped mechanically in a filter—or in your lungs.

The graph on page 7 compares how many cubic feet of clean, smoke-free air the air cleaners can produce per minute when operating at their top speed. An air cleaner with an effective cleaning rate of less than 12 cubic feet per minute is almost useless, especially if some-

one is currently smoking in the room.

Most effective against smoke is an ion generator, the **Energaire,** and two models that use both an ion generator and a fan/filter system, the **Bionaire 1000** and the **Pollenex 1801.** Those models can clear about 50 to 60 cubic feet of air per minute (that's about as much clean air as a small exhaust fan would provide). The next eight models in the graph also clean air at a high enough rate to have a noticeable effect.

Not surprisingly, the small, inexpensive air cleaners—those with a small fan that draws relatively little air through a small, flat filter—are the least effective at removing smoke.

Dust and pollen

An air cleaner can catch dust and pollen more easily than smoke because the particles are much larger—up to 100 microns (.004 inch) in diameter. But pollen and large dust particles are heavy enough that they don't remain airborne for long. They settle onto the floor and furnishings of a room until air currents or people stir them up. An air cleaner can remove dust and pollen only when they're in the air. It won't eliminate the need for dusting and vacuuming.

To many people airborne dust is the problem, causing sneezing, wheezing,

and other allergic reactions. A good air cleaner can be useful in treating dust allergy, especially if it's used during sleep. An air cleaner can also help those allergic to mold spores.

For people allergic to pollen, however, an air conditioner is probably better than an air cleaner. Unlike dust and mold spores, pollen originates outside the house, generally in weather warm enough for windows to be open. An air conditioner lets you circulate air without introducing pollen.

Air cleaners that remove dust the best are the **Norelco HB9000** and the **Space-**

Gard, fan/filter models whose pleated filters have a large surface area. Those two models also have relatively strong fans, moving a lot of dust-laden air through the filter before the dust has a chance to settle.

Two models almost as good at removing dust—the **Bionaire 1000** and the **Pollenex 1801**—are also top-rated for removing smoke. In dust removal, it's their effective combination of fan and filter, much more than their production of negative ions, that works so well. Ions seem to have little effect on dust, judging by the poor performance of three models tested that are solely ion generators, without fans or filters.

Most of the small fan/filter models aren't good dust removers either way. The fan in those models may actually have more of an effect on lowering dust levels than the filter does.

Living with an air cleaner

Effectiveness isn't the only important consideration in an air cleaner. Other factors are:

Noise. Most air cleaners have a fan, and most of those fans can run at more than one speed. At the highest fan setting, some air cleaners barely whisper. Others are so noisy that they would be extremely distracting.

The Ratings show scores for the sound levels measured. For models with more than one fan speed, there's a high and a low score. The best score means the machine is virtually silent at that speed. A model with an average score is about as noisy as a small air conditioner set on "low." The Sears Catalog Nos. 7302 and 7305 (also sold as the **Bionaire 500**) are admirably quiet and fairly effective.

Dirt deposits. The three ion generators that don't have a fan are silent. But they have a problem serious enough to place them at the bottom of our Ratings. Though they are effective in reducing smoke in the air, they cause sooty particles on walls and other surfaces in the room.

The dirt-deposit problem is solved when an ion generator is combined with a fan/filter system. The three hybrid air cleaners don't dirty the wall, though they may collect some dust around them on the table.

Lemon-fresh air. Some air cleaners allow you to inject a citrus scent into the air. In a model that isn't effective in cleaning the air, the scent covers up the smell for a while. Some people may like the scent; others may consider it another form of air pollution.

Several models have a "scent control" switch that can turn off the scent if you don't want it. Some manufacturers sell replacement filters with and without scent. A few models offer no choice.

The cost of cleaned air. Air cleaners use very little electricity, no more than a 100-watt light bulb does. The fanless ion generators use less than one watt. Typical usage is more like 15 to 30 watts.

A greater ongoing expense with most air cleaners is the cost of replacement filters. Manufacturers claim that filters last anywhere from three to six months. But the life of a filter depends on two very variable things: how much the air cleaner is used, and how dirty the air is. You have to judge the filter life by how dirty the filter looks.

Generally, the larger and more effective the filter, the more it costs. The filters for the small, inexpensive air cleaners cost about $4 to $6. Large filters may cost $15 to $20.

A guide to the three types of air cleaner

Fan/filter systems. Most air cleaners on the market use a fan to draw air through a filter. Fans' ability to move air varies a lot.

The filters vary, too. Granular materials—activated carbon, silica gel, a proprietary resin bead—are often part of the filter cartridge in an air cleaner. Such materials are supposed to interact with and trap various gases. But none of the models CU tested use enough of the material to produce much effect.

Another type of filter is a web of synthetic or glass fiber. It works like a strainer, catching particles that pass through it. A fibrous filter can be made more efficient by increasing its surface area—typically, by folding it into accordionlike pleats. In the CU tests, most models with pleated filters did quite well. A "high-efficiency particulate air filter," or "**HEPA** filter," pleated and made of glass fibers, is the ultimate in fibrous filters; it has been used since World War II to filter air in hospitals and laboratories.

Another way to increase the efficiency of a fibrous filter is to include fibers that have an electrical charge. The "electret" filter does this. Many particles in the air have a weak electrical charge, especially when the heating system has dried things out and the air is full of static electricity. An electret filter uses static electricity to catch small charged particles that otherwise would pass through. That sort of filter is used in many top-rated models.

Electrostatic precipitators. Air cleaners using this technology draw in air with a fan past an electrode that gives airborne particles a relatively high electrical charge. Then the air passes a collector plate that has the opposite electrical charge, causing the dust and other charged particles to stick to it.

An electrostatic precipitator can be bought as a component to be built into a forced-air heating system (a method well worth considering if you want to clean the air in an entire house). Manufacturers such as Honeywell and Emerson Electric also sell room-size electrostatic precipitators similar in size and price to room air conditioners. Such models have been around for years, often purchased on an allergist's prescription. New, smaller models have been coming onto the market. CU tested one from Oster and found it works fairly well.

Negative-ion generators. Some twenty years ago negative-ion generators were sold as a miracle cure for just about any ailment until the Food and Drug Administration called them quackery and halted their sale.

Now the negative-ion generator has been born again—as an air cleaner. Negative-ion generators can be truly effective at removing smoke.

These devices spew a stream of electrons into the air, turning air molecules into negative ions. The ions apparently give airborne particles a negative charge. The particles then drift to grounded surfaces such as walls and ceilings, where they stick.

Which machine cleans the most air?

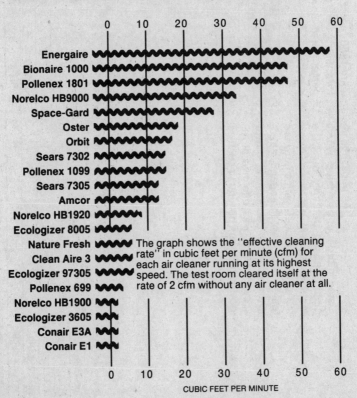

The graph shows the "effective cleaning rate" in cubic feet per minute (cfm) for each air cleaner running at its highest speed. The test room cleared itself at the rate of 2 cfm without any air cleaner at all.

CUBIC FEET PER MINUTE

Ratings of air cleaners

(As published in a January 1985 report.)

Listed in order of estimated overall quality, based primarily on CU's tests for removal of tobacco smoke from the air. Models with a fan were tested at highest fan speed; effectiveness would decrease at slower speeds. None was effective at removing gaseous pollutants such as formaldehyde. Prices are suggested retail. + indicates shipping is extra.

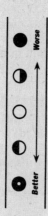

Better ● ◐ ○ ◐ ● Worse

Brand and model	Price (replacement filter)	Type [1]	No. of fan speeds	Dimensions (HxWxD)	Weight	Smoke removal	Smoke removal (room)	Dust removal (desk)	Noise (high/low)	Airflow (high/low) [2]
BIONAIRE 1000	$299(16)	ION/PF	3	9x14x8 in.	13 lb.	●	●	◐	○◐	60/40
POLLENEX IONIZER 1801	100(15)	ION/FF	4	11x13x9	8	●	●	●	◐◐	100/30
NORELCO HB9000	100(20)	PF	3	7x16x9	8	◐	◐	⊙	◐◐	80/50
SPACE-GARD 2275	140(12)	HE	2	14x12x12	12	◐	◐	⊙	○◐	90/70
OSTER 402-06	158(7)	EP	2	6x14x11	8	◐	◐	○	○◐	50/40
SEARS Cat. No. 7302	60+(10+)	PF	2	5x14x11	9	○	○	○	○◐	40/20

Model	Cfm	Type		Dimensions		Ratings					
POLLENEX 1099	60(15)	FF	4	11x13x9	7	○	○	○	◐	●	100/30
SEARS IONIZER Cat. No. 7305	100+(7+)	ION/FF	2	5x11x7	4	○	◐	○	◐	◐/	30/20
NORELCO HB1920	36(6)	FF	2	7x7x7	3	◐	○	●	◐	◐/	40/20
ECOLOGIZER 8005	100(20)	HEPA	[3]	11x13x10	9	○	○	○	○	○/	100/80
NATURE FRESH AP30B1	45+(5+)	PF	2	4x7x11	2	◐	◐	○	◐	◐/	20/10
CLEAN AIRE 3	15(41)	FF	1	17x6 [4]	3	○	●	◐	○	○	20
ECOLOGIZER 97305	29(5)	FF	1	9x5 [4]	2	◐	●	●	●	◐	20
POLLENEX 699	20(6)	FF	2	7x10x8	2	◐	●	●	○	◐/	20/10
NORELCO HB1900	20(4)	FF	1	8x5x5	2	●	●	◐	○	◐	10
ECOLOGIZER 3605	40(5)	FF	2	8x7x9	3	●	●	●	◐	○/○	20/20
CONAIR E3A	30(4)	FF	2	7x7x7	3	●	●	◐	◐	◐/○	30/20
CONAIR E1	22(4)	FF	1	8x5x5	2	◐	●	●	◐	◐	20

■ The following models, all fanless ion generators, were severely downrated because they badly soiled walls in CU's smoke-removal test.

Model	Cfm	Type		Dimensions						
ENERGAIRE	50	ION	0	9x3 [4]	1	●	●	●	●	—
ORBIT	79	ION/C	0	8x3 [4]	1	○	◐	●	●	—
AMCOR FRESHENAIRE	44	ION	0	3x6x4	2	○	◐	●	●	—

[1] ION = ion generator; PF = pleated filter; HE = high efficiency; FF = flat filter; HEPA = high-efficiency particulate air filter designed to filter radioactive airborne contaminants; EP = electrostatic precipitator; C = collector.

[2] Cubic feet per minute.

[3] Continuously variable.

[4] Diameter; model is round.

SPECIFICATIONS AND FEATURES
Except as noted, all: ● Have a fan. ● Have plastic cabinet. ● Have on/off switch.

Negative-ion generators are also sometimes claimed to have a beneficial effect on people's moods. Some manufacturers mention the "revitalizing" effect of negative ions or the refreshing negative-ion charged air you find around waterfalls. Such claims have yet to be proven in well-controlled scientific studies.

Recommendations

The best way to clear the air of a smoker's lair is to open a window a few inches. A fan in or near the window will speed the process. The heat lost during an hour's airing will cost no more than a few cents, even in cold weather.

But an air cleaner can help when opening the window isn't feasible. If you live in a house with forced-air heating, consider a built-in air cleaner for your heating system ($300 to $400 plus installation) before you consider a small model.

We suggest you limit your choice to the top eight models, since they are the only ones likely to work. If you've already bought one of the cheap fan/filter models, you can turn it into a reasonable tabletop fan by throwing away the filter cartridge.

Air conditioners and fans

Air conditioners. These are unquestionably the most effective coolers, for they chill the air and dehumidify it. They do exact a price in high operating costs, but some models are more efficient and versatile than others.

Ceiling fans. With their slow-moving blades, they can do an effective and quiet job of stirring the air in a room. Some of the more ornate ceiling fans may also add a certain amount of charm to a room.

Portable box fans. A movable fan isn't glamorous, but it can ventilate a room effectively or whip up a breeze. It is also about the cheapest summer cooler around.

Whole-house attic fans. One large fan is not a substitute for air conditioning, but it can provide a lot of comfort.

Air conditioners

A room air conditioner is the most effective summer cooler. It's also about the most expensive to buy and run. You can, however, reduce its impact on your electricity bill by choosing an energy-saving unit, one with an energy efficiency rating (EER) of 7.5 or higher.

The EER is the ratio of a conditioner's cooling capacity to the electricity it uses. For example, a machine that consumes 1,000 watts of electricity to produce 10,000 British thermal units (Btu's) of cooling per hour has an EER of 10. Use the Cooling-Load Estimate Form on page 14–15 to find out how many Btu's you need for the space you intend to cool. The Ratings on pages 17–19 cover units with capacities of about 8,000 Btu's. That's probably about right for daytime cooling of a medium-sized upstairs bedroom. If you have a room that's hard to heat in winter, you may want to consider a "reverse-cycle," or heat pump model: it can cool in summer and heat in winter. For more about this feature, see page 16.

You can't, however, choose an air conditioner solely for its high efficiency. How the air conditioner performs on the hottest, muggiest days and how well it distributes its cooling through the room are also important considerations.

Cooling

Any properly sized unit should keep you at least reasonably cool in moderately hot weather. But in a heat wave, electric companies often lower the voltage they supply. If you live in an area subject to heat and brownouts, you'd be wise to choose a machine that can cope with brownouts.

Automatic setting. At an air conditioner's regular setting, the fan runs all the time to keep air moving through the unit. That way, the thermostat keeps in touch with changes in room temperature. A number of models also have an "automatic" or "energy saver" setting that turns the fan off whenever the air conditioner isn't actively cooling.

Over a cooling season, automatic operation can chop a few dollars off your electric bill. It should also help keep room humidity down (moving air would otherwise pick up moisture that had condensed on the cooling coils).

On the other hand, there's a concealed problem: all running noise stops when the fan cycles off. Sleepers may wake when the fan goes on again. In the morning, if the unit is silent, you may forget that it's on, leaving it to cool your room wastefully all day.

Temperature uniformity. Louvers control side-to-side and up-and-down airflow. With most models, you will get even air distribution when you point one set of vertical louvers left and the other set right, and point horizontal ones a bit upward.

Directional control. If your air conditioner must be installed in a corner of the room, vertical louvers that can direct air effectively to one side or the other assume real importance. And adjustable horizontal louvers are handy for pushing air farther outward into a room—or downward, if your machine must be installed high in your living space. Louvers that focus airflow well also let you spot-cool a bed or a favorite armchair.

Moisture disposal. One of an air conditioner's by-products is moisture, which is drawn from the air as it's cooled. The air conditioner should disperse this condensed water as fine mist.

Dehumidification. High-efficiency air conditioners make room air travel a relatively long path over the machine's cooling coils. Accordingly, the coils tend to become a trifle warmer than they do in a regular air conditioner, raising the possibility that a high-efficiency unit might not dehumidify as well as a regular one. However, the manufacturers' own dehumidification specifications show little significant difference between the figures for a brand's regular and high-efficiency machines of comparable output.

However, be aware that dehumidification can differ markedly between brands. So if you live in a humid climate, give strong preference to a model that dehumidifies well.

Buttons

The controls on most units are usually quite simple and straightforward: One dial lets you set the machine for cooling or fan-only operation, automatic or regular cycling, and fan speed. A second dial controls the temperature.

Setting the thermostat is often a matter of trial and error, especially in the beginning. To find a setting you like, set the dial at its coolest position. When the room is cool enough, turn the dial until the compressor cycles off. If your unit is in a bedroom, you may want to do that again at night to find the right temperature for sleeping.

Once you've found those dial positions, you can shift between them readily if your control is well marked. Most dials have numbered settings.

"Hi cool" and "lo cool" settings merely change the fan's speed, with some effect on dehumidifying but not on actual cooling. Still, a high speed can provide a little extra breeze, while a slow speed reduces noise and energy consumption.

A number of machines have "fan only" settings that move room air a bit without cooling it. To get the same

effect on units that lack the setting, just turn the thermostat to its warmest position.

Generally, there's a setting to "exhaust," or remove, room air. Some machines will also "ventilate," or draw in air from outside. Such provisions are only minimally effective, at best. Indeed, they won't even clear a room of cigarette smoke very fast. They also shouldn't be used when you need maximum cooling, since the warm air they bring in imposes an extra load on the machine and your electricity bill.

Noise. A raucous machine can be especially annoying in a bedroom or a study. Outdoors, an air conditioner's noise could be irritating near a patio or a neighboring bedroom. So try to choose a model that's noted for its quietness.

Filter cleaning. Plastic-foam filters inside each unit retain dust from air moving through the unit. These filters need only a rinse now and then. To prevent a mess at the sink, they can also be vacuum-cleaned. It's best to unplug the unit before cleaning the filter.

Mounting and maintaining

If winters are cold where you live, an air conditioner sitting in a window can cause a loss of heat. The mounting panels at the machine's sides are poor insulators and are apt to leak heat. Cold air can also blow through the machine. It makes sense to remove the unit each

autumn and put up a storm window.

That's a challenge that you won't want to meet without a helper, however. Most models are bulky, awkward heavyweights.

The majority of machines are designed for double-hung windows; but some have a slide-out chassis that also allows through-the-wall installation, and a few will fit a casement window.

In window mountings, the typical first step for installation is to attach a bracket or two to the windowsill to support the machine. Most machines come with those brackets.

With models that have a slide-out chassis, you mount the empty cabinet first, then slip in the heavy chassis only when everything is secure.

To keep rain or condensate from dripping back into your room, mount the air conditioner so that it slopes slightly toward the outside (a slope of one-quarter inch per foot is about right). Most models have a leveling adjustment.

With the machine in place, your next step is usually to expand a pair of pleated panels to seal the gaps at the unit's sides, finishing off with a small bead of caulk at each panel's top and bottom.

The condenser fins facing outdoors may eventually clog with dust and grime, particularly if mounted in a window that faces traffic, or if you live in a particularly sooty area. You should clean them each spring, when you put the air conditioner back in the window; the dusting brush of a canister vacuum serves well. Use the vacuum as a blower for hard-to-reach spots. Some localities offer specialized services for cleaning condenser fins that are especially dirty or that are far too difficult for you to clean.

Some fan motors require periodic oiling, perhaps once a season. A slide-out chassis makes the job easier.

All but very powerful machines will run on any 15-ampere household circuit (your local electrical code, however, may require a separate circuit). Be sure the circuit's wall outlet is grounded. Don't use a toaster, iron, or other high-draw appliance on the same circuit while the air conditioner is running. If you must use an extension cord, use only a heavy-duty one sold specifically for air conditioners.

Recommendations

The Cooling Load Estimate Form on pages 14–15 will help you decide how powerful an air conditioner you need. And the work sheet on page 17 will help you estimate the cost of running the unit.

If you live in an especially muggy area, consider a machine with better dehumidification, but you may have to sacrifice something in all-around performance.

Cooling load estimate form

Adapted from a form published by the Association of Home Appliance Manufacturers in Standard RAC-1. **Note:** If you will use an air conditioner only for night cooling, make these adjustments: The factor for item 1 is 200. Disregard item 2. In item 4, the factor for heavy construction is 20; for all others, 30. In item 5, the factors are 5, 3, 7, 4, and 3.

HEAT GAIN FROM	QUANTITY		FACTOR		LOAD (QUANTITY X FACTOR)

1. DOORS AND ARCHES. Multiply the factor by the total width (linear feet) of any continually open doors or arches between room to be cooled and an uncooled space. Consider rooms connected by a door or arch more than 5 feet wide as a single large room for this and following calculations.

_____ ft.   X   300   =

2. SUN THROUGH WINDOWS. Multiply window area for each exposure by applicable factor. For windows with inside shades or blinds, use factor for "inside shades." For windows with outside awnings (with or without shades or blinds), use factor for "outside awnings." Factors given are for single glass only. For glass block, multiply factors by 0.5; for double glass or storm windows, multiply factors by 0.8

		No shades		Inside shades		Outside awnings	
Facing northeast	sq. ft	X	60	or	25	or	20
Facing east	sq. ft	X	80	or	40	or	25
Facing southeast	sq. ft	X	75	or	30	or	20
Facing south	sq. ft	X	75	or	35	or	20
Facing southwest	sq. ft.	X	110	or	45	or	30
Facing west	sq. ft.	X	150	or	65	or	45
Facing northwest	sq. ft.	X	120	or	50	or	35
Facing north	sq. ft.	X	0	or	0	or	0

Enter at right only the largest figure from the column below

3. CONDUCTION THROUGH WINDOWS. Multiply total square feet of all windows in room by the applicable factor.

Single glass	sq. ft. X	14 =
Double glass or glass block	sq. ft. X	7 =

4. WALLS. Multiply total length (linear feet) of all walls exposed to the outside by the applicable factor. Consider doors as part of the wall. Consider walls shaded by adjacent structures, but not by trees or shrubbery, as having "north exposure." "Light construction" means an uninsulated frame wall or a masonry wall 8 inches thick or less; "heavy construction" means insulated frame or masonry thicker than 8 inches. Also make computation for inside walls between uncooled spaces.

		Light construction		Heavy construction
a. Outside walls				
North exposure	ft. X	30	or	20 =
Other than north exposure	ft. X	60	or	30 =
b. Inside walls (between conditioned and unconditioned spaces only)	ft. X	30	or	30 =

5. CEILING. Multiply total ceiling area by the factor for ceiling construction that most nearly matches your ceiling.

Enter one figure only

a. Uninsulated with no space above	sq. ft. X	19 =
b. Insulation 1 inch or more, no space above	sq. ft. X	8 =
c. Uninsulated with attic space above	sq. ft. X	12 =
d. Insulated with attic space above	sq. ft. X	5 =
e. Occupied space above	sq. ft. X	3 =

6. FLOOR. Multiply the factor by the total floor area. (Disregard this item if floor is directly on ground or over a basement.)

sq. ft. X   3 =

7. SUBTOTAL. Add the loads calculated above.

SUBTOTAL =

8. CLIMATE CORRECTION. Multiply item 7 by correction factor for your locality, selected from map above.

(Item 7) X _____ (Factor from map) =

9. PEOPLE. Multiply by 600 the number of people who will normally occupy cooled space. Use minimum of two people.

X   600 =

10. ELECTRICAL EQUIPMENT. Determine total wattage for lights and electrical equipment in the cooled area (except the air conditioner itself) that will be in use when the conditioner is operating. Many appliances may give wattage on their nameplates; if not, multiply the nameplate amperage by the voltage for an estimate. Multiply the total wattage by the factor.

_____ watts   X   3 =

11. TOTAL COOLING LOAD. Add items 8, 9, and 10. The result, in Btu per hour, should be matched fairly closely by an air conditioner's Btu per hour rating.

TOTAL BTU PER HOUR =

Air conditioners that heat

In effect, an air conditioner cools your room by heating the outdoors—it extracts heat from air and pumps it outside. Engineering design makes it possible to reverse the machine's operation to heat your room.

Air conditioners that also heat are nothing new, especially in the South. But the usual approach has been to use resistance heating elements, just like the ones in an ordinary electric heater. That's a costly way to provide heat.

More recently, manufacturers introduced "reverse-cycle" air conditioners. As the name implies, those machines can blow heat indoors as well as out, and they are apt to be more cost effective than an air conditioner with heating elements.

In its heating role, a reverse-cycle air conditioner is a space heater. Like any space heater, it allows you to lower the thermostat setting of a central heating system while keeping the room you're in at a comfortable level.

It may be hard to think of chilly air as containing heat, but it does. That heat is a form of energy, there to be extracted by the air conditioner.

One measure of the air conditioner's abilities is its coefficient of performance (COP), the ratio of energy output to energy input. A COP of 1 means you break even; anything higher and you're ahead. If a unit delivers 5,000 Btu per hour in heat and consumes 586 watts (2,000 Btu) per hour in electricity, its COP is 2.5. In other words, you get back 2.5 watts as heat for every watt used as electricity.

The COP, however, isn't a fixed number. As the outside temperature drops, it becomes harder for a reverse-cycle air conditioner to extract energy from the air, so its capacity and efficiency are reduced. Some manufacturers say their units should not be run when outside air is colder than 45°F. (Standard performance tests are run at 47°.)

Should you consider buying a reverse-cycle air conditioner? Yes, if you live where winters are mild (usually not below 35° or so) and summers are hot enough to justify air conditioning.

Estimating the running cost

To estimate the cost of running a room air conditioner in your area, collect the following data:
● The cooling-season electric rate in cents per kilowatt-hour and the average number of hours of cooling required in a season (you can get both figures from your electric power company). ● Your cooling load (from page 15). ● The labeled EER of the unit you're considering. Then make your calculation.

Calculation	Example*	Your room
1. Electric rate × hours of cooling:	7.75×700 = 5425	_____ × _____ = _____
2. Item 1 × your cooling load:	5425×8000 = 43400000	_____ × _____ = _____
3. Drop last five digits from item 2:	434	_____
4. Annual operating cost (item 3 divided by EER):	434 ÷ 8.8 = 49.32	_____ ÷ _____ = _____

* Assuming an electric rate of 7.75¢/kwh, a 700-hour cooling season, a room with an 8,000 Btu/hr. cooling load, and an air conditioner with an 8.8 EER.

Ratings of air conditioners

(As published in a July 1982 report.)

Listed in order of estimated overall quality. Except where separated by bold rules, closely ranked models differ little in overall quality. Except as noted, all were judged easy to install in a double-hung window. Dimensions are rounded to next higher ¼ in. Prices, rounded to nearest dollar, are average (and range) of retail prices quoted to CU shoppers in 11 cities.

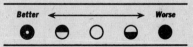

Brand and model	Price	Btu/hr.	EER	Dehumidification (pt./hr.)	Regular conditions	Under extreme conditions	Temperature setting	Automatic setting	Temperature uniformity	Directional control	Moisture disposal	Indoors, at high	Indoors, at low	Outdoors, at high	Outdoors, at low
							Thermostat performance						**Relative quiet**		
FRIEDRICH SS08F10	$543 ($449 to $615)	8000	9.0	1.5	○	●	◑	◑	●	◑	●	●	◑	●	●
FRIEDRICH YS09F10 [1]	753 (625 to 839)	8800	9.3	1.6	●	●	◑	◑	◑	◑	◑	◑	◑	●	◑
CARRIER 51DXA0081	468 (348 to 560)	8200	9.2	2.2	—	●	◑	●	●	◑	○	◑	○	◑	◑
WARDS 5161	502 (450 to 580)	8300	9.0	2.9	○	●	◑	◑	◑	◑	◑	○	◑	◑	◑
AMANA ES92MS	455 (390 to 529)	8500	9.3	1.9	◑	●	●	◑	◑	●	●	○	◑	◑	◑
J.C. PENNEY 1178	416 (336 to 430)	7800	8.7	2.6	—	●	●	●	⦿	●	●	○	◑	○	◑
EMERSON QUIET KOOL 8JS7E	383 (349 to 430)	8200	9.4	2.5	○	●	◑	◑	◑	●	●	○	◑	◑	◑
GIBSON AM08B6ELBB	384 (344 to 429)	7800	8.7	2.6	○	●	●	●	●	●	●	○	◑	○	◑
SEARS 71089	404 (350 to 440)	7800	8.7	2.6	◑	●	●	●	⦿	●	●	○	◑	○	◑
HOTPOINT KM908F	437 (375 to 519)	8100	8.8	2.0	◑	○	◑	◑	◑	●	◑	○	◑	◑	◑
GENERAL ELECTRIC AQ908AA	422 (386 to 499)	8000	8.5	2.4	◑	◑	●	◑	⦿	◑	◑	○	◑	◑	◑
KELVINATOR MH308T1Q	404 (359 to 449)	7800	8.7	2.6	—	●	◑	◑	◑	○	○	○	◑	◑	◑
FEDDERS AST08F2HK	419 (359 to 504)	7900	9.4	1.9	◑	●	◑	◑	◑	◑	◑	○	◑	◑	●
WHIRLPOOL AHFP802	422 (365 to 499)	8000	8.8	2.3	◑	○	●	◑	⦿	●	○	●	○	○	○

[1] Reverse-cycle air conditioner; see box, page 16.

Brand and model	Weight (lb.)	H x W x D (in.)	Window widths (in.)	Inside projection (in.)	Cooling settings	Advantages	Disadvantages	Comments
FRIEDRICH SS08F10	124	16x26x26$^{3}/_{4}$	26$^{1}/_{4}$ to 42	3$^{1}/_{4}$ to 6	5	A,B	—	*B,C,D*
FRIEDRICH YS09F10	135	43726x26$^{3}/_{4}$	26$^{1}/_{4}$ to 42	3$^{1}/_{4}$ to 6	5	A,B	—	*B,C,D*
CARRIER 51DXA0081	113	16$^{1}/_{2}$x25x24$^{3}/_{4}$	27 to 41	5$^{1}/_{2}$	3	A,C,F	a,g	*B,C,D*
WARDS 5161	105	14$^{1}/_{2}$x24$^{1}/_{4}$x23$^{3}/_{4}$	28 to 38	4	3	G,I	m	—
AMANA ES92MS	127	15$^{3}/_{4}$x24$^{1}/_{2}$x23$^{1}/_{2}$	28 to 40	3	2	F	c,h,i,j	*B,E*
J.C. PENNEY 1178	102	15$^{3}/_{4}$x23$^{1}/_{4}$x19$^{3}/_{4}$	27 to 36	4	3	L	a,d,e,l,n	*D*
EMERSON QUIET KOOL 8JS7E	79	14x25$^{1}/_{4}$x20$^{1}/_{4}$	29$^{1}/_{2}$ to 40	4$^{1}/_{2}$	3	K,L	f,k	—
GIBSON AM08B6ELBB	104	16x23$^{1}/_{4}$x20$^{3}/_{4}$	27 to 37$^{1}/_{2}$	5	3	L	l,n	*A,D*
SEARS 71089	103	15$^{3}/_{4}$x23$^{1}/_{4}$x20$^{1}/_{2}$	27 to 36	4$^{3}/_{4}$	3	E,L	l,n	*D*
HOTPOINT KM908F	131	18$^{3}/_{4}$x26$^{1}/_{4}$x26$^{1}/_{4}$	30 to 39	6$^{1}/_{2}$	3	A,D	e	*H,E*
GENERAL ELECTRIC AQ908AA	85	16$^{1}/_{4}$x23$^{1}/_{2}$x21$^{3}/_{4}$	26 to 38	3	3	D,F,H	—	—
KELVINATOR MH308T1Q	101	15$^{3}/_{4}$x23$^{1}/_{4}$x18$^{1}/_{4}$	27 to 39	2$^{1}/_{2}$	3	E,F	a,d,i,j,l,n	*D*
FEDDERS AST08F2HK	94	15$^{1}/_{4}$x24x24	26$^{1}/_{2}$ to 39	3$^{3}/_{4}$	3	D,H,J	—	—
WHIRLPOOL AHFP802	102	15$^{1}/_{4}$x22$^{3}/_{4}$x23	26 to 38	3$^{3}/_{4}$	3	—	b,i,k,m	*D*

SPECIFICATIONS AND FEATURES

All: • Can be run, local codes permitting, on ordinary 15-amp branch circuit protected by circuit breaker or time-delay fuse. • Should be used only with grounded outlet. • Are rated at 7.5 to 8.5 amps but can draw more under extreme conditions. • Should be carried and installed by more than one person. • Have air-exhaust provision, judged minimally effective.

Except as noted, all have: • An automatic setting that turns fan off with compressor. • Barrier to prevent accidental contact with fan blade when cleaning filter. • Adjustable vertical and horizontal louvers. • Convenient controls. • Leveling provision, 2 sill brackets, and expanding side panels that allow air leakage unless well sealed. • Grille on outdoor side to protect cooling fins.

KEY TO ADVANTAGES

A – Has slide-out chassis. Can be installed through the wall.
B – Horizontal louvers can be adjusted downward.
C – Horizontal louvers can be adjusted downward, but they must be inverted (a simple operation).
D – Horizontal louvers can be adjusted slightly downward.
E – Lower half of horizontal louvers can be closed for more powerful airflow.
F – Control panel not covered: access judged more convenient than on most.
G – Has electronic controls; judged very convenient but may require some study of instruction book at first.
H – Control markings easier to read than on most.
I – On automatic setting, fan changes speed with temperature but never shuts off.
J – Expanding side panels have metal frame, judged stronger than plastic frame on most others.
K – Has pull-out filter; does not require removal of front panel.

L – Filter has plastic frame; judged somewhat easier to clean than most.

KEY TO DISADVANTAGES

a – Lacks automatic setting.
b – Fixed horizontal louvers.
c – Vertical louvers move in direction opposite to control knobs.
d – Vertical louver knobs interfere with movement of horizontal louver.
e – Has spring-closed cover over controls, a small nuisance.
f – Control markings hard to see.
g – Thermostat control, though marked, lacks reference numbers.
h – Fan blade exposed while filter is being removed.
i – Filter, retained by elastic band, judged more difficult than most to remove and replace for cleaning.
j – Screwdriver required to remove front panel.
k – Has no sill bracket or leveling provision; judged least secure of tested models during installation or removal.
l – Has only 1 sill bracket.
m – Lacks grille on outside.
n – Had moderate false-start problem.

KEY TO COMMENTS

A – Has motorized "air sweep" louvers; feature did not significantly improve temperature distribution; works only on cooling settings, making adjustment of vertical louvers difficult on fan-only settings.
B – Has ventilation as well as exhaust setting.
C – Side panels must be cut to fit; less convenient than expanding panels but provide a better seal.
D – Fan motor must be oiled.
E – Has special provision for disposing of condensate: drain plug for $1/2$-in. drain line **(Amana)** or optional overflow kit **(Hotpoint)**.

Ceiling fans

Fan manufacturers know that charm sells: Potted palms and characters in white linen suits are the centerpieces in ads for some ceiling fans. Indeed, a ceiling fan can conjure up exotic feeling even in a raised ranch house in the suburbs. But people who use ceiling fans know that fans aren't just for show. On top of their genteel charm, they are an effective way to stir up a breeze so that a room feels cooler.

Large blades allow a ceiling fan to move lots of air while running sedately and quietly. By comparison, a portable fan is noisy and blows up a storm. Also, a properly installed ceiling fan is out of reach—and thus safer if children are around.

A ceiling fan, however, isn't a substitute for a portable box fan or an air conditioner. Unlike a portable fan, a ceiling fan can't bring cool outside air into a room; it can only circulate the air that's already there. And you can't move a ceiling fan from room to room. Nor can you expect a fan—either one on the ceiling or one on the floor—to dehumidify; that's an air conditioner's job.

The lowest-priced ceiling fans generally have a plain painted motor housing, which might be appropriate for a kitchen or for a room with contemporary or casual decor. At the high end of the price range are expensive models with brass finish or baroque trim. In many cases, a model is available in several styles and finishes and in several price lines, so you can let your taste and your budget guide you.

The size of a fan is defined by the diameter, or sweep, of its blades. The most common size is 52 inches. But there are smaller units, 36- to 38-inch models, suitable for a small room. All the fans give you a choice of at least two speeds.

Blowing

Typically, a relatively vigorous column of air, a foot or two wider than the diameter of the blades, descends beneath the fan. The air in the center of the column is usually a bit calmer. Nearing the floor, the air disperses toward the walls and is then swept upward and back toward the fan. Besides that main pattern, there are gentler currents and eddies in the room.

At their maximum speed, most fans generate a 2- or 3-mph breeze in the column of air below the blades. Beyond that column, out to about 16 feet, there's a fairly consistent downward flow of about 1/2 mph.

But airspeed alone indicates only how breezy a fan is. Some manufacturers describe their fan's performance according to the total amount of air moved in a given time, usually in terms of cubic feet per minute (cfm). Claims generally run from about 3,500 cfm for small models up to about 8,500 cfm for large ones.

Air-moving ability. If you have a large room to cool and if the climate in your area is very hot, you'll want a powerful fan.

However, the most powerful fan isn't always the best. Any fan will move enough air to keep you cool in a moderate-sized room in most parts of the country.

Performance range. Suppose you want a fan that can churn up a very strong breeze during the heat of the afternoon and stir up the air gently in the quiet of the evening. A fan with a control that provides three speeds or infinitely variable speed should give you a satisfactory choice.

Minimum speed. One of the charms of the old-fashioned ceiling fan was the lazy way in which its blades swept by— and that's where "minimum speed" is important. A minimum speed of about 50 revolutions per minute (rpm's) or less should satisfy anyone to whom that ambience of tranquillity is important.

Quiet

Ceiling fans are characteristically much quieter than other types of fans. Some models, in fact, are virtually silent when running at their lowest speed.

When a fan is running at less than 100 rpm, airflow is relatively low and air noise is still insignificant.

The mounting hardware of most ceil-ing fans includes a rubber block or grommets to prevent mechanical noise and vibration from being transmitted to the ceiling. Though such flexible hard-ware is useful, it allows an out-of-bal-ance fan to wobble. In most fans that wobble will be barely perceptible.

Operating cost

Operating costs for ceiling fans vary considerably, but even the most profli-gate model will cost only pennies a day to run.

Evaluating the controls

Most of the fans we tested have a con-venient pull chain to turn them on and shut them off. The pull chain on the **Mistralaire,** however, controls a built-in fluorescent lighting fixture. A rotary switch on the motor housing turns the fan itself on and off. The switch is less convenient than a pull chain—and ob-viously impractical if the fan is above arm's reach.

On the two- and three-speed models, the pull chain also provides speed con-trol. Successive pulls cycle the switch through its various speeds and the off position.

Variable-speed models have a rotary control for changing the speed. On most such models, the control knob is only an inch or two below the blades; making adjustments while the fan is running is hazardous, and setting the speed by the trial-and-error process of turning the fan off and on repeatedly is inconve-nient. The **Mistralaire's** variable-speed knob, however, is a reasonable distance from the blades. And the **Leslie-Locke** has a pair of pull chains that allow you to rotate its speed control remotely, as if pulling a cord over a pulley.

Assembly and installation

To insure proper installation, you should hire an electrician to put up a ceiling fan. Nevertheless, the instruc-tions that came with our fans are obvi-ously aimed at do-it-yourselfers.

Ceiling fans are shipped disassem-

Can a ceiling fan help in winter?

Many manufacturers claim that a ceiling fan can blow hot as well as cold—that it can save energy in the wintertime. Their premise is that a fan can move the warm air near the ceiling down to people level, where it's needed. That in turn permits a lower thermostat setting for the same level of comfort.

Most manufacturers say that the fan's motion should be reversed for such use, so the blades pull the air upward rather than pushing it down. Many fans on the market are reversible. Consumers Union tested a couple of large ceiling fans in a staffer's home, in a large room with a 10-foot ceiling. No matter whether a fan was running in forward or reverse, and no matter what the room temperature was, the occupants always felt cooler when the fan was running than when it was off. A subsequent check of temperature at various levels showed why: the air temperature at the ceiling was only a few degrees warmer than at the floor. The cooling effect of the air movement outweighed any improvement in temperature distribution, even when the fan ran at low speed.

You shouldn't rule out the possibility that a fan might reduce heating bills in some very special situations—for example, in a room with a very high ceiling or with a localized heat source such as a wood stove. But, under most circumstances, a ceiling fan isn't likely to be much of an advantage in the winter.

bled. Putting one together isn't a difficult job—just a matter of paying attention to detail. Before you begin, make sure that wooden blades aren't warped; if they are, they won't run true. The large **Hunter** came with a warped blade. Putting things right meant replacement with a complete set of four matched blades (something your dealer should do at no charge). The blades of our **Patton** and the **Penneys** had mislocated mounting holes. We had to enlarge some holes to attach the blades.

For safe clearance, the ceiling should be at least eight feet high so that the blades will be at least seven feet from

the floor. That precludes the use of the **Mistralaire** unless you're tall enough to reach the control, or unless you wire it to be controlled by a wall switch. Take special care in areas of the house where you might raise your arms while getting dressed or where you might playfully raise a child overhead. Watch out for curtains that could get tangled in the blades. The optional lighting kits that attach to the fan's motor housing also reduce head clearance under the fan.

An easy way to install a ceiling fan is in place of an existing light fixture, since the wiring is already in place. An alternative would be to use a decorative

Ratings of ceiling fans

(As published in a July 1982 report.)

Listed by size. Large models are listed by groups in order of estimated overall quality; within groups, listed alphabetically. Small models are listed in order of estimated overall quality. Prices are suggested retail for fan and blades, where more than one type of blade is available, model number of blade kit is given in parentheses. Optional light kits, which attach to fan housing, are generally available at an extra charge; + indicates that shipping is extra.

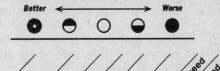

Better ← → Worse
● ◐ ○ ◑ ●

Brand and model	Price	Blade diameter (in.)	Speeds	Air-moving ability	Performance range	Minimum speed (rpm)	Relative quiet at max. speed	Relative quiet at min. speed	Efficiency factor at max. speed
Large models									
CASABLANCA CB4222 (B35)	$429	52	1	◐	●	20	◑	●	○
FASCO A952 (9852)	387	52	2	○	◐	45	◑	●	◐
PATTON 1015W, *A Best Buy*	250	52	1	◐	●	35	◑	●	◐
SEARS Cat. No. 9066, *A Best Buy*	250+	52	3	○	◐	50	◐	●	◐
HUNTER 22272 (22234)	400	52	2	●	◑	110	○	○	◐
PANASONIC F525A	270	52	3	◑	◐	55	◐	○	◑
CODEP 600 J [3]	360	52	1	◐	◐	75	◑	◑	○
LESLIE-LOCKE DF522	300	52	1	◐	◑	70	◑	◑	●
MISTRALAIRE MAS104	150	48	1	◐	◑	50	◐	◑	◐
NUTONE PFM52 BE	270	52	3	◐	○	70	●	◐	◐
J.C. PENNEY Cat. No. 5115	240+	52	1	◐	◐	55	○	◑	○
Small models									
HOMESTEAD HP30 (BO38)	299	38	1	◑	○	100	○	●	◐
NUTONE PFL36 BE	156	36	3	◑	◐	95	○	●	◐
EMERSON CF363A	150	36	2	◑	◐	95	◐	◑	◐
HUNTER 22270 (22271)	290	36	2	◑	◑	190	◑	◐	○

1. *Speed is infinitely variable between high-est and lowest settings.*
2. *Speed on low is determined by a separate control and can be preset at any point between highest and lowest settings.*

3. *According to the company, a later designation of this model was **1600**.*

chain to camouflage a wire that is draped along the ceiling and down to a wall outlet. Be sure the hardware holding the fan is secure; it should be able to support at least 45 to 50 pounds.

Most of these fans have sealed or pre-lubricated bearings that don't require

servicing. The **Hunter** models have an oil reservoir whose level should be checked once a year. That might seem like a bother, but in our opinion, the reservoir should prolong the life of the motor.

Recommendations

Among the large fans, the models in the first Ratings group (see page 24) have a slight edge over the others. Within that group, the **Sears** and the **Patton** are Best Buys.

Models that are lower in the Ratings may also warrant consideration. The large **Hunter,** for example, is ruggedly constructed and powerful. Its major shortcoming is its vigorous performance

when it's set to run at its lowest speed. But in an outsized room or in extreme heat, its whirlwind performance could be an advantage.

A 36- or 38-inch model would be best where space is tight—in a room up to about 10 by 10 feet or in an alcove. Of the four we tested, the **Homestead** and the **Nutone** are the fans of choice.

Portable box fans

The box fan, or "breeze box," has been a household fixture for decades. Inexpensive and versatile, the box fan can be used to focus on a particular area or to act as a stand-in for a window fan, pushing warm air from a room to make way for cool air from outdoors. Since it is portable, you can use it in the kitchen during the day or in the bedroom at night.

True, a box fan is a long way from being an air conditioner. And it can't always provide the ventilation of a properly mounted window fan. But breeze boxes are more popular than any

other type of portable fan.

The more air a fan moves in a given time, the better its cooling effect. With a box fan, you can use the air movement to ventilate—by putting the fan in or near a window to exhaust warm indoor air while sucking in cool air from outside through another window. Or you can use the movement to circulate—by putting a fan on a table or the floor to stir up the air and speed up evaporation from your skin. Fans with a 20-inch blade sweep are the best compromise between performance and portability.

Operating cost

Run for just an hour, most fans have pretty much the same operating cost—a penny or so, at the national average electricity rate of 7.75 cents per kilowatt-hour.

At their speediest, when displacing the most air, just about any fan will produce the same mild roar. At the lowest speed, when moving much less air, a fan will, of course, be quieter.

Features and convenience

When it comes to controls and features, simplicity is a fan's keynote. Legs, a handle, and a speed selector switch are all that most have—or need. Only a few have special features, and only a few depart from the norm with their standard features.

Legs. Some are "outrigger" legs, designed to protect against accidental tip-overs, or to help keep a fan stable on a high-pile rug. The legs on most models have padded ends or are made of plastic so they're unlikely to scratch floors. Be wary of fans with metal legs that can scratch a bare floor.

Thermostat. A thermostat to shut off a fan when the temperature falls below a certain point can be a blessing when a chilly morning follows a hot night. But thermostat controls are not marked in degrees—you can't set them for a given temperature.

If you use a thermostat-equipped fan near a window, remember to switch the fan off in the morning; otherwise, the fan will start again as the temperature rises, bringing hot daytime air into the house.

Ease of cleaning. You may want to remove a fan's grilles to clean blades and other surfaces after a season or two of cooling. That usually entails removing several screws, or prying out an easily removable grille.

Reversible motor. An electrically reversible motor lets you shift from intake to exhaust without turning the entire fan around. That's handy in a window-mounted fan, but it's not much of an advantage with a free-standing box fan.

Safety

The whirling blades of any fan can be dangerous. All fans have a grille that should provide reasonable protection against accidental contact with those blades—but no fan can ever be considered safe for a small child's fingers. In that regard, a fan set on a table or windowsill is likely to be less accessible than a fan set on the floor.

Current leakage and short circuits could result if rain gets into the motor of a window-mounted fan. So don't operate a fan in or even near a window when it's raining.

Some manufacturers specifically warn against using the fan in a window, and some models are labeled or marked to that effect. Such a warning, in a readily visible location, is required for any fan that carries an Underwriters Laboratories listing without having met UL requirements for rain resistance.

Government records show that fan-related fires account for a relatively large share of reported accidents. Sometimes the cause of the accident was obvious: a motor overheating when the airflow was blocked, or when a fan

Ratings of portable box fans

(As published in a July 1982 report.)

Listed by size; within size groups, listed in order of estimated overall quality. Models judged approximately equal are bracketed and listed alphabetically. Except as noted, all models are box fans with metal frame and have 3 speeds. Prices are manufacturer's suggested retail rounded to nearest dollar; + indicates that shipping is extra.

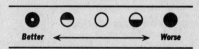

Better ← → Worse

Brand and model	Price	Performance		Noise		Weight (lb.)	Advantages	Disadvantages	Comments
		Ventilation	Circulation	At high speed	At low speed				
20-inch models									
LAKEWOOD P83	$70	◉	◉	●	◉	22	A,B,E	b,d	B,F,G
FRIGID P20B	50	◑	○	●	◑	18	E,G	—	—
EDISON 204007B	40	◑	○	●	◑	15	E,F	f,g	—
EAGLE AIRE EA203	30	○	○	◑	◑	14	E	e,f	—
EDISON 204011B	37	○	○	●	○	11	E,F	g	—
GALAXY 3713	27	○	○	●	○	13	—	f	—
GALAXY 3714	24	○	○	●	○	13	—	a,f	—
LAKEWOOD P223	26	○	○	●	◑	13	E	d	—
SEARS Cat. No. 8120	23+	○	○	●	◑	13	—	a,f	A
SUPERLECTRIC 2066T	40	◑	○	◑	◑	13	E	i	A
GALAXY 3746	30	○	○	●	◑	13	A	c,d,f,h	B,E
LASKO 4520	30	◑	◑	●	◉	8	—	g	B,C
SEARS Cat. No. 8121	30+	○	○	●	◑	13	A	c,d,f,h	A
SEARS Cat. No. 8139	30+	◑	◑	●	◉	8	—	g	A,C
SUPERLECTRIC 2023B	33	◑	◑	●	◑	12	E	i	—
SUPERLECTRIC 2012B	33	●	◑	●	◑	12	E	a,i	A
EDISON 204014A	70	●	◑	●	●	18	D,E	f,g	A
12-inch model									
GYRO AIRE GA1670	80	●	○	●	◉	9	C,D,I	—	A,D

KEY TO ADVANTAGES

A – Thermostat.
B – Indicator light warns when thermostat shuts off fan.
C – Variable speed control.
D – Air-deflecting vanes may be set to oscillate **(Edison)** or rotate automatically **(Gyro Aire)**
E – Has feature that protects against electrical overload and resets automatically.
F – Judged more stable than most.
G – Pry-out grilles for easy cleaning.

KEY TO DISADVANTAGES

a – Has only 2 speeds.
b – Thermostat control is not marked to show which way to turn knob for higher or lower temperature setting.
c – Thermostat reference markings not labeled.
d – Judged slightly less stable than most.
e – Outrigger legs judged easy to bend; when bent, legs made fan less stable than any tested.

f – Frame has pointed corners.
g – Access to interior for cleaning judged less convenient than with most.
h – Uncomfortable handle.
i – Unshielded metal legs may scratch bare floors.

KEY TO COMMENTS

A – According to manufacturer, fan is not for use in a window.
B – Shape of frame unsuitable for steadying with a lowered window sash.
C – Has round frame made of plastic.
D – Plastic frame.
E – Controls face the outside when fan is set in window for exhaust.
F – Motor is electrically reversible; fan can move air in either direction.
G – Model discontinued at the time the article was originally published in *Consumer Reports*. The test information has been retained here, however, for its use as a guide to buying.

---

blade tangled with a drape and stalled. Sometimes the cause was hard to pinpoint, except as spontaneous failure of the fan's motor.

A fan equipped with an automatically resetting overload protector can cope with a short-lived problem—temporary blockage of the airflow, say. But if the protector trips for no apparent reason, it's a sign that something is wrong internally. At that point, have the fan repaired or replaced. Always disconnect a fan before trying to clean or service it. And beware of surprise start-ups by a fan equipped with a thermostat.

Recommendations

A 20-inch fan should be able to handle the ventilation or air circulation in a modest-sized room, or even be powerful enough to ventilate one floor of an average-sized house.

If you're planning to use a box fan for ventilation only, consider a window fan instead. It will be out of the way, and the draft-tight seal provided by the mounting panels permits good air intake.

Whole-house fans

It's easy to spend $400 or so for a medium-sized room air conditioner. And it's easy to pay $40 over a summer to run one—even a high-efficiency model—at typical electricity rates. Payoff: one cool, comfortable room.

Now tally up the account of a whole-house fan. A good 36-inch model can be bought for close to $250. Running it for a 1,000-hour summer season would probably cost $41, where electricity is billed at the national average rate (7.75 cents a kilowatt-hour). Payoff: a more comfortable house.

Some fringe benefits: A house that's fan-ventilated at night will retain some of the cooling effect by day—at least for a while—if you turn off the fan, close windows, and draw shades. When you lose that coolness, you may choose to open things up and run the fan. The breeze may make you feel more comfortable despite the heat. And, by helping to keep attic temperatures down during the day, a fan will cool off a house quicker again in the evening.

Any liabilities? Several—and fairly serious ones at that. An air conditioner dehumidifies air; a fan doesn't. Stricly speaking, a fan doesn't even *cool* air. For a fan to cool a house, the outside temperature has to be lower than the inside temperature. So, while a fan may serve well in the milder temperatures of evening after a sweltering day, it may not keep you comfortable at noon. Finally, a fan imposes costs for installation, usually in the attic, and for such accessories as louvers, shutters, and remote switches. Those installation and accessory costs may run anywhere from $200 to $500.

On balance, then, a whole-house fan isn't a substitute for air conditioning, nor is it as cheap as one might first suppose. But it does have enough advantages to make it a reasonable cooling alternative for some people in some locations.

Note: Whole-house fans were once commonly called attic fans. However, the term "whole-house" has come into use to distinguish these large fans from smaller units, properly called attic fans, that are sold for ventilating only attics.

The right-size fan

Open your windows in the cool of evening, and your house will cool off by itself—though not very fast. The outdoor temperature may drop from 85°F to 75° in two hours, but the still air in an ordinary house will take about twice as long to cool off those 10°. With a whole-house fan of the right capacity, though, the air indoors will be replaced with outdoor air every minute or two. So rooms in the house will cool off almost as fast as the garden.

Indeed, you may feel cooler before your thermometer confirms that you really are. Some people feel several degrees cooler when air moves past them as slowly as 100 feet per minute. A whole-house fan can easily produce that mild breeze.

The key to getting the best from a fan is to choose one of the right capacity for your house.

Before you settle on a specific brand or size, you need to calculate how powerful the fan should be. To determine the proper fan capacity for your house, calculate the volume of space (in cubic feet) to be ventilated. Multiply length by width by height of each room, hallway, and stairwell to be ventilated. You may not need or want to ventilate all the rooms in the house, or there may be out-of-the-way rooms that would be impractical to ventilate. In any case, don't include closets, storage rooms, or the attic itself. In especially warm climates, the volume you wish to ventilate should

closely match the fan's capacity, as measured in cubic feet of air per minute. Where summer temperatures are reasonably moderate, divide the volume by two to find the proper fan capacity.

Note that, in the case of one- or two-speed fans, you'll want a fairly close match between house volume and fan capacity at maximum speed. If a variable-speed model has somewhat more capacity than your house warrants, you can always turn it down to a speed that suits your needs.

How can you tell capacity?

Most fans carry a manufacturer's rating in cubic feet per minute. That cfm number represents "free air delivery," or performance when nothing restricts air flow. Such ratings aren't very helpful, since something usually does restrict the air. A house's size and layout and the fan's location all contribute to the restriction.

Accordingly, you'll usually find a second rating on these fans. That one represents performance when airflow is not totally free. It's derived from the amount of air that can be delivered by a fan working against a standard air-pressure load.

But that figure may not give an absolute yardstick for comparing fans either, we found. Various testing organizations certify the performance of these fans, and the test results may not be directly

comparable from fan to fan. Test factors other than pressure load can also affect performance.

Once you've decided on a fan with a capacity that's an appropriate match for your house, choose the next larger capacity available. That will help to offset any possible underrating by the manufacturer. Furthermore, even if you've selected a model that's slightly oversized, it should have the reserve capacity needed for the hottest weather of the year but it won't cost much more to operate.

A larger-capacity fan will also come in handy, considering that it may have to buck the resistance of "automatic" shutters on the exhaust opening, which then float on the breeze and slow the passage of the air.

Where to mount it

The fan can perch horizontally over a ceiling opening to the attic. It can stand vertically near the opening, enclosed in a suction box that directs air into the fan. Or it can be affixed to a wall so that the entire attic serves as a suction box. If your attic is well ventilated, you need

a fan of slightly larger capacity for a wall installation than for its alternatives.

The simplest and cheapest fan shutters are the ones designated as "automatic." They open purely by the fan's suction or air blast. But automatic shutters tend to reduce air flow more than shutters that open and close mechanically or electrically; the reduction is often as much as 50 percent when the fan is turning at low speed. For some installations, manually operated or motorized shutters might be needed, especially at whichever opening is farther from the fan.

Automatic shutters list at about $50 to $150.

Most shutters have louvers with a felt edge meant to reduce air leakage when the shutters are closed. Even so, don't rely on the felt to seal a ceiling shutter during the heating season. You'd be wiser to cover the shutter opening in winter, perhaps with plastic sheeting.

Installing the fan

No matter how you install the fan, two openings will be required—one between the living space and the attic, a second between the attic and the outdoors. Normally, both those openings would be cut, framed, and shuttered. You can eliminate some of that carpentry, though at the expense of convenience. For example, you may be able to use an existing door to the attic as the intake area and install the fan in an attic wall. But then you'd have to open the door each time you use the fan. If your attic has windows, they could serve as the exhaust openings for a ceiling-mounted fan—if you are willing to make regular trips to the attic to open and close the windows.

There are other things to consider, and potential hazards to avoid, when installing an attic fan. These include the following.

Vent openings. The vent opening nearest the fan is generally the same size as the fan: a 3 × 3-foot shutter for a 36-inch fan, for example. The other required opening should be somewhat larger. To decide on its size, divide the fan's actual capacity (cfm) by 750 (moving air should not exceed 750 feet per minute). For example, if the fan delivers 8,500 cfm, you'll need a little more than 11 square feet of unobstructed exhaust area (8,500 ÷ 750 = 11.3). If the opening is to have insect screening, you'll have to double the size of the opening.

Other grilles or louvers that obstruct the opening will also add to the required size of the exhaust area.

Noise. Compared with other motor-driven appliances, whole-house fans are not particularly noisy. The amount of sound you hear will depend on where the fan is installed (they're generally louder when mounted on the ceiling than when mounted on the wall) and how the unit is installed. If the fan chassis is attached directly to wall studs or ceiling joists, the hum, drone, and rumble of the fan will be transmitted into the living area. Some models are supplied with resilient pads that can be used to make ceiling installation quieter.

Whole-house fans aren't particularly noisy. But a poorly planned installation can magnify a fan's sound disturbingly. Wall studs or ceiling joists with a fan attached directly to them will carry the fan's hums and rumbles to your living space, perhaps amplified by wall or ceiling panels. And a ceiling-mounted fan is apt to be more audible than one at the end of the attic, if only because it's nearer to you.

The fan's on/off switch (or speed control) should be mounted in a convenient location in the living area. It's easiest and least expensive to wire a fan into an existing circuit. But that may not always be practical, either because your local electrical code prohibits it or because existing circuits don't have the capacity to handle this kind of appliance. Most fans put a brief but taxing load on an electric line when the motor is turned on; it's not unusual for a fan to draw 30 to 40 amps for a second or two on start-up.

Fire safety. Should a fire break out, an operating fan will literally fan the flames. An answer to that threat is a fusible-metal control element, or link, that melts in the heat of fire, shutting down the fan and closing motorized or manually operated shutters.

If you own a smoke detector, you may want to relocate it when a fan is installed. It should be located in more or less neutral air—out of the path of fan-blown air, but out of corner dead spots, too.

Blade safety. Some sort of shielding must be provided if a fan is to be installed in an area accessible to people or animals. A homemade blade guard or cage can be constructed of wood framing and coarse screening.

Self-service. Safety shielding should not be impenetrable, however. Once a season, you'll need to check the fan's drive belt and adjust its tension if it's slack. Check the action of the shutters, and oil the fan if need be. If a fan's motor housing provides oil ports—and there are no lubrication instructions to the contrary—you may assume that the motor bearings will benefit, once a sea-

---

An alternative to a whole-house fan

If you are discouraged by the price and bother of having a contractor install a whole-house fan, there may be a cheaper alternative. Buy several simple box or window fans (pages 25–28). You may lose something in convenience, but you certainly save on carpentry.

A 20-inch fan can move 1,800 cfm of air with an electrical consumption of 265 watts. Three of them might have a theoretical delivery capacity of 5,400 cfm with a power consumption of 495 watts. Those totals are similar to the cost and power figures for some whole-house fans.

So much for arithmetic; there are practical matters to consider. Fans create a suction behind them. Three fans working in a house might therefore interfere with each other in a way that reduces their total output. But trial-and-error positioning of the fans should reduce that effect. In a two-story house, one fan or two upstairs and one or two down might do, especially if you could close doors between floors.

Moving several fans about, of course, is less convenient than controlling one whole-house fan with the flip of a switch. You could expect their output to be cut significantly if they have to blow through screened windows. They raise a home-security question, too, since they must work in open windows.

son, from a few drops of SAE 20 oil.

Be sure that the fan won't start up as you service it. (A dormant fan isn't necessarily off; its overload protector may have merely idled the blades temporarily.) An off switch installed at the fan is your best safeguard; otherwise, interrupt the circuit at the fuse or circuit-breaker box.

Ratings of whole-house fans

(As published in a June 1981 report.)

Listed, except as noted, in order of estimated overall quality. Models judged approximately equal in overall quality are bracketed and listed alphabetically. Air flow and wattage are as measured by CU with fan mounted upright and with automatic wall shutter, working against a pressure load judged representative of household installation. Prices are suggested retail; + indicates shipping is extra.

Excellent	Very good	Good	Fair	Poor
◉	◖	○	◗	●

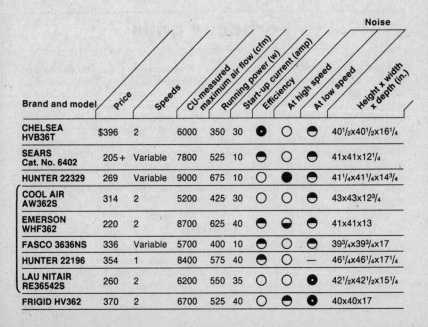

Brand and model	Price	Speeds	CU-measured maximum air flow (cfm)	Running power (w)	Start-up current (amp)	Efficiency	Noise: At high speed	At low speed	Height x width x depth (in.)
CHELSEA HVB36T	$396	2	6000	350	30	◉	○	◐	40½x40½x16¼
SEARS Cat. No. 6402	205 +	Variable	7800	525	10	◐	○	◐	41x41x12¼
HUNTER 22329	269	Variable	9000	675	10	○	●	◐	41¼x41¼x14¾
COOL AIR AW362S	314	2	5200	425	30	○	○	◐	43x43x12¾
EMERSON WHF362	220	2	8700	625	40	◐	◖	◐	41x41x13
FASCO 3636NS	336	Variable	5700	400	10	◐	○	◐	39¾x39¾x17
HUNTER 22196	354	1	8400	575	40	◐	○	—	46¼x46¼x17¼
LAU NITAIR RE36542S	260	2	6200	550	35	○	○	◉	42½x42½x15¼
FRIGID HV362	370	2	6700	525	40	○	◖	◉	40x40x17

Recommendations

The first step in choosing an attic fan is to measure the volume of the house, in order to determine how many cubic feet of air per minute the fan should deliver. Keep in mind that manufacturers' air-delivery figures should be discounted roughly 25 percent.

The air delivery you need won't automatically lock you into buying a fan of a given size. For example, a 36-inch, 1/3-horsepower model might provide about the same performance as a smaller fan powered by a 1/2-horsepower motor. But the larger the fan, the quieter and more efficient it tends to be. A higher price doesn't necessarily mean a larger capacity.

Once you've determined your house's air-delivery requirements, you have to select wall and ceiling shutters and decide whether you want a single-speed or a multiple-speed fan. Consumers Union favors multiple-speed models, combined with controlled shutters. That combination adds to the initial cost but offers slightly better fan efficiency and greater convenience.

You must also decide whether to mount the fan in the ceiling or in an attic wall. A wall installation tends to be quieter than a ceiling mount but might require a slightly more powerful fan if your attic is not tight.

Wherever the fan is installed, it may require a separate electrical circuit, either because local codes require one or because the existing wiring won't accommodate the added load.

Barbecue grills

A gas grill is simple enough to operate. You turn the gas control, ignite the burner, and relax for a few minutes while the grill preheats. That's a lot easier than getting a charcoal fire started.

But what about that great barbecue flavor? you may ask. No problem. That special barbecue taste comes from the meat fat and juices that drip onto the hot coals—or, in a gas grill, onto the hot lava bricks—and cause flaming and smoke.

Most portable gas grills use liquefied petroleum (LP) gas. There are some that use natural gas; but they haven't been big sellers, probably because they are expensive to install. The portables aren't exactly cheap, though—$180 to $455.

Outdoor cooking should be fun, not a chore, so a grill should be convenient to light, easy to adjust and move around. It should also be thrifty with a tank of fuel.

Grilling

Most any grill should be able to cook a thick steak well done on the outside, rare or raw in the middle. A steak cooked that way may not be what you would want for eating, but a grill that can cook meat that way can also produce other degrees of doneness by cooking longer with a lower flame.

If you don't need a grill's entire cooking area, you can save fuel by using only one burner. Even with two steaks cooked on one burner, the results should be close to perfect.

Chicken is a more sensitive food to cook. It contains lots of fat and juices; too high a temperature makes the fat flare up and burn the outside of the bird. Also, chicken must be cooked through to the bone.

A grill should be able to produce chicken with a crispy, brown skin and a fully cooked inside, if you overlook an occasional charred spot on the bone side, where there's little edible meat.

Some grills may char the chicken skin before the inside is fully cooked. With that kind of grill you have to bake the bird—lighting one side of the grill and placing the chicken on the other side. That way, the fat and juices won't drip onto the hot bricks, and the heat will be distributed more evenly.

Self-cleaning ability. With maximum temperatures as high as 1,000°F, you might expect a gas grill to be self-cleaning.

If you run a grill for about 15 minutes at the highest heat setting—after cooking is done—most units will look reasonably clean but still need some brushing—especially if the residue is heavy.

Costs and convenience

The cost of operating a gas grill depends on the grill's efficiency as well as on the price of LP refills. It cost Consumers Union's engineers $8 to refill a 20-pound tank. (When you first put a tank into service, you should have it purged of air before it's filled. That should cost about $2.)

Tank duration. With both burners on high, one 20-pound tankful of LP gas will last 19 to 24 hours in some units; others only 12 to 16 hours. With only one burner on, most will run 50 to 80 hours at the low setting.

Convenience of controls. It's convenient to have controls mounted on a panel in front of the firebox. Less convenient are controls mounted on the post; to work these controls, you have to stoop or kneel.

Portability. Grills differ widely in weight. The lightest weighs about 50 to 60 pounds without fuel; the heaviest, about 90 pounds. (A full fuel tank adds about 38 pounds.) However, weight isn't a major factor in the ease of pushing or pulling. Nor are the outside dimensions important; most grills are quite similar in overall size.

Far more important is the design of the grill's handle. For example, the handle may be too close to the firebox, and a shelf for trays and utensils makes it hard to grip the handle comfortably. Perhaps there's a tripod base with wheels on two of the legs. The leg without a wheel can get in the way of your feet as you move the grill. It's inconvenient to have a handle at the rear of the grill, opposite from where you'd stand when you needed to move the grill; for proper balance, you would tend to rest the grill against your forearm, so you'd be unable to move the grill when it's hot.

Special features

A number of features make some grills more convenient than others.

Igniter. An igniter is easier to use and somewhat safer than matches. Typically, you turn on the gas and push a button repeatedly to make the igniter produce a spark. An igniter button on the front panel is more convenient than one on the post. An automatic solid-state igniter is a nice convenience, and offers a slight safety advantage. However, it requires a 9-volt battery.

Fuel gauge. It would be frustrating, to say the least, to run out of gas in the middle of a barbecue. Several grills have a fuel gauge to help avoid that problem.

A few have a gauge that protrudes from the top of the fuel tank itself. Those gauges work well—especially when the tank is near empty, when a gauge is most needed.

Some grill gauges work like a spring scale, weighing the tank as it rests on a platform; the more gas you use, the lighter the tank becomes, and the closer the needle moves to empty.

If the grill you buy doesn't have a fuel gauge, you might want to buy an extra fuel tank, about $25.

Cooking grids. Typically, cooking grids are made of steel rods. In some models, a porcelain coating may help resist corrosion and may make cleaning easier. Some models have grids made of plain cast iron or cast iron with a porcelain coating.

On some grills, you can raise or lower the cooking grids to adjust the broiling intensity. That might be useful in a

A complete meal on a grill

In hot summer weather, cooking an entire meal outdoors is an appealing way to keep the heat out of the kitchen. But will the food be appealing? To find out, Consumers Union testers ran some informal tests, cooking vegetables and even dessert on a gas grill.

They didn't use fancy cookware. They plunked some foods onto the grates in their natural state, and they used aluminum foil or disposable foil pans for others. They placed some foods on the grill before others, depending on cooking time. The results should hearten those who haven't ventured beyond hamburgers, hot dogs, or an occasional serving of ribs.

Potatoes. Scrubbed potatoes, pricked with a fork, were put on the warming rack, and turned several times as they cooked. The four- or five-ounce spuds needed about 30 minutes to cook on medium heat with the hood closed. The potato skins browned nicely. For softer skins, rub the potatoes with margarine and wrap them in foil.

Corn. The testers soaked several ears of corn—still in the husk— in cold water for about 30 minutes. Then they shook off the excess water and placed the ears around the edges of the main cooking area. They cooked the corn on medium heat for about 25 minutes, turning it occasionally, then peeled away the browned husks. If you prefer, you can remove the husks and silk before cooking and wrap the corn in heavy foil. Add a bit of water before folding the foil around the corn.

Mixed vegetables. A concoction of carrots, green peppers, and onions was easy to prepare. Since carrots ordinarily take longer to cook than peppers and onions, the staffers cut them into thin strips three or four inches long to speed up the cooking time. They quartered two medium onions and cut several peppers into large chunks. All the pieces went into a pouch made of heavy aluminum foil. (If you use lighter foil, double it. Or use an aluminum-foil pan and cover the food with foil.) Before closing the pouch they added spices and a sprinkling of water. Twenty-five minutes of cooking time was plenty.

Cooking on a grill is hard to time precisely. If you're cooking food wrapped in foil, you should carefully unwrap it from time to time and peek at the food. Be careful of steam when uncovering food cooked in or under foil. Open the foil away from yourself to avoid burns.

Grilled herbed tomatoes. The testers cut six medium tomatoes in half crosswise, crushed the sliced portions with melted margarine (you can use butter), and spooned on a mix of two tablespoons of melted margarine, one-half cup of bread crumbs, and one-half teaspoon each of basil, rosemary, and salt. They put some of the tomatoes in a foil pan that was placed uncovered on the warming rack; the rest of the tomatoes went directly onto the warming rack. The tomatoes took about eight minutes to cook; all held their shape without becoming mushy.

Baked apples. Cored Granny Smith apples (cored without cutting through to the bottom) were filled with pieces of pecan, some margarine, a few drops of lemon juice, and a sugar/spice mixture. (For six apples, combine one-third cup of sugar, one teaspoon of cinnamon, and a dash of nutmeg; spoon about two teaspoons of the mix into each apple.)

The apples were spread out in a foil pie pan, covered with foil, and the pan placed on the warming rack as they started cooking the potatoes. The apples also could have been put on the rack much later—just as people started eating the meal.

An indoor barbecue pit on a countertop

Countertop broiling and barbecuing need not be done in an enclosed appliance. Farberware markets an "open-hearth" broiler to handle those functions. The **Farberware 455N** is not inexpensive ($90). It's not particularly versatile. But what it does, it does well.

The **455N** is essentially a rectangular pan, open at the bottom and supported by a pair of plastic handle/leg assemblies. A single heating element nestles below the lift-out broiling rack. Food drippings go right past the element and into a separate collection pan perched underneath. The entire unit comes apart for cleaning; much of it is made of stainless steel and can go right into a dishwasher. (You would still probably have to scour burnt-on residues by hand, though.) A pair of removable, eight-position side pieces supports the rotisserie motor and plastic-handled skewer. The rotisserie parts remove easily from the pan and can be stored independently.

The **455N** takes up $7\frac{3}{4}$ x $21\frac{1}{2}$ x $10\frac{3}{4}$ inches without its supports—about as much space as a typical broiler oven displaces. Its rack will hold eight hamburgers at a time. The rotisserie can handle meat up to 12 inches in diameter, much more buxom than either of the rotisserie-equipped broiler ovens we tested.

In Consumers Union's tests, the **Farberware** broiled better than the broilers did, and its rotisserie yielded an excellent barbecued chicken. In general, it turned out foods within its scope at least as well as a full-sized oven. On the other hand, it was not without drawbacks. It has no on/off switches or other controls—you plug in separate power cords for its rotisserie motor and its heating element. That monopolizes two electrical outlets. The simplicity of its design limits the kinds of chores you can do. It was on the slow side: it took $1\frac{3}{4}$ hours to handle a chicken of a size the other two rotisserie models could cook in an hour. And when broiling items larger than an inch thick, it took half again as long as the broiler ovens, and used more energy as well. Its open design also raises the possibility of splatters with fatty foods such as goose.

All in all, a **Farberware** is a useful auxiliary to your regular range—if the ability to roast relatively large food items is important to you.

charcoal grill—but, in a gas grill, the gas control knobs let you adjust the cooking temperature more easily and precisely. Thus, such height adjustment isn't a significant advantage.

Drip-catchers. Not all the fat and juices that drip onto the lava bricks flare up and burn away, especially on the low heat setting. Most grills have some provision for catching the drippings—generally, in an empty tin can that you position under an opening at the lowest part of the firebox. A variety of clamps and brackets are used to hold the can in place. Most allow you to attach the can securely without much fuss.

Replacing the fuel tank. After you have the fuel tank refilled, mounting it on the cart is generally a straightforward job. Most grills have a clamp that goes around some part of the tank.

Some models come with a handle that slips over the fuel fitting to make connecting the gas line easier. On others, you must use a wrench.

Assembly. Very few grills come fully asembled. Many require something under two hours to do the job.

With most grills, leave yourself up to four hours for assembly and arm yourself with a selection of screwdrivers and wrenches. A socket-wrench set would come in handy. You may also need Teflon tape to seal gas-line fittings.

Safety precautions

Careless handling of a gas grill's fuel tank could cause a fire or an explosion. An LP tank has a device that vents gas automatically if the pressure builds up excessively, as it might during hot weather or when the tank is overfilled. LP gas is much heavier than air. If it's not dispersed by a breeze, it flows like an invisible stream to the lowest point and collects there. You should never keep the fuel tank indoors, even if you bring the grill itself inside for storage.

When you're bringing the tank home after a refill, prop it up securely in your car so it can't tip. Drive with a window open. You should especially avoid tipping the tank when it's connected to the regulator on the grill; a spurt of liquid fuel could damage the regulator. After attaching the tank to the regulator, check the connection with a few drops of detergent solution; bubbles indicate a leak.

Another hazard is the firebox, which can become hot enough to ignite flammable materials. Keep the grill away from walls and from under low ceilings on patios and porches. Use sensible protection such as long-sleeved oven mitts and long-handled tongs. Avoid loose-fitting clothing.

The hood on most models has one or two handles on the side for convenient raising and lowering when you need to check the food. With grills having a handle on the front or top of the hood, it's important to use an oven mitt to avoid burns.

Recommendations

Choose a model that meets your price objective and that has the features that suit your cooking and living style. Then enjoy alfresco cooking as well as dining.

Blow dryers

Where once men slicked back their hair with lotion and women painstakingly set theirs in curlers or rollers, both sexes now blow-dry a hairstyle in minutes.

There are two basic types of blow dryers. Standard dryers are big, offer a lot of settings, and advertise a high-wattage heat output. There are also small, lightweight dryers that have a lower labeled wattage; these can be used with both 120-volt and 240-volt electricity and are perfect for traveling.

Drying

Consumers Union's testers checked drying speed by wetting a wide hank of hair and drying it for fixed periods with the dryer at its highest heat and fastest airflow. By weighing the hair before and after the test, they could tell how much water each dryer had removed from the hair in the set amount of time.

The fastest dryers in this test were about twice as fast as the slowest ones. But, unless you have very long or very thick hair, you shouldn't have to spend more than about five minutes with any model.

You can't judge drying speed from a dryer's labeled wattage. A high wattage on the label doesn't always mean a faster dryer. To begin with, the power actually used in the dryer can be considerably less than the rated wattage and still satisfy the industry's standard labeling requirements. For example, there are models rated at 1,500 and 1,400 watts that actually use less than 1,300 watts, yet meet the labeling requirements.

Dryers judged above average or average in CU's drying-speed tests were rated at 1,500, 1,400, 1,350, and even 1,200 watts.

Heat

There are people who use a blow dryer on high heat, controlling the temperature by moving the dryer farther away from the hair; hairstylists often use that method. But not everyone likes high heat and fast airflow. You may prefer hot and fast at the beginning of the drying, then cool and slow for styling. Or you may prefer to do the whole thing at a high airflow, but a comfortable medium heat. Short-haired people may be happy with medium heat and airflow.

Most dryers have from one to three switches that provide at least three combinations of heat and airflow. Some provide as many as six combinations, more than you really need.

Convenience

A good dryer should be easy to handle and operate, and not too noisy. It should be comfortable to hold, and light enough so you don't feel your arm is going to fall off before you finish drying your hair.

Travel dryers and some other models are relatively light—under three-quarters of a pound. You might want a light-weight dryer if you have thick hair and must dry your hair for several minutes—but a larger, heavier model does tend to be faster, so, despite its weight, you'll likely be holding it for a shorter period.

A number of dryers have a large handle and could prove difficult to hold for someone with small hands.

Some dryers are easier to operate than others. The switches vary in number (from one to as many as four) and in location (usually on the left side of the handle, but sometimes on the front or rear).

A single switch is generally easier to get used to. While it usually offers only two or three settings, those should be adequate for most people. A switch located on the front or rear of the handle is easier to manipulate with either hand than one on the left-hand side.

Some models have their switch-setting legend on one side of the handle and the switches themselves on the other; that can be inconvenient.

On a few models with only a single switch, the "off" is between the fast and slow settings. That's inconvenient; every time you want to switch from fast to slow or vice versa, you have to shut off the machine.

Most dryers have a cord about five to six feet long. Some of the cords are coiled, and that's not always a convenience. If your electrical outlet is not close to your mirror, the pull of the coiled cord can be a hindrance.

Many dryers come with a concentrator, an attachment that fits over the end of the nozzle and supposedly makes styling easier by directing the air to a smaller section of hair. These concentrators may not work because very few of them concentrate the air.

Travel dryers

Most of these have only one heat setting when used on 240 volts. But that isn't a very serious problem.

A bit more serious is the fact that a travel dryer may suck in hair through its intake openings and entangle the hair around its fan. That can hurt, of course;

it can also cause the dryer to malfunction.

Such machines pose a problem for people who have long, straight hair, but not for people whose hair is short or curly.

Safety

Over the years, changes in design have made blow dryers safer. Today none uses an asbestos heat shield. And all have a dual safety device, as required by Underwriters Laboratories, to shut off power in case of overheating.

Nonetheless, a hair dryer can be dangerous. Every year a number of small children—and a few adults—are electrocuted when a hair dryer that is plugged in but turned off falls into (or is

pulled into) a bathtub that has water in it. Accidents can happen around sinks and showers, too.

Even if a dryer is turned off, the cord itself is "live" where it connects with the switch.

Blow dryers carry a tag warning of the danger, and their instructions advise that the dryer be used in the bedroom, not the bathroom.

Ratings of blow dryers

(As published in a June 1984 report.)

Listed in order of estimated overall quality; differences between closely ranked models were slight. Prices are approximate retail; + indicates shipping is extra.

Better → Worse ◉ ◑ ○ ●

Standard models

Brand and model	Price	Size	Drying speed	Temp./airflow combinations [1]					Convenience			Comments
				Cold	Cool	Warm	Hot	Very hot	Noise	Holding	Switch	
CLAIROL SALON POWER 1500 IPD2	$24	Large	◉	—	—	M,H	M,H	—	○	○	○	F,S
CONAIR QUIET TONE 1500 086CC	34	Large	◉	M	M,H	H	—	—	◑	◑	◑	F,J,S,T
GENERAL ELECTRIC DIAL POWER Pro 22	18	Large	◑	L	L,H	M	—	—	○	◑	◑	E,G
NORELCO CHIC 1400 CPG14	22	Large	◉	L,H	H	L	H	—	○	◑	●	G,R
SEARS 8780	20	Large	◉	L	L,H	H	—	—	○	◑	○	G,H,O,S
OSTER PROFESSIONAL STYLE 33207	24	Large	◉	M	M	M,H	H	—	○	◑	◑	G,S
CONAIR PRO STYLIST SUPER MINI 092	27	Medium	◑	—	L,H	—	—	—	◑	◑	◐	O
GILLETTE PROMAX 1500 9100	22	Large	◉	—	L,H	L,H	H	—	◑	◑	◑	G,S
SEARS Cat. No. 8765	10+	Medium	◑	—	L,H	L	H	—	○	○	○	G,H

Brand and model		Size										Comments
SUNBEAM PRO-AIRE 1350 52471	19	Medium	◐	—	L,H	L	H	—	○	◑	○	G,Q
WINDMERE PRO 1250 P12 T/575	20	Large	○	L	L	M	L	M	◐	○	○	G
CONAIR PRO STYLE 1250 085D	30	Large	◐	L	L,H	H	—	—	○	◐	◐	G
WINDMERE THE BOSS 1500 PLUS	32	Large	◐	L	H	H	L	H	◐	◐	◐	G,O,S
NORTHERN PRO PISTOL 1250 18783	20	Medium	◐	L	L	H	—	—	○	◑	○	G,B
VIDAL SASSOON 1500 VS207	19	Large	○	L,M	L,M	M	L,M	L	◐	◑	◐	F,H,J,S
GENERAL ELECTRIC SUPER PRO 6	29	Large	◐	M,H	M,H	H	—	H	○	●	●	G,O,R
CLAIROL SON OF A GUN TD2Z	26	Medium	○	L	L,M	L,M	—	M	◐	◐	◐	R

■ The following 2 wall-mounted models were downrated because they sucked hair into their fan.

ANDIS HANG UP HD1	24	V. small	○	—	—	M	—	M	○	◐	○	E,I,J,K,L,M,U,V
NORELCO WALL MOUNTED CWD10	35	Medium	●	—	VL	L	—	L	○	◐	○	E,I,J,K,T,V

Travel models

■ The following 9 travel models were downrated because they sucked hair into their fan.

BRAUN TRAVELAIR DV1400	25	Small	○	—	—	M	—	M	○	●	●	C,L,P,R,T
GENERAL ELECTRIC GRAN TOUR 1200 Pro 17	24	Small	○	—	—	L	—	M	○	●	●	D,L,P,T
CLAIROL RAPIDE 1250 MT1	18	V. small	○	—	L	—	H	—	○	●	●	E,L,M,T
SUNBEAM POWER BREEZE 52491	14	Small	◐	—	VL	L	—	M	●	●	●	E,L,M,T
WINDMERE RUNABOUT 1200 MD12/652	19	V. small	◐	—	VL	—	—	M	◐	◐	●	L,M,T
SEARS MICRO MINI Cat. No. 8745	13+	V. small	○	—	L	—	—	M	◐	◐	●	E,L,M,U
NORELCO CHIC 1200 CFD34	22	Small	◐	—	L	—	—	M	●	●	●	L,M,N,T
VIDAL SASSOON FOLDING DRYER VS222	13	Small	●	—	VL	—	—	M	○	◐		E,L,M,N,U
CONAIR VAGABOND 1250 125C	20	V. small	●	—	VL	L	—	M	○	◐		A,E,L,M,N,P,T

1 Average air temperature and air speed as measured by CU. VL = very low air speed; L = low; M = medium; H = high.

SPECIFICATIONS AND FEATURES

Except as noted, all: ● Have a concentrator. ● Have side-mounted switches. ● Weigh about 1/2 to 3/4 lb. ● Have a non-coiled cord about 5 to 6 ft. long. ● Operate on 110-125 v. electricity only. ● Are rated at approx. 1200 watts.

KEY TO COMMENTS

A – Has relatively low sound level, but has loud whistle.
B – Vibrated slightly.
C – Concentrator judged effective.
D – Concentrator judged slightly effective.
E – Has no concentrator.
F – Heavier than most, 1 to 1 1/4 lb.
G – Somewhat heavier than many, 3/4 to 1 lb.
H – Has 7-ft. cord.
I – Cord length from cradle to plug about 32 in.
J – Has coiled cord.
K – Comes with cradle for wall mounting.
L – Can be used on 240 v. electricity.
M – Only 1 setting on 240 v. electricity.
N – Handle folds for compact storage.
O – Comes with accessories: stand, waving brush, curler, curler nozzle.
P – Comes with storage case or bag.
Q – Rated at 1350 watts.
R – Rated at 1400 watts.
S – Rated at 1500 watts.
T – Switch is on front or rear.
U – Switch is on left side.
V – Has additional switch on cradle.

How to use a dryer

To get some tips on the proper use of a blow dryer, Consumers Union asked a few leading questions of a theatrical hair designer.

Q: Is there a *best* way to use a blow dryer?

A: Not really. Everyone has his or her own method, and most methods work fine. It's really just good common sense.

Q: What do you mean?

A: There are some things you *shouldn't* do with a blow dryer, but you find them out pretty fast.

Q: Such as?

A: You shouldn't keep the flow of hot air on one part of your hair for more than a few seconds. Of course, once you burn your scalp, you get the idea. And you should keep the flow of air moving over the hair, or you risk overdrying it.

Q: How far away from your hair should you hold the dryer?

A: Well, no one uses a ruler when drying hair, but 6 to 12 inches is about right. But people should do what feels comfortable. If you're burning your scalp, chances are you're holding the dryer too close to your head. Common sense is the key; if it feels right, it's probably right.

Q: Is fast drying better?

A: Not always. A slow airflow and cooler setting makes styling easier. But most people are in a hurry and want to use the hotter settings. I say take the extra two minutes.

Q: Can you blow-dry your hair every day without damaging it?

A: In most cases, yes. Just be careful not to over-dry your hair. Everyone's hair is different, but all hair needs moisture to be pliable. Your hair should still be slightly damp even after it's been styled and dried.

Q: Should you blow-dry hair that has a perm?

A: No. The whole point of a permanent is that you let your hair dry naturally. If you have a perm and must blow-dry, use one of the cooler settings on the dryer.

Q: Should you blow-dry hair that's been bleached or dyed?

A: Sure. But you must also be careful to condition the hair regularly, and remember not to overdo it with high heat as you blow-dry.

Q: Is it okay to blow-dry damaged hair?

A: Only if you use the blow dryer on a cooler setting. Damaged hair should not be subjected to heat.

Recommendations

No single blow dryer is likely to be just right for everyone.

If you have short or curly hair, you can also look among the travel models. All of those are light in weight and either small or very small. They permit operation at 120 to 240 volts. Though they won't dry as quickly as the larger models, the fastest may be fast enough for people with short hair.

Before choosing any blow dryer, see if you can try out a sample in the store. If you can plug it in, check that the dry-er's fast airflow is fast enough and the slow one slow enough for your needs. Make sure the dryer isn't too noisy and that the temperature of the heat settings suits you. If you can't plug it in, you can at least check that the switches are conveniently located and easy to set. You can also make sure that the dry-er feels comfortable to hold, that its barrel isn't too long for your reach or its handle too fat (or thin) for your hand, and that it isn't too heavy to maneuver.

Clock radios

Any up-to-date clock radio comes back on schedule to give you a second wake-up call—and plenty more.

Gone is the traditional clock face with its two hands; gone, too, for the most part, are the mechanical-leaf digital displays that once seemed so modern. Electronic display has taken over, with some "screens" a virtuoso display of programmed information.

For a clock radio with a list price around $50, you'll get both AM and FM reception and, most often, a choice of waking up to the radio or to a buzzer or beeper. You'll also get a drowse switch and frequently you will get additional features that range from a double alarm (so your mate can be awakened at another time) to a hidden calendar (day and date appear at the touch of a button).

Features

The radio should be sensitive and selective enough to pick up stations you want to listen to regularly. Sound quality won't be high fidelity but it should be pleasant enough for bedside listening.

Every radio should have a power-failure indicator and an indoor built-in antenna.

Several other features appear on some radios:

Choice of alarm sound. This gives you the option of waking to the sound of the radio or a buzzer or beeper. The buzzer volume can be controlled on a few models.

Alarm indicators. Often, a light goes on when the alarm is set to operate, and the alarm setting is displayed when you press a button. On many models, when you press the button, the alarm setting temporarily replaces the time setting on the display panel. But the alarm setting may have a separate display, so the time setting and the alarm setting are visible at the same time. Or there may not be an on-off light for the alarm; the only way to tell that the alarm is set to operate is to look at the position of the alarm switch.

On some models, you can bring several kinds of information to the display panel at the touch of a button, such as the alarm setting, the drowse period, and the automatic-shutoff period.

Alarm reset. Most models have a push button or a touch plate that performs two functions: It turns the alarm off and resets it for the next day. On several models, you have to work a slide switch to turn the alarm off, then remember to turn it on again for the next day.

Double alarm. You can set some alarms to go off at two different times—at seven and at eight o'clock, say. That's handy if an early riser and a late riser use the same clock.

Drowse or snooze switch. If you want to sleep a little longer after the radio comes on or the alarm sounds, you can hit the drowse control and the radio will be silent for a period, then come back on. You can repeat this cycle several times.

Many models have a drowse period that is fixed at about ten minutes.

Automatic shutoff switch. If you want to go to sleep with the radio on, you can set it to turn off automatically.

On most models, the shutoff period can be set to run from one to about 60 minutes.

Reversible clock-set. You usually set the clock and your alarm time by pressing a button until the right digits come up. On many models, the time runs forward only; if you overshoot the mark, you have to run through a whole day's series again. But on some models, you have the useful option of running the digits forward or backward.

Protection from missetting. With most models, it's pretty hard to change the settings accidentally. In some cases you can change a setting only by pressing two buttons at once. In others, you avoid missetting by placing the time-set switch in a neutral position.

Battery backup. A digital clock doesn't remember information when the household current goes off for even a few seconds; the time and alarm settings disappear. To prevent that, many clock radios have a backup power system—usually a small, 9-volt battery. If the household current goes off, the displays go dark but the clock will keep correct time and hold the alarm setting for several hours. The clock may lose or gain time when it's operating on the battery.

Unless the household current stays off for several hours, or unless you leave the unit unplugged for a long time with the battery installed, the battery may last about a year. On most models that have battery backup, you can check the battery strength by disconnecting the unit from the power line for a few minutes. If the display flashes and shows the wrong time when you plug the unit in again, that means you need a new battery.

A model without battery backup should at least let you know there's been a power failure. When the power comes back on, the display should flash.

Brightness control. On most models, you can adjust the display brightness.

Illuminated radio dial. Most models have an illuminated radio dial. A few have only an illuminated pointer; you can see the pointer move in the dark, but you can't see the numbers.

Special features

A few models have extra features:
- Terminals for connecting an external FM antenna
- An earphone jack
- A tone control
- A hidden calendar, displaying the month and the date when you press a button
- The ability to display the seconds part of the current time when you press a button
- Provision for wall mounting

Ratings of clock radios

(As published in an August 1983 report.)

Listed in groups in order of estimated quality based on FM performance; within groups, listed in order of useful features. The prices are suggested retail; + indicates shipping is extra. In the feature columns, ✓ means the model has the feature; x means it doesn't.

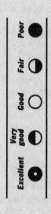

Legend: ⊙ Excellent ◕ Very good ○ Good ◑ Fair ● Poor

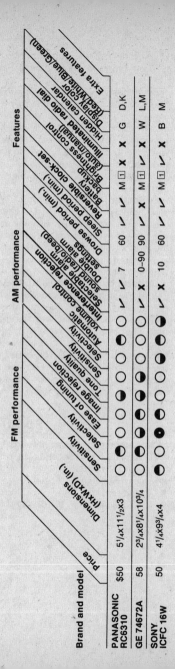

Brand and model	Price	Dimensions (HxWxD) (in.)	FM performance — Sensitivity	Selectivity	Ease of tuning	Image rejection	Tone quality	AM performance — Sensitivity	Selectivity	Automatic Volume control	Interference rejection	Selectable alarm sound (radio/beep)	Double alarm settings	Browse period (min.)	Sleep period (min.)	Reversible clock-set	Battery backup clock-set	Brightness control	Auto/manual illuminated radio dial	Hidden calendar	Display color (Red/White/Blue/Green)	Extra features
PANASONIC RC6310	$50	5¼x11½x3	○	○	○	◑	○	○	◑	○	○	✓	7		60	✓	✓	✓	M☐	x	x	G	D,K
GE 74672A	58	2⅜x8¼x10¾	○	○	◑	◑	○	○	○	○	○	✓	x	0-90	90	✓	x	✓	M☐	✓	x	W	L,M
SONY ICFC 16W	50	4¼x9¾x4	◑	◑	○	○	◑	⊙	◑	○	◑	✓	✓	10	60	✓	✓	✓	M☐	✓	x	B	M

Brand / Model	Price	Dimensions (in)	Ratings																		Extra features
EMERSON REL5200	50	3x10³/₄x5¹/₂	○	○	◐	◐	◐	◐	◐	✓	9	X	60	X	X	M	X	X	✓	X	R	B,L,N
ZENITH R451W	50	3³/₄x10¹/₂x7¹/₄	◐	◐	◐	◐	◐	◐	◐	X	9	X	60	X	X	X	X	X	✓	X	R	C,N
SEARS Cat. No. 23270	50+	3¹/₂x11¹/₂x6	◐	○	○	○	◐	○	○	✓	9	✓	60	✓	✓	A	✓	✓	X		R	D,F,G,L
LLOYDS J266B	40	3¹/₄x11³/₄x6	◐	○	◐	◐	◐	○	◐	✓	9	✓	60	X	✓	M	✓	✓	✓		R	—
SANYO RM6100	37	2¹/₄x10¹/₄x6¹/₄	◐	◐	●	●	◐	●	◐	X	9	✓	60	X	✓	M	✓	✓	X		R	A,N
MAGNAVOX D3210	54	2¹/₄x9¹/₂x6¹/₄	○	◐	◐	◐	●	●	◐	X	9	✓	60	X	✓	A	✓	✓	X		W	M
SOUNDESIGN 3728	53	3x10³/₄x8	○	○	◐	●	●	●	◐	✓	9	✓	60	X	X	M	X	✓	✓		R	K,N
TOSHIBA RCK1	50	2³/₄x10¹/₂x5¹/₂	○	◐	◐	◐	◐	◐	◐	✓	4	X	64	✓	X	M[1]	✓	✓	X		B	E,H,M
HITACHI KC659H	55	3¹/₄x11¹/₄x5¹/₂	○	◐	◐	◐	◐	◐	○	✓	9	X	60	X	X	M	✓	✓	X		G	J,N
WESTCLOX 80168	48	2³/₄x8x5	◐	◐	◐	◐	◐	◐	◐	✓	9	✓	60	X	✓	x[1]	X	V	R		B	
RADIO SHACK 121541	40	2¹/₂x10¹/₄x5	◐	◐	●	●	●	◐	◐	X	9	X	120	X	✓	M	X	X	X		R	A,D,I,N

SPECIFICATIONS AND FEATURES

All have: ● AM/FM reception. ● Built-in antenna. ● Power-failure indicator. ● Push-button or touch-plate drowse switch. ● Protection from accidental reset. ● Simple alarm reset switch. *Except as noted, all have:* Digits approximately half-inch high.

[1] *Digit display may be too bright.* (GE also shows bright light underneath.)

KEY TO EXTRA FEATURES

A – Has earphone jack.
B – Has extra-large digits.
C – Has terminals for external FM antenna.
D – Alarm works on backup battery.
E – Shows time and alarm settings at same time.
F – Has hidden time display during power failure; it's visible when you push a button.
G – If radio signal is too weak, buzzer sounds at wake-up time.
H – Has convenient push-button array for setting time and alarm.
I – Has strength indicator for backup battery.
J – Has tone control.
K – Has hidden seconds display; it's visible when you push a button.
L – Buzzer volume is controllable.
M – Has illuminated dial pointer.
N – Has slide switch for alarm reset.

Clothes dryers

A clothes dryer has to be adaptable to the needs of a variety of fabrics. Most cottons are sturdy and make modest demands. They can tolerate high drying temperatures, and they take the return to room temperature in stride.

Permanent-press items call for more pampering. To rid them of unwanted wrinkles requires a drying temperature of at least 150°F. But heating them to much more than 180° may damage the fibers or finish. Moreover, permanent-press fabrics insist on being cooled as they tumble-dry; otherwise they can start to wrinkle all over again. When the dryer stops and you remove them from the drum, permanent-press fabrics should be close to room temperature.

Delicate fabrics and synthetic knits are even more demanding. They need some heat, but 140° or under is best.

Plastic and rubberized items can't take any significant rise in temperature. You can speed up their drying time, however, by tumbling them in a dryer with the heat off. That may also serve to "air fluff" quilts and pillows.

A good dryer will be able to meet all those temperature demands. It will also be able to get a wash load precisely as dry as desired. Years ago, you had to guess how long a wash load would take to dry, set the dryer's timer accordingly, and hope for the best.

Modern dryers have a timer you can set, but they also have automatic cycles that should take the guesswork out of drying. The cycles usually work by means of a thermostat that measures moisture indirectly, by measuring the air temperature in or around the dryer's drum. When the laundry is dry enough for the air to rise to a certain temperature, the thermostat cycles the heating element or burner off. The dryer's controls advance through the automatic cycle only when the heater is off. After the control has registered a predetermined amount of heat-off time, it reaches the end of the heating portion of the cycle.

But some dryers measure moisture directly, inside the drum, with an electric sensor. A full-sized dryer should be able to handle a large laundry load, about 12 pounds when dry. The most deluxe machines have at least two maximum drying temperatures and a number of controls for various temperature settings and drying cycles.

Some of those controls are there just for show. For instance, some models have separate controls for different types of fabric. Though you might think each control delivers a "customized" drying condition suitable for each fabric, the controls often produce the same temperatures.

Performing

Consumers Union's tests measured performance in several ways.

Drying speed. Using a large load (10½ pounds dry, 16½ pounds wet) of cotton/polyester fabrics, a dryer should do the job in about 30 minutes, on average. Most electrics draw about 5,400 watts in operation. Gas dryers' fuel consumption, as stated by the manufacturers, ranges from 15,000 British thermal units per hour to 22,000 Btu/hour. Usually, the higher the Btu rating, the speedier the dryer—though only by minutes.

Though gas and electric models are equally fast on average, gas models generally cost only one-third as much to run. Computing the cost of electricity at 7.75 cents per kilowatt-hour, the national average, the cost for an average load would be a little more than 20 cents with any electric.

Figuring the cost of natural gas at 61.7 cents per therm (100,000 Btu, or about 100 cubic feet of gas) and using 7.75 cents per kwh for the dryer's electric motor and controls, the drying cost for a gas model should be about 7 cents per load.

Those figures are based on national averages, and your local energy rates may well be different. If so, you can figure out which type of dryer is cheaper to operate by remembering that gas and electric dryers cost the same to run when the cost per therm of gas is about 25 times the cost per kwh. If the cost per therm is less, gas drying is cheaper; if more, electricity is cheaper.

Permanent-press cool-down. Operated at the manufacturer's recommended setting, a modern dryer should take the wrinkles out of permanent-press fabrics. If your dryer doesn't, try operating it at a lower heat setting than that suggested by the manufacturer.

The dryers most effective at preventing wrinkles during cool-down are those that leave the fabrics at room temperature when the cycle is over. Those judged least effective leave fabrics warm to the touch at the end of the cycle. Inadequate cool-down may be a problem with large loads, which often need a longer-than-average cooling period.

Dryers generally follow the permanent-press cycle with a heatless tumble cycle. That's desirable, especially with models that provide a short cool-down, since extra tumbling improves the cooling of large loads and reduces the likelihood of wrinkling.

A dryer should let you operate the permanent-press cycle at any temperature you want. That's useful for fabrics that demand low-temperature drying, since it lets you use the added tumble cycle.

Dryness range. You shouldn't expect a dryer, however complex its automation, to produce precisely the desired dryness in load after load of wash. Too many variables come into play—different load sizes and different fabrics within a load, for instance. You may want a load just a touch damp for ironing or—with corduroy and quilted materials—to avoid puckering and shrinkage.

To deal with such variables, some models have controls that let you work within a moisture range at one or more automatic cycles.

Convenience

Most dryers have a rotary dial to control timed cycles. Different segments of the dial usually govern the regular and permanent-press cycles and other functions. The most convenient dials are those whose markings and dial positions are clear and unambiguous.

Instead of a rotary dial, some machines have an array of "membrane" switches. The switches hardly move at the press of a finger; a small light goes on above the pressed switch to show the cycle you've chosen. In some ways, membrane controls are less convenient to use than rotary dials. A rotary dial is continuously adjustable; it lets you set a drying cycle of any length you wish, up to the machine's maximum. With a membrane control, you may be unable to determine if the drying cycle is adequate if the lights above the membrane switches merely show the cycle chosen, not its progression.

An "air-fluff" or no-heat setting is useful for rubberized fabrics, fabrics that contain foam, or fabrics grown musty from storage. You use this setting in conjunction with timed drying. If operated at the no-heat setting on the automatic drying cycle, many dryers will keep running until you remember to turn them off.

Other factors contributing to convenience include:

Lint filter. Cleaning the filter after each load helps keep the dryer working at peak efficiency. A few machines have a whistle to warn of lint buildup. Some have a top-mounted filter—convenient, but when you pull it out, you can easily spill lint on top of the dryer. Other dryers have the filter in the doorway, forcing you to stoop.

Drum opening and door. Some dryers have a D-shaped drum opening, with the flat side of the D at the bottom; others have an O-shaped opening. The D shape supposedly keeps clothes from falling out when you unload the dryer.

Consumers Union's testers found that items occasionally dropped out of all drums, regardless of drum shape.

A door that opens downward into a horizontal position makes a handy shelf for unloading the wash but can get in the way when you're reaching into the rear of the drum.

A side-opening door should swing a full 180°. Otherwise you may find it getting in your way.

Lights. A light inside the drum is useful even in a well-lit laundry room.

Rack. A number of dryers have a rack for drying items unsuitable for tumble-drying, such as sneakers; the rack fits inside the dryer drum, but doesn't rotate.

On/off power switch. On all models, the cycle stops when you open the door, though it may take a few seconds, during which some of the laundry may fall out onto the floor if you open the door immediately.

Finish. A porcelain cabinet top is more resistant to scratches than the baked-enamel finish of most models, unfortunately.

Recommendations

First, you need to choose between electric and gas. You must buy an electric dryer if natural gas is not available to you.

You also need to make sure that the dryer you choose can be vented in the direction that suits your laundry room. (The hot, moist air is generally vented outside the house.)

If you have access to a supply of natural gas and if the dryer you're buying is not a replacement for an electric dryer, a gas dryer is probably a better choice than an electric model. Gas dryers tend to be higher priced by about $40 because they're a bit more complicated

than electrics as machines; even so, the savings in energy should easily make up for the price difference in the first year of ownership. If you are replacing an electric dryer, it may be most economical to buy another electric model, particularly if you would need to do extensive gas plumbing before buying the dryer.

An important factor to consider when choosing a dryer is a brand's Frequency-of-Repair record, which is based on reader response to Consumers Union's Annual Questionnaire. The repair record of most good brands is average or better.

Ratings of clothes dryers

(As published in an October 1984 report.)

Listed by types; within types, listed in order of estimated overall quality. Differences in quality between closely ranked models were slight. Prices are the average and range of retail prices quoted to CU shoppers in a 13-city survey for models in white.

Better ◐ → Worse ●

Dryness range

Brand and model	Average price	Price range	Dimensions (HxWxD), in.	Depth, door open, in.	Drying speed	Permanent-press	Cool-down effectiveness	Delicate-fabric temperature	Regular fabric	Permanent-press	Added tumble time, min.	Maximum timed cycle, min.	Control convenience	Brand Frequency of Repair	Advantages	Disadvantages	Comments
Electric models																	
WHIRLPOOL LE7800XM	$378	$329-410	43¼x29x28	41½	◐	◐	●	○	○	◐	40	60	○	I	F,J,L		B,E,F,H,K
SEARS KENMORE 65921	416	379-480	43x29x28	41¾	◐	●	○	○	○	○	40	50	○	A,C,F,J,K,L	—		A,B,E,F,H
WHIRLPOOL LE9800XM	447	380-500	43¼x29x28	41¾	◐	◐	○	○	○	◐	150	80	◐	E,F,J,L	h		A,B,D,E,F,H,K
GENERAL ELECTRIC DDE9200D	410	370-495	43½x27x28¼	53½	◐	◐	○	○	○	○	15	70	○	B,G,J	e,j,m		A,C,I,K
HOTPOINT DLB2880D	362	310-409	42½x27x28½	53½	◐	●	●	○	○	○	60	50	○	2 A,G,J	e,f,m		C,I,K

Brand and model	Average price	Price range	Dimensions (HxWxD), in.	Depth, door open, in.	Drying speed, door open, in.	permanent-press	cool-down effectiveness	delicate-fabric temperature	regular fabric	permanent-press	Added tumble time, min.	Maximum timed cycle, min.	Control convenience	Brand Frequency of Repair	Advantages	Disadvantages	Comments
SEARS KENMORE 65741	391	330-469	43x29x27¾	41¾	○	●	◐	○	●	○	30	50	◐	A,C,L	g,q		E,F,H
GENERAL ELECTRIC DDE8200D	385	350-449	43x27x28¼	53½	○	◐	◐	◐	●	◐	15	70	○	G,J	j,m		C,I,K
MAYTAG DE612	429	399-489	43x28½x28	47	○	◐	◐	●	●	◐	24	60	○	H,J	d,q		G,I
AMANA TEA800	392	348-465	43x27x29	47½	○	◐	◐	◐	●	◐	16	70	—	A,F,J	j,o,q		—
SPEED QUEEN HE7003	376	329-420	43x27x29	47½	○	◐	◐	◐	●	◐	16	70	—	A,F,J	j,o,q		—
FRIGIDAIRE DECIM	368	329-409	44x28¼x27¼	53	○	◐	◐	○	●	○	72	100	○	J	c,e,p		C
WARDS 7841	407	379-460	44x31x28	54	○	◐	◐	◐	●	○	60	110	◐	D,F,I	a,f,j		C,J
NORGE LDE9120	360	310-465	43x31x27¾	54	○	◐	◐	◐	●	◔	60	110	○	—	a,f,i,p,q		C,J
MAGIC CHEF YE20CN4	338	280-400	43x31x28	54	○	◐	◐	◐	●	◔	60	120	○	—	a,f,i,p,q		C,J
MAYTAG DE712	454	415-500	43x28½x28	47	○	◐	◐	●	●	◐	40	60	◐	H,J	d,q		A,G,I
WHITE-WESTINGHOUSE DE800E	365	315-445	43¾x27x26	48	○	◐	◐	○	●	○	30	60	◐	A,G,J	k,l,n		—
KELVINATOR DEA800A	351	300-439	43½x28¼x27¼	53	○	◐	◐	○	●	○	15	80	○	B,J	e,i		C
GIBSON DE28A6WP	341	299-375	43x28¼x27½	53	○	○	◐	○	●	○	0	90	○	—	b,c,e,i		C
ADMIRAL DE20B6	331	289-389	43x31x28	54	○	◐	◐	◔	●	○	50	110	—	—	a,f,i,p,q		C,J

Gas models

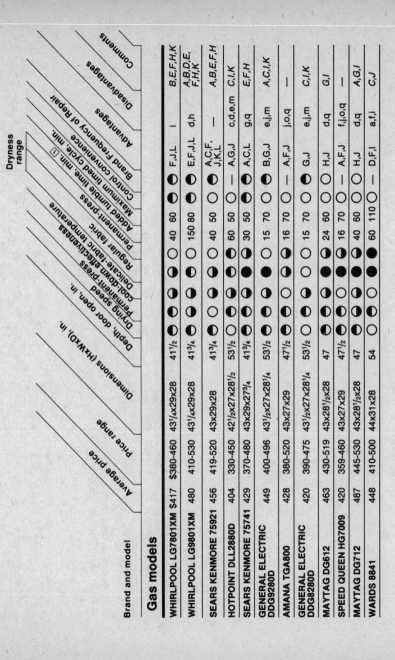

Brand and model	Average price	Price range	Dimensions (HxWxD), in.	Depth, door open, in.	Drying speed: Regular fabric	Drying speed: Permanent-press	Cool-down permanent-press	Delicate-fabric effectiveness	Regular fabric	Permanent-press	Added tumble time, min.	Maximum timed cycle, min. [1]	Control convenience	Brand Frequency of Repair	Advantages	Disadvantages	Comments
WHIRLPOOL LG7801XM	$417	$380–460	43¼x29x28	41½	●	◑	◑	◑	◑	○	40	60	○	◑	F,J,L	i	B,E,F,H,K
WHIRLPOOL LG9801XM	480	410–530	43¼x29x28	43¾	●	◑	◑	◑	◑	○	150	80	◑	◑	E,F,J,L	d,h	A,B,D,E,F,H,K
SEARS KENMORE 75921	456	419–520	43x29x28	43¾	○	◑	◑	○	○	○	40	50	○	○	A,C,F,J,K,L	—	A,B,E,F,H
HOTPOINT DLL2880D	404	330–450	42½x27x28½	53½	◑	◑	◑	◑	●	◑	60	50	—	◑	A,G,J	c,d,e,m	C,I,K
SEARS KENMORE 75741	429	370–480	43x29x27¾	41¾	◑	◑	◑	●	◑	●	30	50	●	◑	A,C,L	g,q	E,F,H
GENERAL ELECTRIC DDG9280D	449	400–496	43½x27x28¼	53½	◑	◑	●	◑	●	◑	15	70	○	◑	B,G,J	e,j,m	A,C,I,K
AMANA TGA800	428	380–520	43x27x29	47½	○	◑	◑	○	○	○	16	70	—	◑	A,F,J	j,o,q	—
GENERAL ELECTRIC DDG8280D	420	390–475	43½x27x28¼	53½	◑	◑	●	◑	●	○	15	70	○	◑	G,J	e,j,m	C,I,K
MAYTAG DG612	463	430–519	43x28½x28	47	◑	●	●	●	●	○	24	60	○	○	H,J	d,q	G,I
SPEED QUEEN HG7009	420	359–460	43x27x29	47½	○	◑	◑	●	◑	○	16	70	—	○	A,F,J	f,j,o,q	—
MAYTAG DG712	487	445–530	43x28½x28	47	◑	●	●	●	●	○	40	60	○	◑	H,J	d,q	A,G,I
WARDS 8841	448	410–500	44x31x28	54	○	○	◑	●	◑	○	60	110	—	●	D,F,I	a,f,i	C,J

Dryness range

Brand and model	Average price	Price range	Dimensions (HxWxD), in.	Depth, door open, in.	Drying speed	Permanent-press cool-down effectiveness	Delicate-fabric temperature	Regular fabric	Permanent-press	Added tumble-press	Maximum timed cycle, min.	Control convenience, min.	Brand Frequency of Repair	Dryness range	Advantages	Disadvantages	Comments
WHITE-WESTINGHOUSE DG800EX	399	354-498	43³/₄x27x26	48	◐	○	●	●	◐	◐	30	60	—	—	A,G,J	k,l,n	—
FRIGIDAIRE DGCIM	404	359-455	44x28¹/₄x27¹/₄	53	◐	◐	●	●	◐	○	72	100	J	—	f,p	C
NORGE LDG912	381	350-412	43x31x27³/₄	54	○	●	●	●	●	○	60	110	—	—	a,f,i,p,q	C,J
MAGIC CHEF YG20CN4	376	315-440	43x31x28	54	○	●	●	●	◐	○	60	120	—	—	a,f,i,p,q	C,J
KELVINATOR DGA800	410	350-475	43¹/₂x28¹/₄x27¹/₄	53	◐	◐	●	◐	◐	○	15	80	B,J	—	e,f,i	C
GIBSON DG28A6WPFA	375	354-409	43x28¹/₄x27¹/₂	53	◐	◐	◐	◐	○	◐	0	90	J	—	b,f,i	C
ADMIRAL DG20B6	365	329-412	43x31x28	54	○	◐	●	●	●	○	50	110	—	—	a,f,i,p,q	C,J

[1] At end of permanent-press automatic cycle.

[2] Brand was a ○ but fell to a ● in last two yr.

SPECIFICATIONS AND FEATURES

All have: ● Automatic drying cycles. ● 4 adjustable leveling legs.

Except as noted, all have: ● Continuously adjustable timed cycle. ● Satisfactory permanent-press cycle. ● "Air-fluff" or no-heat setting that coincides with maximum timed cycle. ● 3 drying temperatures. ● End-of-cycle signal (buzzer, beep or bell) that is not adjustable for loudness and/or does not have an on/off switch. ● Drum light. ● Round drum opening. ● Door that opens to right. ● Venting from rear, back of both sides, and bottom.

KEY TO ADVANTAGES

A – End-of-cycle signal is adjustable for volume.
B – End-of-cycle signal may be turned on or off.
C – Whistle indicates when lint filter is full.
D – Has lighted dial.
E – Has on/off power switch to stop drum before opening door (but flipping switch off will cancel cycle).
F – Comes with separate rack for drying items without tumbling.
G – Has porcelain-coated drum.
H – Has porcelain-coated cabinet top.
I – Has no-tumble control.
J – Automatic permanent-press cycle can run at any temperature.
K – Has hanger rack.
L – Drum has large oval opening.

KEY TO DISADVANTAGES

a – Has no low-temperature automatic cycle; must use timed cycle.
b – Has no added heatless tumble.
c – With permanent-press load, drying temperature judged somewhat high unless on low setting.
d – With small permanent-press load, drying temperature judged somewhat high unless on low setting.
e – With regular load, drying temperature judged somewhat high unless on lower setting.
f – If run on automatic cycle at no-heat setting, dryer will not shut off automatically.
g – Maximum time for no-heat setting only about 25 min., too short for some fabrics.
h – Timed cycle can be set only for 20, 30, 40, 60, or 80 min.
i – Has no drum light.
j – Drum light was very dim.
k – Door may be blocked by basket.
l – Door does not open a full 180 degrees.
m – Lint filter may fall out when clothes are loaded or unloaded.
n – Lint filter is not removable for cleaning.
o – Lint filter can be put in incorrectly, which prevents door from closing.
p – Cycle-selector dial turns only in 1 direction.
q – Has only 2 drying temperatures.

KEY TO COMMENTS

A – Moisture-sensing model.
B – During added tumble, drum turns for about 10 sec. every 5 min.
C – Drum opening has "D" shape.
D – Has solid-state (membrane) switches.
E – Lint filter accessible from top.
F – Door opens downward.
G – Door opens to left.
H – Vents from rear only.
I – Vents from rear, left, and bottom.
J – Vents from rear, middle right, and bottom.
K – Manufacturer offers special help for do-it-yourself repairs.

Coffee grinders

Once you open a can of vacuum-packed coffee, the contents begin to go stale in only a few days. The solution is to grind your own, as needed.

You can, of course, crush your coffee beans with a mortar and pestle, or you can use one of the old-fashioned hand-cranked grinders. You can even pulverize the beans in a blender, though it's not well suited to that task. But most people interested in grinding their own coffee turn to an electrical appliance designed just for the purpose.

Home coffee grinders come in two basic styles. One is typically cylindrical, with a pair of whirling blades in a little covered cup. Some manufacturers call this kind of grinder a mill, but it's really a "chopper." Grinders that send the beans between a pair of toothy disks are more aptly termed "mills."

Grinding

Different ways of making coffee demand grinds of different coarseness. If you boil your coffee, cowboy-style, the grind can be fairly coarse. You need progressively finer grinds for percolator, drip, and espresso coffee makers. Obviously, a good home coffee grinder should be able to grind either coarse or fine, depending on your needs.

Coffee labeled "drip grind" is supposed to be considerably finer than a percolator grind, but drip grinds are apt to be relatively coarse, and only slightly finer than the perc grinds.

Coarse-cut. Just about any grinder can produce a suitably coarse grind. So if you boil or percolate your coffee, the choices are broad.

Fine grind. A broad range of fineness allows you to tailor the grind to your particular brewing method and even to your own coffeepot.

Neither type of grinder—chopper nor mill—is superior in terms of fine grinding. But individual units do excel at one or another type of grind.

The finer a coffee bean is ground, the more flavor it can yield. You can use a smaller amount of a finer-than-usual grind without giving up flavor and aroma.

In tests using a good automatic drip coffee maker, and finely ground coffee, Consumers Union concluded that the coffee quality remained good until the recipe was reduced to below 4 grams (two level teaspoons) of grounds per cup. A very fine grind, then, can indeed make a pound of coffee go further.

When using less-than-normal amounts of coarser grinds, coffee quality will suffer. Can you save by regrinding already-ground coffee? Perhaps, but don't count on it.

Temperature

Brewed coffee owes its aroma and flavor to highly volatile substances that start to vaporize as soon as the beans are roasted. Heat accelerates those losses, so a grinder shouldn't warm the beans too much in the grinding process.

In general, a mill adds less heat than a chopper.

Convenience

A good mill is more convenient than a chopper.

A mill has a hopper that can hold and even store at least a quarter-pound of beans, or enough for about 20 cups. Most choppers can handle about 1½ to 3 ounces of beans at once, or enough for 8 to 15 cups of brew; the choppers can't be used to store beans. It's best to store only a small amount of beans in a grinder. Keep the rest of the beans in an airtight container in the freezer, so they stay fresh. Grind only enough beans for immediate use.

Mills have a control that adjusts the fineness of the grind. Choppers don't. To grind coffee in a typical mill, you pour in the beans, set the fineness control, and turn on the machine. The grounds are delivered directly into a removable receptacle.

A chopper will run only as long as you press down on its cover or on/off button. Up to a point, the longer a chopper runs, the finer it grinds. You should time the grinding in order to get the same degree of fineness batch after batch—though it's also possible to gauge the fineness of grind by listening for a change in pitch as the chopper whirrs away. And you must upend a chopper to pour out the ground coffee.

If you expect a good cup of coffee, keep your grinder clear of oily ground-coffee residues. Coffee oils go stale after only a few days' exposure to air and will turn rancid not long thereafter.

A grinder isn't especially easy to clean. A chopper, at least, presents no important obstacles once you get past the blades. Its chopping bowl is relatively shallow, with metal-faced surfaces that are reasonably easy to brush or wipe.

Safety

The on/off switch on a chopper is designed with an interlock: The machine won't run unless the cover is in place—an effective protection against contact with the spinning blades. With the safest units, it's especially hard to defeat the interlock, since two steps are needed to work them—you twist on the cover and flip the switch, or hold down the cover and flip the switch.

Hand-cranked coffee mills

These days, the hand-cranked coffee mill is usually kept around only for show. Still, manual grinders are capable of pulverizing coffee beans.

The hopper at the top of the **Catalina 84160** ($17) holds the beans. Ground coffee falls into the little drawer at the base (the drawer holds grounds for about eight cups of coffee).

The **Catalina** produces grounds finer than most brands of ground coffee, though coarser than an "extrafine" brand. And the hand-cranking barely raises the temperature of the ground coffee.

But grinding a drawerful of beans takes two minutes and leaves you tired. Adjusting the fineness setting is a guesswork procedure that involves turning a nut atop the grinder to change the spacing of the cutting disks inside. Cleaning the **Catalina** is none too easy, either.

If the **Catalina** were included in the Ratings on page 62, it would land just below the **Salton**. The hand-cranked mill is perfectly adequate as a coffee grinder, provided you have the stamina to operate it. Otherwise, you'd better use an electric.

Recommendations

Grinding beans yourself should give you a better cup of coffee—and can make a pound of coffee go further. Most coffee drinkers seem to brew by the drip method, which requires a fine grind. If you use a percolator, almost any grinder will serve.

A guide to coffee beans

The reason you grind your own beans is to improve the brew through the use of fresher grounds or by locating a particular bean or blend of beans that best suits your taste.

Coffee trees grow best in hilly regions in the world's tropical areas. The beans they produce are usually one of two main species—**coffea arabica** and **coffea robusta,** each with a characteristic shape, color, flavor, and aroma. (The beans, by the way, are virtually all gathered by hand; advertisements that boast of "hand picked" beans are meaningless.)

The robusta variety, grown mainly in Africa, India, and Indonesia, tends to have a harsh flavor. Robustas are used mainly in instant coffees and in blends of ready-ground coffees. They have about twice the caffeine content of arabicas. Many decaffeinated coffees are made from robusta beans, since the processing rids the bean of much of its natural harshness while providing a large harvest of caffeine for separate sale to makers of soft drinks and drug products, and to other commercial users.

Most beans found in specialty stores are arabicas. There are two main kinds, brazils and milds.

Brazils. As you might expect, these comprise any Brazil-grown beans. There are dozens of varieties, but the main ones take their names from the ports of export. **Santos** beans have a sweet, clear flavor. **Rios** are pungent. Other important kinds are called **Victorias** and **Paranas.**

Milds are those arabica coffees not grown in Brazil. They come from Central and South America, the Caribbean, the Middle and Far East, and parts of Africa. Among the varieties:

Arabian. Only Arabian-grown coffees bear the designation **Mocha.** A fine Mocha is heavy-bodied, with a smooth and delicious flavor and an unusual acid character.

Colombian. Experts consider these winy in body and fine in flavor and aroma. Colombian coffees you may encounter include **Supremo,** termed rich, mellow, full-bodied, and mild in aroma; **Colombian Espresso,** a deep, dark roast like Italian espresso; and **Colombian Amaretto, Cinnamon,** and **Swiss Chocolate Almond,** varieties of flavored coffee.

Haitian. These coffees are described as mellow and rich, with a heavy body and fairly high acidity.

Hawaiian. The principal Hawaiian bean is **Kona,** a variety of excellent quality, with mild acidity, distinct sharpness, and medium body.

Indonesian. Some robusta coffees are grown on the island of Java, but those sold in the U.S. as **Java** are traditionally arabicas. Another Indonesian, **Mandheling,** is grown in a district on Sumatra's west coast. Some experts term it one of the finest coffees in the world.

Tanzanian. Tanzania produces both arabicas and robustas. One popular arabica is **Tanzanian Kilimanjaro,** grown on the southern slopes of the Kilimanjaro range. The coffee is mild, aromatic, and rich, with a distinct sharpness.

In the store, you're likely to find blends of two types of beans, as well as one or more of the individual beans mentioned above. One of the most popular blends is **Mocha-Java**—the **Mocha** for flavor, the **Java** for body and aroma.

Ratings of coffee grinders

(As published in a January 1981 report.)

Listed in order of estimated overall quality; models judged approximately equal in overall quality are bracketed and listed alphabetically. Except as noted, all choppers can grind about 1½ to 3 oz. of beans at a time, enough (at 1 tbsp. per cup) for about 8 to 15 cups of coffee; all mills can hold and grind 4 oz. of beans, enough for 20 cups of coffee. Unless otherwise indicated, all grinders have a line cord about 3 to 4½ ft. long. Prices are suggested retail; + indicates shipping is extra.

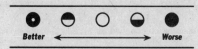

Better ←————————→ Worse

Brand and model	Price	Type	Fineness of grind	Convenience	Ease of cleaning	Advantages	Disadvantages	Comments
BRAUN KMM10	$55	Mill	◐	◐	○	A,B,D,E	—	A
BRAUN KSM2	24	Chopper	◐	○	○	—	—	—
KRUPS 202	26	Chopper	○	○	○	—	—	—
WARING 11CM10	24	Chopper	○	○	○	—	—	A,B
ZABAR'S Z75 ①	18+	Chopper	○	○	○	—	—	C
KRUPS 223	55	Mill	◐	◐	●	A,B	a,b	A,D
WARING 11CG10	40	Mill	◑	○	○	D,E	—	B
KRUPS 203	29	Chopper	◑	○	○	—	—	—
KITCHENAID KCM	60	Mill	○	○	●	C	c,d,e	E
SALTON GC4	22	Chopper	○	◑	○	—	h	A,B,E,F
■ *The following models were judged significantly lower in quality than those preceding.*								
CHEMEX ECG	40	Mill	◑	○	○	D,E	f	B,C
VARCO MX228	25	Chopper	◐	○	○	E	g	A,B

① Zabar's 2245 Broadway,
New York, NY 10024

KEY TO ADVANTAGES
A – Has timer-controlled dispenser with 12-cup maximum setting; dispenses about 4½ gm. (**Braun**) or 5½ gm. (**Krups**) of ground coffee per cup.
B – Hopper can store slightly more than ½ lb. of beans.
C – Hopper can store about 1¼ lb. of beans.
D – One grinding disk is removable to improve access for cleaning.
E – Heated beans less than most grinders of this type.

KEY TO DISADVANTAGES
a – Ground relatively fine at coarsest setting.
b – Fineness control is at back of unit and has hard-to-read markings, minor drawbacks.
c – Lacks shield over grinding mechanism, but depth of hopper makes accidental contact with grinding disks unlikely.
d – Must be upended for access to fineness control, a minor drawback.
e – Outlet chute for grounds, hidden behind hinged plate, tended to clog.

f – Downrated because outlet for ground coffee is large enough to allow a finger to contact moving parts; would otherwise have ranked with **Krups 223** and **Waring 11CG10**.
g – Downrated because cover could be opened about 1¼ in. at one side, exposing spinning blades.
h – Can grind beans for only about 4 cups of coffee at a time. Can hold enough beans for about 8 cups of coffee, but then can produce only a coarse grind.

KEY TO COMMENTS
A – Line cord is about 5 to 6 ft. long.
B – Claimed able to grind dry foods such as grains, seeds, and nuts.
C – Relatively heavy (4 lb.).
D – Comes with cleaning brush and wall bracket.
E – Comes with coffee measure.
F – Requires two hands to operate, a minor inconvenience, but a slight safety advantage.

Coffee makers

Coffee making consists of dissolving the right amount of the roasted and ground coffee bean in the right amount of hot water. Dissolve too much and the coffee will be overextracted; it will taste bitter regardless of the amount of water in the brew. If not enough of the ground coffee dissolves, the taste will be underdeveloped, again regardless of the amount of water.

But people's tastes in coffee are highly variable. There are still those who like their brew boiled in the pot, the way coffee was made on the prairie. That's the way coffee was made just about everywhere until a hundred years ago. By today's standards, the drink was probably too harsh, too bitter, and—if the grounds didn't settle—unpleasantly chewy.

There are many ways to go wrong, of course, on the path toward the proper balance. Overextracted coffee can result from brewing too long. Underdeveloped coffee can be produced by too short a brewing time, or from water that isn't hot enough. Too much or too little water yields coffee that's too watery or too rich. Overcooking coffee that's already brewed produces a burned taste. And, of course, you can vary your results by experimenting with different brands and grinds of coffee.

Electric coffee makers are supposed to take care of all those problems except for the selection of coffee.

The percolator, invented in the last century, bettered the coffee drinker's lot somewhat. The brew could still turn out harsh, bitter, or overcooked, but the percolator kept most of the grounds out of the liquid.

Drip-style brewing refined coffee making even further. The simplest glass drip brewers, such as the **Melitta** and the **Chemex,** consist of nothing more than a cone to hold a paper filter and grounds, and a carafe to catch the brew. You simply boil water and pour it over the grounds. Water and coffee meet only once in a drip brewer, yielding a drink its proponents swear by. And there are other, less widely used coffee-extracting techniques with equally ardent adherents.

Predictably, someone figured out a way to automate drip brewing, so you could put cold water into one end of a machine, flip a switch, and get hot coffee out the other end.

An automatic drip coffee maker isn't necessarily any faster or more convenient than a percolator or a simple cone-and-carafe brewer. Coffee and water must still be measured out, soggy grounds discarded, and various parts cleaned up.

The essential difference between percolator and drip: In a drip coffee maker, hot water passes through the ground coffee only once; a percolator circulates its brew continually through the ground coffee. A heating element surrounding the well at the bottom of a percolator boils the water in the well, causing steam-vapor bubbles that force the hot water up the pump tube. The water seeps through the ground coffee and goes back into the reservoir, where it is reheated in the well and recirculated. The process continues until all the brew reaches a temperature near 200°F. A thermostat then shuts off the main heating element. A second element on most percolators keeps the coffee warm.

Brewing

Which is better: drip or perk? A lot seems to depend on what you're accustomed to. Judging from sales, the drip coffee maker has taken a convincing lead over all others in the marketplace: half of all households own one. Evidently, a lot of people prefer the drip method. For one thing, dripped coffee tastes better than percolated coffee and an automatic drip coffee maker can make coffee fast and keep it hot.

An automatic drip coffee maker—or ADC, as it's abbreviated in the trade and on coffee cans—has six basic components: a tank for fresh water, a device for heating the water, a basket to hold a filter and ground coffee, a carafe to catch the liquid coffee as it drips, a thermostatic control system, and a hot plate to keep the carafe of coffee at drinking temperature. All coffee makers use disposable filters; some can also use a permanent filter. Several have a clock that can be set to start your coffee the next morning.

If you ask a sensory expert to compare coffee that's been perked with coffee made in an automatic drip machine, the expert will most likely confirm that just about any drip maker produces coffee that tastes better than perked or instant coffee. The expert taster will also tell you that the best coffee tends to be very full-bodied and fragrant. However, that doesn't mean the coffee is very strong. Coffee that's too strong is full of sourness, bitterness, and astringency, which overwhelms its more delicate flavors.

Most automatic drip models brew at the rate of about one cup per minute, starting with cold water as recommended by manufacturers. You can't make instant coffee much faster than that.

A few models are a little slower, brewing one cup every 75 to 100 seconds. If a new machine, or your old one, for that matter, doesn't live up to expectations, be patient. Your recipe could need adjustment, particularly if you grind your own beans (see pages 60–61).

Going by the cup marks on the coffee makers, the standard cup is about five ounces, and the manufacturer's recommended recipe is usually one tablespoon of ground coffee per cup (drip-grind, which is a little finer than regular grind). This conflicts with the recipe on most coffee cans—one tablespoon per *six*-ounce cup. You might also keep in mind that when a coffee maker is claimed to have 12-cup capacity, it will probably produce only about 10 six-ounce cups.

Convenience

These coffee makers are variations on the same design theme. The top of the unit holds the water reservoir and the heating elements. The carafe nestles in a niche in the body of the unit and rests on a warming plate. When you shop for a coffee maker, here are the main things to look at:

Adding water. The best models have a large, easy-to-reach opening for the water reservoir. Look for a deep well for the water; CU testers found that shallow water reservoirs tend to cause splattering and are easier to overfill.

On most machines, you can use the coffee carafe as a water pitcher to fill the reservoir because water doesn't begin to flow over the grounds until it's hot.

Most models have a lid for the water

reservoir. It's a nice touch.

The basket. Some baskets that hold ground coffee fit into the mouth of the carafe, and some slide into a holder below the opening where the hot water comes out. With a slide-in basket, you can take the carafe off the hot plate and pour coffee without disturbing the basket.

Most baskets have only one or two parts. Some have three. The fewer the parts the better.

Some models have an extra minibasket for making small amounts of coffee. The idea is to confine the ground coffee to a small space so it will be thoroughly saturated.

The filters. All models can work with a disposable paper filter. Most filters cost about $1 per 100 to 200. Some models require filters that cost about $4 to $5 per 100.

A few models also have a permanent plastic filter. According to the instructions for some, filter baskets can be used with or without an extra paper filter. Coffee made either way will taste equally good. But if you use a finer grind without the paper filter the brew may contain some sediment.

A few drip models come with another kind of filter—an activated-charcoal filter that "screens out impurities for richer-tasting coffee," as one description says. An activated-charcoal filter won't screen out the minerals that make hard water taste bad, and the manufacturers of coffee makers don't claim that it will. However, activated charcoal can remove other substances that affect the taste of water. Such filters work best if the water flows through them very slowly.

The carafe. Most models come with a glass carafe. A few are insulated-glass inside, plastic outside. With the hot plate turned off, an insulated container keeps coffee warm for a few hours.

Some carafes have a bulging shape, so you have to turn them almost upside down to pour the last drop of coffee from the carafe with the lid on.

You can use the carafe to fill the tank, but be cautious because some carafes have volume marks that aren't very accurate.

Head starts. With most machines, you can fill the water reservoir at any time, then flick a switch or plug in the machine when you want coffee. That allows you to set up the coffee maker early (before dinner guests arrive, say), then brew when you're ready. It also allows you to rig up a timer, so you can get the machine ready at night and have fresh coffee waiting for you in the morning.

The hot plate. Most hot plates are recessed to center the carafe on the hot plate. The recess also helps hold the carafe securely in place.

Generally, the hot plates kept coffee between 160° and 190° F, which is satisfactory.

The longer the brewed coffee stays on the hot plate, the harsher its flavor gets. Some experts feel that coffee left on a hot plate for as little as an hour will have deteriorated quite noticeably.

A clock. Several models have a built-in clock that will trigger the brewing process. Most of the clocks are digital and can be set to trigger at exactly the minute you want.

Although a clock is handy, remember that you'll have to fill the tank and the basket ahead of time. And if ground coffee stays in the basket overnight, it loses some of its freshness; the coffee you brew in the morning won't be as flavorful as it could be.

Models with a clock cost about $15 to $25 more than similar models without a clock.

Switches and lights. Nearly all models have an on-off or reset switch and a pilot light to show when the machine is in operation.

Dimensions. Most of the models are between 10 and 13½ inches tall. A few are a little taller—perhaps too tall to fit under an overhanging cupboard.

Cleanup. The carafe and brew basket should be cleaned after every use, because dried-up coffee oils can ruin the taste of even the best blend. Soapy water will usually do the trick.

Cleaning. Coffee-maker housings are usually made of smooth plastic and are easy to wipe with a sponge.

Most models have a lid that completely covers the tank. You can open the lid to fill the tank or to see whether it needs cleaning. On some that don't have a lid, the design of the tank makes it hard to see inside. Because minerals accumulate in the tank and the tubes, it's very

important to clean the inside of your coffee maker every so often, especially if you have hard water. If you live in a hard-water area and use your coffee maker every day, be sure to clean it at least once a month with white vinegar. It's a chore, but it's worth the trouble. Follow the manufacturer's instructions.

Safety. Coffee makers shouldn't present any shock hazard in a normal, dry environment, or even if water is accidentally spilled on the hot plate.

The temperature under the hot plate, at the counter surface, is invariably less than 200°—not likely to harm any common type of countertop material over the short term. It's hard to say what might happen after months or years of use. A heat pad under the unit will prevent any damage.

Recommendations

If you only drink instant or percolated coffee, you'll probably find that coffee produced by an automatic drip machine is a big improvement.

But electric drip coffee makers don't have a monopoly on good-tasting brew. The nonelectric **Melitta** can make coffee better than the brew from most of

the electrics. Any nonelectric drip coffee maker might make a worthwhile alternative to an electric coffee maker, especially if you're short on counter space, or if you want a decent cup of fresh coffee without spending a lot for the brewing machinery.

Percolators

Although an electric percolator does not seem to produce as high a quality brew as an automatic drip coffee maker, a percolator still has some desirable qualities. It's automatic, it's relatively cheap, it's compact, and it can be easily stored when not in use. And if you want a party-size coffee maker, a percolator is the reasonable way to go.

Here are some features to consider:

Repercolation. Be sure the percolator has a keep-warm element. If it

doesn't, you must remove the pump and basket as soon as the initial percolation is over.

Cleanability. Any leftover residue from previous brewings will degrade the coffee's flavor, so a scrupulously clean pot and interior components are a must. The easiest pots to clean are those made of glass or ceramic. Next come stainless-steel pots and components, which are easier to keep clean than aluminum or plastic.

The size of the top opening, the shape and depth of the heating well, the pump and basket design, the presence or absence of residue-holding grooves, abrupt or gently curved shoulders at the top, and the spout design—all those things affect the ease with which a percolator can be cleaned. A spoutless pot or one with a short, wide spout is easier to keep up to snuff than one with a long narrow spout. (For the latter you need a brush designed for the purpose.) If the pot is immersible, that makes cleaning less of a chore.

Hazards. Some metal pots have sharp edges inside the top that could scrape your hand during cleaning. So be careful.

Top locks. Electric percolators are supposed to have a top that will stay on should the pot be knocked over accidentally. Some are not as effective as others.

Strength controls. Many percolators have a "strength control." The controls work by adjusting the thermostat and, thus, the perking time. Set at "strong," the control permits the brew to reach its maximum temperature, dissolving and extracting at the percolator's maximum capability. Turned to the "weak" side, the control lowers the thermostat's shutoff temperature, which reduces the temperature of the final brew and the brewing time. Since percolators sometimes tend to underextract, a condition that would be made worse by reducing the brewing time, start out with the control set to the strongest position. If

that setting turns out to make coffee too strong for your taste, putting less coffee in the basket will reduce its strength.

Signals. Some percolators have signal lights that flash on when the perking is over, an advantage if there is no plastic or glass knob that shows the coffee perking, or if you can't tell by listening.

Sediment and clarity. A clear cup of coffee is preferable to one cloudy with sediment or with a few grounds swirling around as you pour and stir. A few percolators have filters either built into the basket or as separate units that you fit into the spout. Wetting the coffee basket before loading in the coffee helps to prevent dry coffee particles from falling through the basket into the water.

Markings. Cup markings on the pot tell you how much water to add for the desired amount of coffee. Some are easier to read than others; a few are just marks that don't indicate the number of cups. Some percolators have convenient water-level view tubes outside the pot.

If you like to routinely remove the grounds basket as soon as the pot has finished perking, you will appreciate a pot that makes the operation convenient. With some, the basket-pump assembly and the lid lock together, so you can carry lid and all to the sink. Press a release button and the basket drops off so that the lid is free to go back onto the pot. Others have a plastic-tipped pump that shouldn't be too hot to hold as you lift out the pump and basket.

Ratings of automatic drip coffee makers

(As published in a March 1983 report.)

Listed in three groups separated by rules, based on quality of brewed coffee as judged by CU's sensory consultants. Within groups, models are ranked primarily by convenience. Closely ranked models differed little in quality of brewed coffee; models within brackets were judged approximately equal and are listed alphabetically. Prices are suggested retail; + indicates that shipping is extra.

Brand and models	Price	Claimed capacity (cups)	Advantages	Disadvantages	Comments
KRUPS BREWMASTER 261A	$75	10	C	r	A,C
J.C. PENNEY Cat. No. 3543	55+	12	C,F,H	i	G,H,I,J
NORELCO READY BREW II HB5192B	67	12	A,C,F,I	l	I
SEARS COUNTERCRAFT Cat. No. 67552	33+	12	C,F,H	i	G,H,J
SEARS COUNTERCRAFT Cat. No. 67952	50+	12	C,F,H	i	G,H,I,J
GENERAL ELECTRIC BREW STARTER II DCM14	49	10	C	—	C,I
NORELCO DIAL-A-BREW HB5185	39	10	A,F,I	l,o,s	G
BRAUN KF35	80	12	I	a,c,q,r,s	A
GENERAL ELECTRIC DCM10	37	10	C,I	—	—
GENERAL ELECTRIC DIGITAL BREW STARTER DCM50	67	10	C,I	—	I
PROCTOR SILEX BEVERAGE BREWER II A510W	73	12	C,F	—	I
J.C. PENNEY Cat. No. 2966	27+	10	F,H,L	i,l,o	—
KRUPS AROMA SUPER 268	95	12	C,E,G,I	c,r	—
OSTER ROWENTA	93	8	C,D,K	l,r,s	C,F
FARBERWARE 236	90	10	C,L	j	A,C,I
MELITTA ACM10A	45	10	C,I,L	e,f,l,r	—
PROCTOR SILEX A415AL	36	10	I	l,o,s	E
WARDS 45424	22	10	C,E,H	m	C,G

Brand and models	Price	Claimed capacity (cups)	Advantages	Disadvantages	Comments
CHEMEX CA2	72	10	I,J	f,l,m,r	C,D
FARBERWARE 265C	55	12	A	f,t	C
MELITTA AROMA PLUS ACM10B	55	10	C,I	a,e,r	—
MR. COFFEE CM-1	35	10	E,I	k,l	—
MR. COFFEE CMX-1000	64	12	I	k,l	A,I
MR. COFFEE CBS-900	50	10	A,I,L	c,k,l,o,p	G
WEST BEND FLAVO DRIP 5977	50	10	I	f,g	—
KRUPS KAFFEE-TEE AUTOMAT 245	90	8	C,K	a,c,d,r	A,C,J
BUNN POUR-OMATIC GR (B-8)	60	8	B,I	b,e,g,h,l,o	B,J
HAMILTON BEACH 791	34	12	C	c,e,h,n,s	C

SPECIFICATIONS AND FEATURES
Except as noted, all are 10 to 13½ in. high and have: ● Brewing heater that turns off automatically. ● Uninsulated glass carafe. ● Coffee basket that mounts on brewing unit, not on carafe. ● On/off switch. ● Capacity within one 5-oz. cup of maximum amount marked on carafe or tank. ● 2 to 3 cup claimed minimum capacity.

KEY TO ADVANTAGES
A – Slightly faster than most.
B – Fastest of all, if reserve water supply kept hot.
C – Tank has cup markings.
D – Tank is removable for easy filling at sink.
E – Has 1-cup minimum capacity.
F – Has coffee-strength control.
G – Has check-valve to stop drip when carafe is removed.
H – Can be used without paper filters.
 I – One-piece basket or no basket—easy to clean.
J – Hot-plate temperature can be varied.
K – Insulated carafe keeps coffee warm for a few hours.
L – Produced somewhat better balanced or more aromatic coffee than most in the same quality group.

KEY TO DISADVANTAGES
a – Somewhat slower than most.
b – Starting with cold water, takes approx. 30 min. to heat water in tank.
c – Cup markings on tank or carafe not very accurate.
d – Spout in middle makes it hard to fill tank.
e – No cup marks on carafe.
 f – Minimum capacity 4 cups.
g – No on/off switch. **Bunn** switch is for hot plate only.
h – No "on" indicator. **Bunn** light indicates only that hot plate is on.
 i – Basket somewhat inconvenient to assemble. **Penneys 2966** least convenient.
j – Hot-plate temperature too high; close to boiling.
k – No hot-plate recess to center carafe under basket.
 l – Carafe hard to empty completely.
m – Carafe lid has to be removed when pouring; otherwise it falls off.
n – Steamed more than others.
o – Tank not completely covered, hard to see if clean inside.
p – Must be filled at beginning of brew cycle—not before—and then manually switched from "brew" to "warm" at end of cycle.

q – Basket easy to knock out of place.
r – Filters relatively expensive.
s – Coffee quality varies depending how many cups are brewed at a time.
t – Housing difficult to clean.

KEY TO COMMENTS

A – Claimed minimum capacity unclear; tested at 2-cup minimum.
B – Comes with pitcher for filling tank.
C – Basket mounts on or is part of carafe.
D – Manufacturer recommends using regular ground coffee, but tests showed drip and regular grind produced similar coffee.
E – Has red spot on slide switch instead of pilot light.
F – Has reset switch instead of on/off switch.
G – Has extra "mini-brew" coffee basket or adjustable basket for brewing small amounts.
H – Has charcoal filter for water.
I – Has built-in clock which can be set to turn machine on.
J – 14 to 15 in. high.

Slow cookers

For centuries, people have cooked stews and other dishes slowly, over low heat. But the pot has had to be tended. Now along come electric slow crockery cookers, which can supposedly cook in 6 to 12 hours almost any dish that requires liquid—and no attention from the cook necessary.

The modern version of the old pot is quite good. It will safely cook a meal while you're at work or while you sleep. At parties, it can be used to keep food warm. During hot weather, it will cook without adding much heat to your kitchen. A good slow cooker, we think, can be a handy kitchen aid.

But slow cooking sometimes requires some adjustments on the cook's part. With most slow cookers, there's virtually no evaporation during cooking, and sauces or gravies may emerge more watery than you'd prefer. Beef cooked at low temperatures may have a pink cast rather than the "done" look of browned meat. And some other foods—such as milk products, pasta dishes, and soft-flesh fish—just won't stand up to the slow-cook process. Such difficulties can be circumvented. You can thicken a watery sauce by adding flour or by boiling it down in another pot just before serving. You can brown beef before adding it to the cooker. And you can add milk products, pasta, and such during the last stage of cooking.

Still, is there a real need for yet another electric appliance? Isn't it really enough to use a large pot on a range top and turn the heat down? Yes and no. It's not a very good idea to let a pot go unattended for long periods on a gas range because of the possibility that the low flame might blow out, leaving a dangerous gas leak. With an electric range success would depend on just how low the heat can be set. If you have an electric range and a large, tightly sealed heavy pot, it can work. But a good slow cooker is likely to work better since it will give you low heat, a tight seal, and the kind of heat conductivity that helps to ensure against burning.

The typical slow cooker is about nine inches high and about nine inches in diameter—with metal or plastic outer shell and a stoneware liner. But there are a lot of variations. Some covers are transparent, some not. Some cookers are oblong or oval instead of cylindrical. Liners may be of aluminum, glass, or steel rather than stoneware, and may or may not be removable for cleaning. Capacities vary too.

But the basic difference, so far as cooking performance goes, is whether the cooker is a **continuous-heat** unit or a **thermostatic** unit, which cycles the heat on and off during the cooking process.

Cooking

You might think a cooker that lets you regulate heat up or down over a wide range would be more desirable than a model that provides only one or two continuous heats. This doesn't appear to be so, we discovered when we cooked with the cookers.

A beef stew recipe is included in one form or another in virtually every model's recipe book. One thing is certain: a slow cooker can soften up relatively cheap, tough cuts of meat.

Continuous heat. A continuous-heat cooker takes from five to seven and a half hours to turn out beef stew. On low, the same chore takes anywhere from 10 to 15 hours or more. That can put you up against recipes that require an impractical amount of cooking time. That is, unless the cooker has an extra "automatic shift" position that delivers high heat for about two hours, then switches to low.

Thermostatic. These may be very unpredictable. Some models cook considerably faster than their recipe books would suggest and others considerably slower. But almost all can be adjusted to turn out a stew in cooking times comparable to those of the continuous-heat models.

In general, finding the right control settings for the various dishes you want to cook takes some experimentation.

Nutrition. There are claims that slow cooking is better than range-top or oven cooking because the higher temperatures often involved in the latter methods allow a greater loss of nutrients. While it's true that some nutrients are destroyed by high heat, other nutrients can be lost because of the lengthy cooking times often required with slow cookery. And with some foods, nutrient loss can occur even at lower temperatures. Of course, water-soluble nutrients that are "lost" simply pass from the food into the surrounding liquid and are recovered if you consume the liquid along with the food—as you would with stew. But there's no hard evidence that slow cooking always means more nutritious cooking.

Convenience

Recipes. Though slow cooking requires special recipes, the books that come with some models don't offer many recipes, But you can always buy a slow-cookery cookbook.

Plugging them in. If you buy a thermostatic model, be aware that some are rated for electrical draws of about 1,500 watts. That just about monopolizes a 15-ampere branch circuit that powers the cooker—a disadvantage in view of the long cooking time involved.

Shape and size. A problem can occur with some cuts of meat in any slow cooker that's shallow or very wide in diameter: If the liquid doesn't cover enough of the meat, part of the meat may not get done as well as you'd like it unless it is turned over during cooking.

Setup. An important advantage for slow cooking shows up at food-preparation time. When making a stew, for example, ingredients can be made ready and put into the cooker at once. Then, whether the cooker is on low or high, all you have to do is to remember to check the dish for doneness as the expected finishing time nears. The result is that you are freed from having to wait around pot-watching, stirring, and adding ingredients that require less time than the meat to cook.

Serving. Look for a model with a detachable cord if you'll use your cooker as a serving dish at the dinner table. Many of the removable liners can be used alone as serving dishes, but note that their handles get considerably hot-ter than the handles of the outer shells.

Food storage. All-metal and metal/plastic cookers let you put cooked food directly into the refrigerator, cooker and all. When they later emerge from the cold, those cookers can also be turned on immediately. You can't do the same with cookers that have glass or stoneware liners, since too-sudden temperature changes may crack the liners.

Cleaning. Washup should prove easiest with the models whose liners are removable. Those liners can be fully immersed in water or, once any adhering material is loosened, put into a dishwasher. When you're washing most of the other models, you have to be careful not to dunk their bottoms and cords in water. Most of the manufacturers suggest that you avoid abrasive cleaners or steel wool in favor of cloths, sponges, or plastic scrubbers.

Energy

If you cooked stew as a casserole in the oven of an electric range, you'd expend about double the energy required by a continuous-heat cooker, whether set on high or low. But the *top* of an electric range is apt to handle the chore using the same amount of energy—or even less—than that required by most slow cookers. In general, continuous-heat cookers use about one-quarter less energy than the thermostat models.

Safety

The lower a cooking temperature, the longer the time needed to cook food. Low cooking temperatures over a long period can pose a health hazard.

Bacteria grow rapidly in foods held more than three or four hours between 60° and 120°F. Some may still grow, though more slowly, at 120° to 140°. Even if the food eventually gets hot enough to kill the bacteria, the heat won't destroy the toxin some bacteria leave behind, and that toxin could make you ill.

The threat of trichinosis from undercooked pork, or even beef, is also a consideration. A cooking temperature of 140° is needed to kill trichinosis parasites, too. So cookers' heating rates and holding levels have to be high enough to prevent such problems from developing. Continuous-heat units shouldn't cause concern; all heat food to well above 140° in a sufficiently short time and keep it there even at a low setting.

But health problems might arise with a thermostatic model that has keep-warm settings below its lowest slow-cook setting. The thermostat can be tricky to adjust; a very small change in position can produce a rather large change in the cooker's temperature. The result could be a temperature that spurs bacterial growth. In using a thermostatic model, start by following its instructions closely. If you find you need longer cooking times, lower the control setting only a little at a time.

To qualify for listing by Underwriters Laboratories, a cooker must meet specifications concerned with overheating under abnormal conditions (a unit that has run dry and a thermostat that has broken down and kept the heating element on continuously for seven to eight hours).

There is, however, some chance of a fire hazard with certain units, if you misuse them. Some models come with very short cords. That reduces the chance of a child's tugging on the cord

and dumping the hot ingredients. However, a short cord may fail to reach a convenient electric outlet. If you then use an undersized extension cord (not possible with the low-wattage continuous-heat cookers, but very possible with the higher-wattage thermostatic cookers), there's a good chance it will overheat dangerously during cooking.

If you need an extension cord, make sure to get one that has sufficient capacity. To calculate the maximum current capacity you need, divide the cooker's wattage by 120 (a 1,600-watt cooker would require 13$\frac{1}{3}$ amps).

When cooking, most cookers will prove distinctly uncomfortable if you grasp their lid or casings, which heat to at least 130° or so, even at a low setting.

The handles of all the cookers, however, will give you reasonable protection; you can grasp any of them comfortably with your bare hands. You may need a potholder, though, if you lift the cover when the cooker is set at high temperature.

Recommendations

Most people should be pleased with a continuous-heat cooker rather than a thermostatically controlled one. Continuous-heat models offer a choice of two cooking temperatures, which should do for most of your slow-cooking recipes. You can reasonably choose among them on the basis of the capacity

you need, the completeness of their recipe books, or a good discount. Cookers with thermostatic controls often also claim to serve for such chores as roasting, deep frying, or regular cooking. But with those models it can be more difficult to predict cooking times.

A unique electric cooker

The **Ultra Chef** is supposed to be a one-pot, no-stir, electronic appliance for the cook who has everything but time. The rotating glass cylinder, the ads say, browns, sautees, fries. What's more, foods are claimed to retain more of their nutrients and flavor. And clean-up is said to take minutes.

Cooking

With most dishes, you don't have to stir the food; the rotating cylinder does that. But you must still do all the initial preparation—washing, trimming, chopping.

As for speed, beef Stroganoff takes only about twenty minutes, for example.

But the food is not usually gourmet fare. For one thing, recipes in the 208-page **Ultra Chef** cookbook often call for canned stock, soup, or bouillon cubes—all of which can lend a processed-food flavor to the end product.

The food is not always cooked to perfection, either. Of the twenty or so recipes tried by Consumers Union testers, half turned out very good to excellent, half only fair to good.

The beef Stroganoff is judged excellent. The meat is melt-in-the-mouth tender. All ingredients go in at one time, yet the cream is not overcooked.

Stuffed peppers do not fall apart in the rotating cylinder. The peppers are underdone, but the stuffing is nicely cooked.

After the recommended two hours of cooking, a pot roast is fork tender only at its narrow end; the rest is somewhat chewy.

The beef in a stew is a bit chewy. And the stew seems to lack the richness that develops with longer cooking.

Pork chops taste all right—but they cooked to a safe 180°F.

Fried chicken is inconvenient to prepare. You have to heat the oil twice in order to cook a cut-up 2½-pound broiler/fryer in two batches. It takes more than an hour—two and a half times longer than a batch made in an electric cooker/fryer.

Chicken curry, shrimp in garlic butter, asparagus in lemon butter sauce, and corn on the cob will turn out very good. But potatoes au gratin, rotating around in the cylinder, won't develop a golden crust. And cheese fondue may come out stringy.

Vanilla pudding is very good. Chocolate mousse suffers from chunks of unmelted chocolate. Caramels are delicious, but getting the hot goo out of the cylinder is a tricky, two-person job.

Handling

Setting up the **Ultra Chef** is easy. But because of its weight (about seventeen pounds) and bulk, you'll probably want to leave it set up on your countertop all the time.

Filling the cylinder with food and programming it to cook is easy, too. You put in the food, set the cylinder in the base, fit on the metal lid, lock it in place with a swing arm, and insert a temperature probe. You tip the cylinder to its cooking angle, set temperature and time with touch pads on the base, then flick a switch to start the machine, which purrs quietly as it revolves.

When the cooking is done, things get more challenging. The directions say to remove the cumbersome, handleless cylinder from the base before taking off the lid. But the cylinder is very, very hot.

Wearing bulky oven mitts, it's not easy to grasp the half-inch lip of the cylinder. And a slip could mean a big mess.

Next, you must remove the lid. That sometimes requires a firm tug. If vapor lock occurs, removing the lid may be a two-person task—one person to hold down the cylinder, the other to struggle with the lid. Even when the lid is easy to open, food may spatter on the counter. And you have to be careful not to get scalded by escaping steam when opening the lid.

Ladling out soup or stew is easy. But you must really scrape out the last of the stew. For that, again, you may need at least three hands: one mitted pair to tip the cylinder, the third to scrape out the food. There's no spout to help make the job a neat one.

It's easier—and probably safer—to remove the lid from the cylinder and ladle out the food while the cylinder remains upright in the base, although the company says it is better to remove the cylinder from the base before removing the lid.

Cleaning is easy. Once the cylinder cools, you rinse it out and put it on the bottom rack of the dishwasher for washing. Assuming there's no sticky food clinging to the cylinder's lid, it will come clean on the dishwasher's top rack—though you must unscrew the knob to clean out food underneath.

The rest of the unit can be wiped clean with a damp cloth after it cools.

Recommendations

The overall quality of food prepared in this expensive appliance is not likely to be superior. Recipes in the cookbook call for too many processed-food ingredients. Handling the hot, cumbersome appliance is a lot harder than working with ordinary pots and pans.

Dehumidifiers

High humidity indoors makes you uncomfortable, and damages possessions. High humidity warps wood, rusts metal, and makes cold-water pipes drip. It encourages the growth of mildew on shoes, clothes, books, and carpets, and on shower curtains, between bathroom tiles, and elsewhere. Mildew spores are a common cause of allergies. And mildew smells.

Merely improving ventilation can solve many humidity problems, especially if the high humidity comes from the "weather" inside the house rather than from the weather outside. Humidity caused by cooking or by showers and wet towels can be reduced with ventilating fans. Adding a vent or louvered doors to storerooms or closets may help prevent mildew from growing. A poorly vented clothes dryer is another cause of excess humidity that is easily corrected.

Basements or rooms partly below ground level are particularly susceptible to dampness. Condensation of warm, humid air as it enters the relatively cool basement is one cause of a damp basement. When the warm, humid air is cooled, it can no longer hold as much water vapor. The resulting "dew" can make a basement feel—and smell—like a dank cave.

Improving the basement's ventilation by installing a fan or by opening doors and windows can reduce condensation. If dripping cold-water pipes are the problem, wrap them with insulation.

Moisture in the ground is another cause of dampness in a house, one not always easy to fix. In a crawl space or basement with a dirt floor, covering the dirt with plastic sheeting (or installing insulation with a vapor barrier) may help keep moisture out of the house. If you have a comparatively dry basement, a dehumidifier may help it to be completely dry. But if moisture is introduced into the dehumidified space substantially faster than the dehumidifier can take it out, even plugging in the dehumidifier won't help. Major moisture problems are usually caused by leaks or seepage of groundwater and are better dealt with through major measures, such as improved drainage, a sump pump, or basement waterproofing.

Basically, if you've established that the moisture is caused mainly by condensation, and the area you're concerned about is not enormous, a dehumidifier can probably do the job of reducing your excess moisture to a level safer for possessions and more pleasant for people.

Drying out

An electric dehumidifier is a refrigeration machine. The heart of the unit is a coil of chilled pipe. A fan circulates air over the coil, where moisture condenses and runs off into a container or into a hose connected to a drain. Now the air is not only dry but chilled. The absorbed heat gets pumped to the condenser coils. Up to this point, the dehumidifier operates like an air conditioner. But where the air conditioner would now return the dry, chilled air to the room and dump the heat load outdoors, in the dehumidifier that heat is picked up again by the chilled air as it passes over the condenser coil, and it is dry, warm air that returns to the room. In fact, the air is warmer than it was in the first place; heat is given up by the condensing water vapor, and heat dissipated by the compressor and fan motors of the unit itself produce heat.

You may feel more comfortable even though the temperature has not been reduced, because the rate of evaporation of perspiration is increased by drier air. There are situations in which the added heat tends to wipe out the comfort advantage of lowered humidity, though not its mold-inhibiting, rust-preventing benefits. A theoretical situation illustrates the trade-off: In a sealed room 20 × 15 × 7 feet, at 80° F and 60 percent relative humidity, using a dehumidifier to lower humidity to 40 percent would raise the temperature to about 90° F.

Therefore, a dehumidifier does not supplement an air conditioner. Even in situations where a dehumidifier may save an air conditioner from dealing with a portion of the moisture load, it gives the air conditioner an extra heat load to contend with.

What capacity?

Among dehumidifiers, the major variable is capacity: the amount of moisture a given model can take out of the air during a day. Dehumidifiers are sold that claim to extract only 12 pints of water a day, or over 48 pints. Large capacity or small, dehumidifiers are typically the size of a large stereo loudspeaker.

The capacity you need is difficult to predict. It depends on the size of the area to be dehumidified and on the amount of dampness. The Association of Home Appliance Manufacturers (AHAM) says that a model with a capacity of 25 pints a day should be able to handle most dampness problems. It may be to your advantage to consider an even larger model. You can always turn a high-capacity model down if need be, but you can't turn an undersized unit up beyond its capacity.

Wringing out the air

Most manufacturers base the capacity claims for their dehumidifiers on the results of a standardized test recommended by AHAM. The test checks performance when the temperature is 80° F and the relative humidity is 60 percent.

Consumers Union engineers duplicated those conditions in an environmental chamber, a heavily insulated room in which the temperature and humidity can be regulated. Most models extract at least as much moisture as claimed.

The AHAM test doesn't tell you everything about a dehumidifier's performance, however. So CU tried each unit under more typical conditions—70° and 70 percent relative humidity, a likely situation in a basement. With most models, performance under those conditions will drop off a bit.

There was another test at 90° and 50 percent relative humidity, an extreme situation that you might find in, say, the ground floor of a building without a basement. (Don't be misled by that lowered humidity percentage. Because hot air can hold more moisture than cool air, the air in this test is actually considerably more moist than in the other two.)

How about low-temperature operation? Most manufacturers tell you not to run a dehumidifier in an area cooler than about 65°. If you do, the coils may ice over and stop condensing water.

Finally, CU ran the units in 90° heat and simulated a brownout by reducing voltage gradually from the normal 120 volts to 100 volts.

Expensive to run?

Comparative running costs of large-capacity units don't vary greatly from one model to another, as the following examples show.

A dehumidifier is generally needed fewer times of the year in New England, along the Pacific Coast, and in the Plains states. There, an average dehumidifier would cost about $31 a year to run, assuming 700 hours of use and the national average electricity rate of 7.75 cents per kilowatt-hour. The least efficient units would cost about $33 a year to remove as much moisture as the average model: the most efficient units would cost about $29 a year.

In the Gulf states, by contrast, a dehumidifier is a virtual necessity for most of the summer. Annual operating costs will be higher, but differences in operating cost from model to model will still be fairly small. An average unit would need to run about 2,800 hours, for an annual cost of about $124 at the average electricity rate. The least efficient units would cost about $133 a year to run; the most efficient ones, about $117.

In other parts of the country, operating costs will fall between those two extremes. Your actual costs, of course, will depend on your local climate, your electricity rate, and the size of the room to be dehumidified.

How convenient?

On all the machines, a dial at or near the top turns the dehumidifier on and off and adjusts the humidistat, a device that senses humidity much the way that a thermostat senses temperature. With the dial turned all the way up, the machine will run continuously. Set the dial lower and the machine will run only long enough to bring the humidity down to whatever level you've set. At that point, the compressor and fan will cycle off until room humidity builds up again.

A switch on the **Comfort-Aire** and the **Dayton** lets the fan run even when the compressor is off. Air then continues to circulate through the machine, allowing the humidistat to monitor the room's

humidity level more closely. In addition, those two models have more than one fan speed. At lower speed, the units are quieter, although they may be slower at removing moisture from the air.

As each dehumidifier works, moisture from the air condenses on the cooling coils and drips into a water container. When the container is full, the machine shuts off and a signal light prods you to empty it. You'll usually have to do that chore once or twice a day, depending on how humid the air is.

All the containers are plastic. They're generally deep, with a fairly wide opening that makes them easy to clean but not convenient to carry. The container on the **Wards,** the **Comfort-Aire,** and the **Dayton** has a handle and a small opening that prevents water from sloshing out. The container on those three is the easiest to remove, lift, and carry to a sink. The small opening, however, poses problems in cleaning—a necessity now and then to prevent molds and bacteria from breeding.

The container on the **Sears** is hard to remove from the cabinet. (The two **Oasis** models have a somewhat similar failing, compounded by clumsy handles that make the container rather difficult to carry.) The **Sears** is designed to remain off as long as the container is out of the machine. The others can cycle on while you're emptying or cleaning the container. That can leave puddles of water inside the cabinet or on the floor. It's easy to replace all the containers. But several demand a bit of care; if they are put back incorrectly, the switch that prevents the container from overflowing can be bypassed. The consequence would be a messy overflow.

You won't need to cope with the water container if there's a floor drain in the room you want to dehumidify. Four machines have an open base, so you can remove the water container and place the dehumidifier over the drain. And you can connect a hose to any of these units to lead water to a drain. A few models have a hose coupling that you must cut open if you intend to use it. If you don't cut cleanly, the coupling may leak.

None of these dehumidifiers are meant to be positioned against a wall, because it will impede the intake of humid air through the back. The instructions for a few models actually suggest that you put the machine in the center of the room—awkward in a cramped basement. Other recommendations range from about six inches from the wall to some indefinite location in which airflow will not be restricted.

Most models are mounted on three or four casters. Four models are particularly mobile; all their casters swivel. Three units aren't especially easy to move around; they have only two casters.

To maintain the performance of your dehumidifier, vacuum or dust its coils at least once a year. Most of the models also need an annual touch of oil on the bearings of the fan motor.

Recommendations

The **Sears 5040** is among the most efficient. It has only one minor inconvenience, a water container that is hard to remove. The **Whirlpool AD0402XMO** is similar to the **Sears** and performs as well, but it has only two casters, so it's not easy to move about. The $386 **Oasis OD3800L** might be worth considering in really sultry climates.

The $329 **Comfort-Aire FDHD41** and the $364 **Dayton 3H324** have some particularly convenient features, including handy controls and a water container that's easy to remove and replace. But they have poor low-voltage performance. You need not rule those two out, however, if heat waves and brownouts are rare in your area.

Ratings of dehumidifiers

(As published in an August 1984 report.)

Listed in order of estimated overall quality. Models judged approximately equal in overall quality are bracketed and listed in alphabetical order. Performance evaluations were made with each model set to its highest water-extraction rate. Evaluations are based on actual amount of water removed without regard to stated capacity. Dimensions are rounded to nearest 1/2 in. and include any projection of water container. Prices are suggested retail; + indicates shipping is extra.

	Excellent	Very good	Good	Fair	Poor
	●	◖	○	◗	●

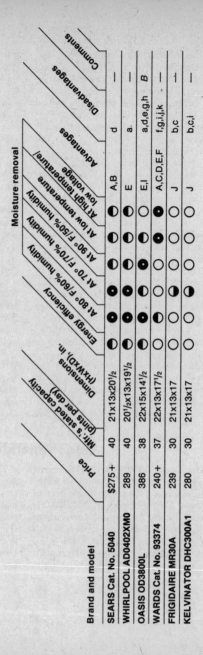

Brand and model	Price	Mfr.'s stated capacity (pints per day)	Dimensions (HxWxD), in.	Energy efficiency	Moisture removal — At 80° F/60% humidity	At 70° F/70% humidity	At 90° F/50% humidity	At low temperature	At high temperature/ low voltage	Advantages	Disadvantages	Comments
SEARS Cat. No. 5040	$275+	40	21x13x20½	◖	●	●	◖	◖	◖	A,B	d	—
WHIRLPOOL AD0402XM0	289	40	20½x13x19½	●	●	◖	◖	◖	◖	E	a	—
OASIS OD3800L	386	38	22x15x14½	◖	●	◖	○	○	○	E,I	a,d,e,g,h	B
WARDS Cat. No. 93374	240+	37	22x13x17½	○	◖	●	●	●	◖	A,C,D,E,F	f,g,i,j,k	*
FRIGIDAIRE MR30A	239	30	21x13x17	○	○	○	○	○	○	J	b,c	†
KELVINATOR DHC300A1	280	30	21x13x17	○	◖	○	○	○	○	J	b,c,i	—

Brand & Model									Advantages	Disadvantages	Comments
J.C. PENNEY Cat. No. 1173	275+	35	20½x13x17	◑	◐	●	○	○	J	b,c,i	—
WHITE WESTINGHOUSE ED358G	310	35	20½x13x17	◑	◐	●	○	○	J	b,c,i	—

■ *The following models were downrated because of poor operation at high temperature/low voltage or at low temperature.*

Brand & Model									Advantages	Disadvantages	Comments
COMFORT-AIRE FDHD41	329	41	23½x13x18½	○	●	●	◐	○	A,C,D,E,F,G,H	f,j,k	A
DAYTON 3H324	364	41	23x13x18½	○	●	●	◐	○	A,C,D,E,F,G,H	f,j,k	A
OASIS OD3800	359	38	22x15x14½	○	◐	●	◐	●	E,I	a,d,e,g,h,i	—

SPECIFICATIONS AND FEATURES

All have: ● Adjustable humidistat. ● Separate water container. ● Provision for disposing of water through a hose. ● Automatic shutoff to prevent overflow. ● 6- to 8-amp current draw, which may increase under extreme conditions. ● Signal light to indicate container is full.

Except as noted, all have: 6-ft. power cord.

KEY TO ADVANTAGES

A – 3 or 4 swiveling casters; unit is especially easy to move about.
B – Unit won't operate unless water container is properly installed; relatively unlikely to drip.
C – Water container more convenient to remove than most.
D – Water container has handle; easy to carry without spilling.
E – Water container easy to position properly.
F – Relatively convenient top-mounted controls.
G – Separate switch lets unit be turned on or off without disturbing humidistat setting.
H – Has 2-speed fan; quieter at low speed.
I – Relatively long (8½-ft.) power cord.
J – Open area in base allows water to flow directly into floor drain; no need to empty container or attach drain hose.

KEY TO DISADVANTAGES

a – Has only 2 casters; more difficult to maneuver than most.
b – Has 3 casters, but only one swivels; somewhat difficult to maneuver.
c – Water container may appear properly positioned when it is actually bypassing automatic shutoff.
d – Water container harder to remove than most.
e – Inconvenient handles made water container somewhat difficult to carry.
f – Water container was difficult to clean.
g – Automatic shutoff operated by plastic float; judged not as durable as most.
h – Signal light not as bright as others.
i – Humidistat not as well marked as others.
j – Sharp edges at top rear of cabinet made unit difficult and hazardous to carry.
k – Hose coupling must be cut open to use; if cut isn't clean, coupling may leak.

KEY TO COMMENTS

A – Fan can run when compressor is off.
B – Separate control at back; used when operating unit at low temperature.

Dishwashers

An automatic dishwashing machine won't necessarily clean your dishes any better than if you wash them carefully by hand, and you shouldn't count on a machine to cut down contagion in your household. What a dishwasher will do is make life easier, and establish a consistent level of dish cleanliness. Nevertheless, in these days of escalating energy costs, you might refrain from dishwasher use. But the fact is that some dishwashing machines can match or better a human dishwasher in saving energy.

If you are accustomed to washing dishes under a steadily flowing stream of hot water, you may be using even more hot water and therefore more energy, than if you wash dishes in a machine.

The first decision to make in buying a dishwasher is whether to buy a model that's built in under the counter, or a freestanding "convertible/portable." Mechanically, the two are generally the same: Convertible/portable models, however, have a finished exterior, wheels, a plug-in power cord, and hoses you attach to the sink faucet each time you use the machine. Convertible/portable models save you the expense of installation (although they can eventually be built-in) and they add to counter space. However, they cost somewhat more than built-ins and aren't as convenient.

Most people choose a built-in. The standard size fits a space that's about two feet wide and two feet deep under the normal counter height of 35 inches.

To install a dishwasher, you'll almost certainly need to hire a plumber. You may also need a carpenter and an electrician. If you're handy and local building codes permit you to do the work, you may be able to install the dishwasher yourself. (Replacing a machine should be easier—and less expensive—than installing one for the first time.)

The next choices you have to make are which brand to buy and how fancy it should be.

How deluxe?

The part that varies the most from dishwasher to dishwasher is the controls. In general, the more expensive a machine in a particular line, the more buttons—and the more cycles—it has. Machines at the very top of many lines have electronic touch switches rather than buttons.

Models at the next level down in the product lines have lots of cycles, too, but they're controlled by a row of push buttons. Buy a cheaper model still, and you get fewer cycles, which you select by pushing a button and turning a dial.

Another part likely to show variation within a line is the racks. All dishwashers have two rolling racks of plastic-coated wire, with room for a flatware basket or two. The more expensive models usually have a more convenient rack arrangement—sometimes an adjustable upper rack, or a provision for double-stacking cups, or an extra flatware basket. Occasionally, a basic model offers one of those conveniences as an option.

Often, however, most of the parts are pretty much the same throughout a manufacturer's line: the same basic cabinet, insulated to muffle noise and to prevent heat loss; the same wash arms, which protrude into the cabinet and spray water on the dishes; the same heating element, which usually heats both the wash water and the drying air; and the same pumps and filters, which circulate water and dispose of the water and soft food residue. Since those are the parts that most affect performance, a manufacturer's deluxe dishwasher may work no better than one of its more basic models. And if you buy the deluxe machine, you may be buying more convenience than you need. The difference in price between a deluxe model and a basic one can be $100 or more.

Cycles

Offering a lot of cycles is one of the main ways a company tries to sell its more expensive models. Those cycles often have names such as "pot scrubber," "soak and scrub," and "china/crystal." The names may lead you to believe that a dishwasher can wash all kinds of dishes, freeing you entirely from the chore. But that's not quite so.

A dishwasher cycle is basically a combination of washes and rinses. The typical "normal" cycle comprises two washes and two or three rinses. Then the heating element in the bottom of the cabinet dries the dishes.

A manufacturer can create other cycles in several ways. To make a "heavy" or "pots and pans" cycle, a manufacturer might extend the wash periods, add a third wash period, make the water hotter, or combine those effects. A "light" cycle typically has one wash period instead of two. In a "china/crystal" cycle, water may shoot through the wash arms less forcefully.

A manufacturer can also increase the number of "cycles" by including a couple of optional phases and counting each cycle more than once. Or it can take a part of the normal sequence of events and call it another cycle. Dishwashers generally have a couple of nonwashing cycles, for instance. One is "rinse and hold," which rinses the dishes and then stops. A "plate warmer" cycle merely turns on the heating element.

Machines with full push-button controls have a button for each cycle, plus an indicator to show what the machine is doing. On machines with dials, you have to press the "normal" button and turn the dial past the "pots and pans" or "heavy" segment for the machine to start its normal cycle. On some machines, you have to dial through various cycles—with the machine starting and stopping—to reach a "light" or "rinse and hold" cycle.

A dishwasher can't always substitute for elbow grease. A machine run on its heavy cycle is likely to clean pots and pans little better than it does on its regular cycle. But on either cycle, the machine will probably still leave the really burned-on stuff for you to scrub off afterward; heated drying may even bake the food on harder. Loading a dishwasher with lots of pots and pans and baking dishes also uses the space inefficiently.

A china/crystal cycle might seem useful for someone who gives a lot of formal dinner parties. But you should think twice before subjecting good crystal or china to possible jostling within the dishwasher.

You should need only two or three cycles—normal, heavy, and perhaps light—to wash all the dishes you can reasonably expect a machine to wash.

Saving energy

Energy-saving features are another way dishwashers are promoted. But the features may not save as much energy as you might expect.

Every dishwasher takes a certain amount of electricity to run the pump, motor, and heating element, but it's not much—less than one kilowatt-hour per cycle. That's less than 7 cents' worth of electricity at the national average rate of 7.75 cents per kwh, or about $20 a year.

Most of the energy a dishwasher uses is in the form of hot water from your water heater. The cost of heating that water depends on the kind of water heater you have, utility rates, and the amount of water the dishwasher uses. Water heated electrically is usually more expensive than water heated with gas or oil.

Some of the dishwashers tested by Consumers Union use 12 gallons of water per normal cycle. At average utility rates, supplying that hot water would cost about $65 a year for electrically heated water, about $25 a year for water heated with gas. But other machines use only 8 gallons per cycle, which means that hot-water costs would drop to about $43 a year (electric) or $17 a year (gas). Choosing the most water-efficient dishwasher, then, can mean a difference of a couple of dollars a month in operating costs—and saves nearly 1,300 gallons of water in the bargain.

Almost all dishwashers allow you to dry the dishes without the heating element—an option that actually saves a little money. Most dishwashers use only about 0.3 kwh to dry a load of dishes—about 2.3 cents' worth of electricity at the average cost per kwh. Over the year, foregoing heated drying would therefore save about $7. (The trade-off is letting your dishes sit overnight to dry

completely. If you want quicker results, open the door a bit after the last rinse; the load should then dry in an hour or two unless the weather is muggy.)

So if you buy the dishwasher that uses water most efficiently and use it without heated drying, you can expect to save roughly $15 to $30 a year, depending on the kind of water you have. But there are still more ways to save on running a dishwasher.

Many people have turned down the thermostat on their water heater from 140° to 120°F to reduce the cost of maintaining hot water. The yearly saving, assuming average utility rates and water usage, is about $30 for an electric water heater and $20 for gas (oil-heated water usually comes straight from the boiler and isn't stored).

But dishwashers generally don't work very well with a water supply that's cooler than 140°. Consequently, some companies have adapted their machines to work with cooler water. The heating element inside the dishwasher heats the water for at least one wash period until it's hot enough to clean dishes. It's cheaper to keep the water heater set for 120° and heat only a few gallons to 140° than to run the dishwasher only on 140° water.

Again, the saving is greatest if you have an electric water heater—2 to 4 cents per cycle or about $6 to $12 a year, plus the $30 or so you'd save on water-storage costs. With a gas water heater, however, the saving made on the hot water used by the dishwasher is likely to be offset by the increased electricity needed to heat the water, so the total saving per year would probably be just the $20 you'd realize by turning down the water heater's thermostat.

If you include the saving on hot-water costs, the amount of money you can save by choosing a particular dishwasher can be quite substantial. Over the life of a machine, the saving would be enough to make an expensive but efficient machine a better buy than a cheaper, less efficient dishwasher.

Consumers Union's tests showed no performance penalty for the saving, either; dishwashers that work with 120° water wash about as well with cooler water as they do with 140° water.

There are still further savings possible. Most dishwashers have a shorter cycle that uses less water. By using one of those cycles all the time, you can save about $10 a year with an electric water heater at the national average electricity rate.

But there's sometimes a connection between how much hot water a dishwasher uses and how well the dishes get washed.

Washing dishes

Consumers Union engineers determined the performance of dishwashers by using a tough test, simulating dirty dishes of a family of ten with hearty appetites. They loaded plates with beef stew, mashed potatoes, spaghetti, spinach, egg, butter, and other foods. They soiled bowls with chicken soup, cereal and milk, or stewed tomatoes. They poured milk, tomato juice, and orange juice into glasses and coffee into cups and saucers, and they dipped spoons, forks, and knives into the various foods. Then they let the dishes sit for an hour, scraped them lightly, and left them in the dishwashers overnight. The machines were turned on the next morning, using the setting the manufacturer had indicated as the everyday setting.

Each machine ran through this test several times; dishwashers designed to cope with 120° water were tested with both 120° and 140° water. Some machines were run at cycles other than the normal one. After each cycle, every item in the dishwasher was examined for cleanliness.

A good modern dishwasher should be able to clean unrinsed dishes without problem (after you've scraped off bones and other hard debris, of course).

Most of the dishwashers did a good job of drying the dishes. Some models have a blower to dispel water vapor more quickly, but it was no guarantee that the dishes would be well dried.

To use a dishwasher most efficiently, experiment with the cycles. If you habitually rinse the dishes or wash them right away, a light or low-energy wash may be all you need.

Living with a dishwasher

One of the main complaints about dishwashers is the noise they make. The way a dishwasher is installed affects the way it resonates. The use of insulating (sound-absorbing) material can make a perceptible difference.

Another important part of living with a dishwasher is its ease of loading and unloading.

Racks should move smoothly even when weighed down by dishes.

Most dishwashers hold 12 full place settings of dishes but a few hold only 11 place settings.

Most dish racks are designed for cups and glasses on top, plates on the bottom. On some machines, you have to watch where you put oversized items to avoid blocking the upper wash arm. A machine should be able to accommodate plates 10½ inches in diameter, and even large platters at the edges of the lower rack.

In some, though, larger dishes in certain positions may get in the way of the wash arm that spins just under the upper basket. If your plates are oversized or oddly shaped, take one along when you shop for the dishwasher to check clearances.

Dishwashers whose upper rack can be raised or lowered an inch or two are best at handling outsized plates or very tall glasses. Some upper racks can also be tilted, which helps water drain from the bottoms of mugs or glasses. Many machines have a rack that folds down or a special arrangement of supports, so you can double-stack cups. Other machines have a removable or adjustable set of props.

A nicety found on some machines is a covered basket or part of a basket for small, light objects—corn-on-the-cob holders or baby bottle nipples—that can be thrown around the cabinet by the strong spray of water.

Another nicety—a rinse-conditioner dispenser—could be a necessity if you live where the water is hard and leaves spots as it dries. A rinse-conditioning agent, which helps water run off glasses, should improve the appearance of glasses washed in hard water.

Every now and then someone starts the dishwasher only to find that the wrong cycle has been selected. You can stop any dishwasher after it's started and restart it, but that's easiest on the less expensive, dial-controlled machines. Usually, all you do is unlatch the door, change the controls, crank the dial to its proper starting point, and latch the door. On machines with push-button controls, you press the "cancel" button and wait anywhere from 15 seconds to a minute or two.

You can interrupt a cycle to load a dish you've overlooked or to rearrange the load if dishes start banging together—simply by unlatching the door and waiting a moment before opening the machine. When you latch the door again, the cycle picks up where it left off.

If you do have to interrupt a cycle, be aware that the heating element under the lower rack can be very hot, especially if the dishwasher was in the heated drying phase. If you do have to reach into that area during a cycle, or after the machine has shut down—to retrieve an item that's dropped to the tub bottom—be sure to let things cool off.

Other features

A feature that's common on dishwashers with electronic controls is a delayed-start timer. That feature probably shouldn't be used to run the dishwasher when you're away. It might come in handy if your household is chronically short of hot water, or it might save you a few dollars a year if your utility company charges a lower electricity rate during off-peak hours.

On some models you can boost the water temperature. On a few machines, this is similar to the automatic boost some machines offer so they can use 120° water as a matter of course. The feature is sometimes called "assured wash temperature," and it ensures that the water used in at least one wash of a cycle is hot enough to clean the dishes. That might be useful if your hot-water supply is unreliable.

Sometimes, however, the feature is

presented as a health benefit, with a name such as "sani-cycle," "sani-steam," or "sani-scrub." On machines with those cycles, the final rinse cycle uses extrahot water. A few brochures talk about "protection you'll appreciate during the flu and cold season" and "ex-tra health protection." Such talk is misleading, and the feature is a waste of money to buy and use. Once you put "sanitized" dishes into the cupboard, they're repopulated with household microbes, which are also on everything else in the house.

Dishwasher reliability

According to the U.S. Department of Energy, the average dishwasher has a life span of about twelve years. Responses to a Consumers Union Annual Questionnaire give you some idea of how reliable a given brand will be during its life, especially in the first few years.

According to the questionnaire's respondents, **GE, Hotpoint,** and **Whirlpool** dishwashers have been much more reliable than average. On an annual basis, only about 10 percent of dishwashers with those brand names needed repair in their first few years.

Kitchen Aid and **Maytag** had only an average repair record. **Sears, Magic Chef,** and **Waste King** models have been worse than average, and **White-Westinghouse** models were much worse than average. About 20 percent of the dishwashers with below-average records needed repair in the first few years.

Recommendations

First of all, try to choose a machine that's known to be excellent. Second, opt for one of the more efficient machines. Sometimes you can find great differences in price in one area, so shop around.

Ratings of dishwashers
(As published in an August 1983 report.)

Listed by groups in order of overall washing ability; within groups, ranked by energy-efficiency and convenience factors. Scores are based on performance with 140°F water and, except as noted, at cycle recommended by manufacturer for normal use. For models that can use 120° water, performance with cooler water was about the same and energy efficiency improved. Price range is as quoted to CU shoppers; + indicates that shipping is extra. ✔ means yes and ✗ means no.

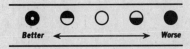

Ratings of dishwashers. Listed in order of estimated quality based on washing ability, energy efficiency, and convenience.

Brand and model	Approximate retail	Range	Washing ability [1]	Energy efficiency [1]	Heated drying [1]	Noise	Time, min. [1]	Water consumption, gal. [1]	Can accept 120° water	Push button / All push button/dial	Cycles [2]	Controls	Advantages	Disadvantages	Comments
GENERAL ELECTRIC GSD 1200T	$525	$480 to $698	●	○	●	79	10½	✓	✗	✓	5	A,H,I,M	a	A,D	
KITCHEN AID KDS-20	650	550 to 759	◐	◐	●	69	8	✓	✓	✗	3	B,D,H,J,O	b,f	A,B,D	
KITCHEN AID KDI-20	570	495 to 639	◐	◐	●	73	9	✓	✓	✗	3	B,P	f	A,G	
MAYTAG WU301	525	479 to 570	③	●	○	66	8½	✗	✓	✓	3	A,G,I,P	—	F,G	
MAYTAG WU901	615	549 to 680	③	●	●	66	8½	✗	✓	✓	3	A,G,I,P	b	B,D,F,I	
WHIRLPOOL DU7900XL	465	390 to 529	○	◐	●	55	9	✗	✓	✗	4	D,H,K,N	—	B,E	
HOTPOINT HDA 865	415	355 to 530	○	○	●	76	9½	✓	✓	✗	3+	—	—	A	
MAGIC CHEF DU45	349	259 to 375	○	○	○	55	11	✗	✓	✗	3	H	—	B	
TAPPAN 61-1341	378	299 to 520	○	○	○	55	11	✗	✓	✗	3	H	—	B	
THERMADOR THD3500	600	510 to 682	○	◐	○	92	12	✓	✓	✗	2	C,E,I,L	c,d,e	A,C,H,I	
WASTE KING WSK 3300	600	485 to 685	○	◐	○	92	12	✓	✓	✗	2	C,E,I,L	c,d,e	A,C,H,I	
WARDS 932	440+	350 to 460	●	◐	○	64	9	✗	✓	✗	3+	D,F,G,I	c,e	B	
FRIGIDAIRE DWU44J	549	360 to 519	◐	○	○	64	9	✗	✓	✗	3+	F,H	c	B	

1. *Measured with 140° water at cycle recommended by manufacturer for normal use.*

2. *Number indicates how many separate washing cycles are available; + indicates that portions of cycles can be selected by advancing dial.*

3. *Performance is based on cycle called "dual wash" by manufacturer, which is closer to cycle other companies designate as normal than the cycle recommended for normal use. When set that way, washing ability was ● that cycle, with only one wash and 2 rinses, redeposited tiny particles of food on dishes throughout the machine.*

SPECIFICATIONS AND FEATURES

All: ● require about 34¾x24x24 in. (HxWxD) for installation and 24 to 26 in. clearance to open door. ● Have at least heavy wash, normal wash, and rinse-hold cycles. ● Have switch to choose no-heat drying with any wash cycle. ● Have heating element under lower rack that can pose burn hazard when energized. *Except as noted, all:* ● Have dial that indicates progress through cycles. ● Have rinse-conditioner dispenser. ● Have 1 full or 2 half-size flatware baskets. ● Have porcelain front panel that allows choice of color. ● Require 120-volt, 15-amp circuit.

KEY TO ADVANTAGES

A – Cup rack has 1 or more fold-down sections for double-stacking of cups.
B – Upper rack has adjustable dividers.
C – Upper rack has movable section.
D – Upper rack can be adjusted up and down to accommodate load.
E – Upper-rack dividers can be folded down to double-stack cups.
F – Upper rack has posts that allow double-stacking of cups.

G – Holds very tall glasses better than most.
H – Holds tall glasses better than others.
I – Flatware basket(s) has covered section(s).
J – Has additional small, covered basket.
K – May be set for delayed start of up to 6 hr.
L – Stainless-steel interior and door liner.
M – Solid-plastic interior and door liner.
N – Solid-plastic door liner.
O – Dual fill valves provide additional protection against accidental overfills and flooding.

KEY TO DISADVANTAGES

a – Timer can't be reset by hand; canceling a cycle takes about 1 min. and wastes all detergent in machine.
b – No timer dial; indicator lights give only rough idea of progress through cycle. Canceling to restart takes about 15 sec. for **Maytag,** about 2 min. for **Kitchen Aid.**
c – Door latch moved stiffly.
d – Upper rack moved stiffly when loaded.
e – Lower rack occasionally snagged on tub gasket and rolled out of tracks.

f – Upper-rack slides dripped a little water when pulled out.

KEY TO COMMENTS

A – If water is not automatically certain temperature for part of cycle, heater switches on.
B – Option provided to make sure water is certain temperature for part of cycle.
C – Option provided to boost water temperature to 175°F for part of cycle.
D – Has no plate-warming setting.
E – Removable flatware rack mounted on door.
F – Plate rack is on top, cup rack is on bottom.
G – Tested with optional rinse-conditioner dispenser.
H – Lacks reversible front panel.
I – Requires 20-amp circuit.

Alternate Dishwasher Models

Most of the dishwashers noted on page 88–89 are loaded with features. The features may make dishwashing a little more convenient, but they don't improve washing performance in any significant way. From studying manufacturers' specifications, CU's engineers compiled a list of some more-basic models whose performance should be the same as that of the models tested. The dishwashers are listed by brand in alphabetical order. Within brands, they're listed in order of approximate retail price.

Brand	Approx. price	Difference from model tested
Frigidaire DWU55J	$689	Push-button controls; blower for drying; adjustable upper rack.
General Electric GSD 1000T	475	Button/dial controls; lacks "light" cycle.
GSD 900T	425	Button/dial controls; lacks "light" and "china/crystal" cycles; different upper rack.
Hotpoint HDA 965	450	"China-crystal" cycle; option to boost water temperature.
Kitchen Aid KDS-60A	690	Convertible/portable version of **KDS-20** tested.
KDP-20	610	Similar to **KDS-20** tested, but lacks option to boost water temperature; small-item basket; upper rack not adjustable.
KDC-20	540	Similar to **KDI-20** tested, but lacks "soak and scrub" cycle.
Magic Chef DU55	415	Push-button controls; "plastic" cycle; blower for drying.
DU35	299	Lacks "heavy wash" cycle, option to boost water temperature, rinse-conditioner dispenser.
Maytag WU701	575	Similar to **WU301** tested, but has option to boost water temperature, rinse-conditioner dispenser; minor differences in controls.
WU501	575	Similar to **WU301** tested, except for minor differences in controls.
WC301	545	Convertible/portable version of **WU301** tested.
Tappan 61.1451	539	Different control layout.
61.1231	449	Lacks "heavy wash" cycle, option to boost water temperature, and tub insulation.
Thermador THD 4500	649	Push-button controls; "short" cycle.
THD 2500	549	Lacks option to boost water temperature.
Wards Cat. No. 042	500+	Electronic controls; extra indicator and warning lights; delayed start option.
982	470+	Convertible/portable version of **932** tested.
Waste King WSK2200	549	Lacks option to boost water temperature.

Brand	Approx. price	Difference from model tested
Whirlpool DU9900XL	730	Electronic controls; "china/crystal" cycle; extra flatware basket.
DU8900XL	580	Push-button controls; "china/crystal" cycle; extra flatware basket.
DP6880XL	470	Convertible/portable version of **DU4000XL**, below.
DU5000XL	425	Lacks "low energy" cycle and delayed start option; different racks.
DU4000XL	405	Lacks "low energy" cycle and option to boost water temperature; different racks and basket.

Electric blankets and mattress pads

Electric blankets and pads contain wires through which electricity flows, generating heat. Some of that heat, along with your own body heat, is trapped by the blanket. The wiring in an electric blanket runs in evenly spaced lines. The wire in electric mattress pads is arranged so that the heat is "zoned," with more heat at your feet than at your torso, and none where the pillow is. (Some pads leave a lot of unheated space at the foot; they may leave tall people with cold feet.)

Since you lie on top of the wires with an electric mattress pad, you might wonder about comfort. It's likely that you will hardly notice the slight ridges formed by the wires amidst the pad's padding.

With either blanket or pad, the amount of heat available to you is set by a control box. Queen and king sizes come with two. Twin sizes, naturally, have only one, and you can get full sizes with one or two. The controls generally set how often the electricity cycles on

and off. The higher you set the control, the longer the electricity stays on. The solid-state controls used with some blankets work differently: they lower the current at times, instead of cycling on and off.

The thermostats, which sense the heat in an electric blanket, may be the most noticeable part of its hardware. Those little bumps sprinkled throughout the blanket are there to keep the blanket from overheating. Thermostats would be impossibly uncomfortable in an electric mattress pad. To guard against overheating, mattress pads are designed so they can't heat up as much as an electric blanket. You should feel as warm, however, because most of the heat from a pad stays where your body is, while some of an electric blanket's heat is dissipated into the room.

An electric blanket will keep you warm, and all the controls will provide a comfortable range of settings. But the prices of electric blankets range widely—from about $50 to more than

$100 for a queen-sized blanket, for example.

What do you get by buying a more expensive blanket? Different blanket material, for one thing. The lower-priced blankets are usually 20 percent acrylic, 80 percent polyester. Slightly more expensive are blankets with a 50-50 blend of those fibers. Often the top-of-the-line fabric is 100 percent acrylic, but some companies also offer a more expensive line whose fabric is 75 percent acrylic, 25 percent wool. The more expensive blankets are usually cut more generously, with a few extra inches of overhang.

In time, an all-acrylic blanket will start to pill and look worn, and may not stand up to years of use and repeated launderings as well as the other fabrics.

Mattress pads are generally composed of a thin polyester filling sandwiched between two layers of material, usually cotton/polyester. Pads are not quilted with stitching, however; a pin-point heat-fused technique is used instead.

Noiseless, solid-state controls are another feature of some expensive blankets and pads. The tick-tick made by the ordinary control as it cycles on and off isn't very loud. But it might keep some people from falling asleep.

Any blanket or pad should have lighted controls that are easy to see in the dark and in the light.

Many blankets and pads come with a control hanger, a plastic clip that you hook onto the bedframe. That keeps the control out of the way but easy to find, although it may not be easy to see. A control hanger can be purchased separately for many of the other lines, too.

A few blankets have snaps at the foot so you can fit the blanket around the end of the mattress. Some people might find that useful for keeping a neat tuck.

Most mattress pads are fitted, in that they have fabric sides to anchor them to the mattress.

Sleeping safety

Sales of electric blankets have never reached the levels once predicted for them, perhaps because a lot of people don't like the thought of going to bed with electricity. In fact, manufacturers often avoid using the word "electric" in their products, calling them "automatic blankets" or "automatic mattress pads."

Yet, as products go, electric blankets are usually safe. Though millions of electric blankets are in use, the Consumer Product Safety Commission said that fewer than 200 accidents associated with electric blankets were reported in hospital emergency rooms in one sample year. None of those accidents was fatal; most involved burns caused by overheating of the electric blanket.

Underwriters Laboratories sets standards that blankets and pads must meet to get the UL label. Those standards were made more stringent for blankets manufactured after 1981. They must have additional thermostats (at least 10 altogether for a twin, 12 for a full or queen, and 20 for a king).

UL conducts a regular test program to make sure manufacturers that use its imprint meet its standards. Consumers Union concluded that if an electric blanket or pad is used according to the manufacturer's instructions, the risk of the blanket's causing a burn or a fire is very small, and the risk of electric shock is even more remote.

Here are the key points in using an electric blanket or pad safely:

■ Don't let someone use one if he or

she can't tell when the bed is too hot or can't use the controls for some reason.

- Make sure the connector between the controls and the blanket or pad is inserted all the way. When the blanket or pad is plugged in, at least one of the pins is electrically live. Some models have a removable shield over the connector. You might also tape the connection shut while the blanket or pad is on the bed.
- Don't fold an electric blanket or pad while it's on. That may cause a section to overheat. An electric blanket should be used with no more than a thin blanket cover or lightweight spread on top. Never tuck the wired part under the mattress. And don't let pets sleep on the bed while

the blanket is on; the spot they're on can overheat, and their claws can damage the blanket.

- To clean a blanket or pad, follow the manufacturer's instructions for laundering (usually a cold or warm wash and low-heat or line drying). Never have one dry-cleaned—the cleaning solvents can damage the wiring.
- A blanket or pad won't become dangerous if the fabric part gets wet. But if the connector is wet, it might be able to cause a shock. For that reason, it's probably not a good idea to use an electric blanket or pad on a water bed.
- It's dangerous to smoke in bed with or without an electric blanket. Blanket fabric won't ignite, but may smolder and eventually burn up.

Recommendations

Some manufacturers of electric blankets and pads try to sell their products as energy savers, telling you how much fuel money you can save by using an electric blanket or pad. Their figures aren't disputable—it's quite possible to save many times the cost of an electric blanket or electric mattress pad. But what saves money is turning the thermostat down at night, not your method of keeping warm. Actually, using an electric blanket or pad is a bit less economical than using a comforter—you can stay just as warm in a cold bedroom with a comforter, which needs no electricity.

But if you prefer electric heat, the electric mattress pad is the better way to go. A pad uses roughly half of the electricity that an electric blanket uses. Since an electric mattress pad is a pure-

ly utilitarian item, you pay just for function, not for a pretty color or a nice fabric. The only drawback to an electric mattress pad is that it probably won't last as long as an electric blanket, especially if you wash it frequently.

Any electric blanket should keep you adequately warm. Unless you're bothered by the noise of the controls cycling on and off, there's no compelling functional reason to buy a blanket from an expensive line.

As with any textile product, the care with which an electric blanket or pad is sewn can vary. It's a good idea to inspect a sample you're about to buy. In addition, a sample's control mechanism can be defective every now and then. If the lowest setting is too hot or the highest setting too cold, return the blanket or pad.

Ratings of electric blankets and mattress pads

(As published in a January 1984 report.)

Listed by types. Within types, listed alphabetically by brand; within brand, listed in order of increasing price. All models tested were judged to work well and safely. "Wear performance" indicates how well blankets kept appearance, how durable pads were. Judgments were based on tests of queen size and one other model in line and are likely to apply to all models in line. Brand-line names in quotation marks are CU's designation. Prices are suggested retail; + indicates shipping is extra.

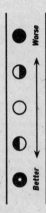

Better ← → Worse

Electric blankets

Brand and line	Wear performance	Control visibility — In dark	In light	Noiseless control	Control hanger [1]	One-control models — Twin	Full	Full	Two-control models — Queen	King	Comments
CANNON REGAL [2]	◐	○	—	—		3705/$40	3706/$50	3707/$58	3708/$68	3709/$93	A,H,O,P
CANNON REGAL SUPREME [2]	○	●	—	—		3720/$45	3721/$55	3722/$65	3723/$75	3724/$99	B,H,J,O
FIELDCREST CORSAIR	●	◐	—	—		10317/$60	16327/$65	16337/$75	16387/$90	16347/$125	A,P,Q
FIELDCREST [3]	○	○	—	—		-13/$75	-23/$80	-33/$95	-83/$105	-43/$135	B

Brand									
FIELDCREST ACCENT	◑	○	○	◐	—	11113/$85	11123/$95	11133/$115	11183/$125	11143/$170	B,H
J.C. PENNEY "2200 Line"	○	○	—	—	✓	2252/$30+	2260/$40+	2278/$50+	2286/$60+	2294/$80+	A,P
J.C. PENNEY "9000 Line" [4]	○	●	—	—	—	9083/$40+	9091/$50+	9109/$60+	9117/$70+	9125/$100+	C
J.C. PENNEY "3100 Line" [4]	◐	○	—	—	✓	3169/$56+	3177/$70+	3750/$80+	3768/$90+	3292/$120+	B
NORTHERN CARLYLE	○	○	—	—	—	4800-900/$49	4801-800/$52	4802-900/$61	4804-900/$74	4803-900/$112	A,Q
NORTHERN FLIP-FLOP	◐	○	—	—	✓	1390/$54	1391/$59	1392/$69	1394/$82	1393/$119	C,I
NORTHERN CAPRI	○	◐	—	—	✓	990-900/$57	991-900/$63	992-900/$74	994-900/$86	993-900/$126	B
ST. MARY'S FAIRLANE	○	◐	—	—	—	04617/$30	04627/$34	04637/$39	04687/$51	04647/$71	A,P,Q
ST. MARY'S PERFORMER	○	◐	—	—	—	06516/$37	06526/$42	06536/$50	06586/$57	06546/$80	C,L
ST. MARY'S LOVING TOUCH [5]	○	●	—	—	—	07513/$45	07523/$50	07533/$60	07583/$75	07543/$105	B
SEARS HAMPTON	○	○	—	—	—	7010/$30+	7011/$40+	7012/$50+	7013/$60+	7014/$80+	A,P,Q
SEARS PRELUDE	◐	○	—	—	✓	7040/$35+	7041/$45+	7042/$55+	7043/$65+	7044/$85+	A,J
SEARS COLORMATE	◐	◐	—	—	✓	7050/$55+	7051/$65+	7052/$76+	7053/$88+	7054/$114+	B,J
SEARS SHOWOFF [6]	○	○	—	—	—	70601/$70	70611/$90	70621/$115	7063/$140	7064/$150	D,J
SUNBEAM RIVIERA	○	◐	—	—	✓	969-900/$63	961-900/$73	962-900/$86	964-900/$100	963-900/$142	B,K

Electric mattress pads

Brand and line	Wear performance	In dark	In light	Noiseless control	Control hanger [1]	Twin	Full	Full	Queen	King	Comments
		Control visibility				One-control models		Two-control models			
FIELDCREST THE SLEEPER	◐	○		—	—	16219/$60	16229/$75	16239/$80	16289/$95	16249/$120	E,J,M,Q,S
J.C. PENNEY "3000/4000 Line"	○	◐		—	—	3094/$36+	3102/$45+	[7]	3110/$55+	4068/$70+	E,J,N,Q
NORTHERN THE BED WARMER	◐	○		—	—	2140-2/$37	2141-2/$49	2142-2/$56	2144-2/$65	2143-2/$63	F,M,P,R
ST. MARY'S THE WARMUP	○	○		—	—	08519/$30	08529/$40	08539/$46	08589/$54	08559/$70	G,M,Q
SEARS HARMONY [6]	○	◐		—	—	7270/$36	7221/$46	7222/$53	7223/$59	7224/$77	E,J,M,Q
SUNBEAM FITTED BED WARMER	○	◐		—	—	2300/$45	2301/$58	2302/$64	2304/$75	2303/$96	E,J,N,Q

[1] *Hanger comes with model; may be purchased for some others.*

[2] *These Cannon blankets were sold only through Wards. They may be available in some Wards retail stores. Prices are approximate.*

[3] *Fieldcrest markets the same line of blankets under 6 names. First 3 digits of model numbers vary according to name: Allure, 170.; Moderne, 173.; Regency, 170.; Temptation, 170.; Tenderness, 168.; Tranquility, 173.*

[4] *Models are no longer listed in catalog.*

[5] *A later designation of this model is Ultima; later model numbers begin with 077 instead of 075.*

[6] *Model numbers are no longer listed in catalog, but models may be available in some Sears retail stores. Prices are approximate.*

[7] *Not available in this size.*

SPECIFICATIONS AND FEATURES

All: ● Would fit size of bed they are intended for, though nominal measurements vary somewhat and actual measurements may differ slightly from nominal measurements. (Nominal measurements for most brands: Twin, 84 × 60 in.; Full, 84 × 72; Queen, 90 × 84; King, 90 × 100.) ● Have control with separate on/off switch. ● Have cord from wall plug to control about 6 ft. long and from control to blanket or pad 9 to 12 ft. long.
Except as noted, all: ● Have unshielded connector or connector with removable shield. ● Come with 5-yr. warranty.
All electric blankets: ● Have built-in safety thermostats. ● Have nylon-bound upper edge.
Except as noted, all electric blankets: ● Allow 3-5 in. to be tucked in at bottom.

KEY TO COMMENTS

A – Fabric: 20% acrylic/80% polyester.
B – Fabric: 100% acrylic.
C – Fabric: 50% acrylic/50% polyester.
D – Fabric: 75% acrylic/25% wool.
E – Fabric: 50% cotton/50% polyester.
F – Fabric: 100% polyester.
G – Fabric: 50% cotton/50% polyester for top; 100% olefin for bottom.

H – Blanket material thicker and wattage lower than most.
I – Blanket is supposed to be reversible but position of control connector allows tucking in from one side only.
J – Fitted model: Blankets have snaps at bottom; pads have fabric sides (unfitted pads have elastic anchor straps).
K – Brightness of on/off light adjustable.
L – Allows 7 in. to be tucked in at bottom.
M – Heated to within about 6 in. of foot of bed.
N – Pad is heated to within 11 in. of foot of bed; may keep feet less warm than others.
O – Connector is permanently shielded, providing best protection against accidental shock but making it hard to see whether pins are pushed in all the way.
P – Blanket or pad material somewhat less flammable than most.
Q – Comes with 2-yr. warranty.
R – Pad comes with a 1-yr. warranty.
S – Nonfitted model available at lower cost, with model numbers prefixed **160**; also available in California King size (69x81 in.), model **16259**, $120.

Electric heaters

Electric heaters draw a lot of wattage, and electricity is expensive in most parts of the country. Yet a judiciously used portable heater can help you cut heating costs.

Your central heating system is the greediest energy-consumer in the house. You can save energy and money by turning the central heating system on later than usual in the autumn, and off again as soon as possible in the spring. Even during the heating season, you can save money by keeping the room you are in warm while lowering the thermostat setting for the whole house.

You needn't, for instance, warm the whole house while you dress or have breakfast; the portable will provide economical comfort. If you want to stay up at night to watch TV after you've turned down the thermostat, use the portable to keep you warm.

Running a typical electric heater at its maximum output of 1,500 watts for an hour costs only about 12 cents at the national average electricity rate of 7.75 cents per kilowatt-hour.

The most popular brands and types on the market include radiant heaters, which use a reflector to direct their heat. Quartz heaters fall into this category. Convection heaters often have a fan to direct warm air into the room; radiant/convection hybrids have both reflector and fan.

Some heaters are boxy "uprights"; others are of a low-slung "baseboard" style. Most draw a maximum 1,500 watts.

A convection heater slowly warms the air around it. A radiant model uses a reflector to concentrate its heat on objects. Radiant/convection models heat up a room more quickly because they beam their heat directly at objects. Though radiant/convection heaters work faster than convection models, both types can heat a room evenly.

Convenience

Some portable heaters are easier to live with than others.

Thermostat and on/off switches. The ideal on/off switch on a heater is one that is separate from the thermostat, so you can turn the unit on and off without disturbing your thermostat setting. Most of the models, however, still combine the two functions, an arrangement we consider a disadvantage. Those models are called out in the Ratings.

Setting a heater's thermostat is a hot-or-miss proposition. All the thermostats lack actual temperature indications. Instead, most have numbers or simply marks for the purposes of setting. Some lack any guide, which makes them difficult to set and reset.

Signal lights. We especially like heaters that have one light to remind you the unit is turned on and another to indicate that the unit's thermostat has cycled on. Five models offer that combination: the **Patton FL15A,** the **Slant/Fin AQ1500,** the **Embassy RAHX1500,** and the **Markel 57TN** and **484T.**

A unit that has only a thermostat light can be misleading. When the thermostat light is not lit, you might leave home thinking the unit is turned off, when in fact it had simply cycled off.

The **Titan T800** also has a light that indicates the unit is plugged in. We think such a light is a good reminder to unplug an electric heater when it's not in use. That model's thermostat light is faint, however, and it is difficult to tell when it has cycled on.

Portability. The word "portable" is sometimes used by appliance manufacturers to describe anything that isn't bolted down. Most of the heaters in our test weigh less than 12 pounds and are quite portable. The most unwieldy models, however, are the six-foot-long baseboard heaters. The radiator-style heaters are the heaviest models in CU's test, though they can be moved about on their wheels.

Some heaters are less portable as they heat up. When hot, the radiator heaters cannot be maneuvered easily with their single handle. The radiator-shaped **Thermal 695** has screws in the handle that become too hot to touch. The **Dimplex 9215,** which is an upright, flat panel, is difficult to move when hot because it has no handle.

Stability. A portable heater should also be stable enough to stay upright if a toddler or an enthusiastic pet bumps into it or if someone tugs on the electrical cord. The heaters that are more stable than most are noted in the Ratings.

Noise. Most are quiet. A few produce some creaks and groans during warm-up and when cycling on and off. In general, the noisiest models are far from irritating, but the culprits are noted in the Ratings.

Defects. Quite a few of Consumers Union's test heaters showed defects from manufacturing or shipping. They had such problems as broken and jammed parts, damaged cases, and rubbing fan blades, broken casters, and one almost unmovable on/off switch. It's a good idea to try out a heater as soon as you get it home.

Safety

A fire hazard occurs when a portable heater, especially a radiant or radiant/convection heater contacts or comes near curtains or bed linens.

Portable heaters can be tipped over easily. On many heaters, a tip-over switch automatically turns off the unit if it falls forward or backward. Some models have an even safer tip-over switch that operates no matter which way the heater falls.

A tip-over switch is sometimes supplemented by an overheat sensor to protect your floor by limiting heat output if a tip-over occurs.

Burns are always a danger with heaters, especially with radiant models, whose surface can get quite hot. The high heat emanating from the unit is usually enough to warn away an adult, but it might not turn away a small child.

Convection units pose less burn hazard because their surface temperatures are generally lower. Radiator-style convection heaters are an exception. Those can become too hot to touch or to carry safely once they warm up.

On some heaters, another potential for shock exists if a child pokes his or her fingers or a metallic object through the grille. If there are toddlers in your house, it would be well to avoid a model that isn't well guarded against probing.

Recommendations

Radiant and radiant/convection models have an edge in overall performance. But a convection-only model is safer to use, and the best of them performs well at whole-room heating.

For spot-heating, your best choice is a radiant-only quartz heater. But a radiant heater requires special care for safe operation. The beam of heat that is so effective for spot-heating may produce scorching of fabric accidentally draped over the grille.

Once you get a heater home, you should, of course, read the instructions and check it over for defects. Then, by following a few guidelines, you can keep warm safely: keep flammables, combustibles, furniture, and curtains away from the heater. Make sure the element on a radiant heater is free of flammable material, such as dust balls or hair. Cleaning the heater's reflector with a vacuum cleaner will improve its efficiency.

Keep the heater away from water. If it tips over near water, don't stand it up again; first disconnect the power.

If you must use an extension cord, use one that can handle 1,500 watts or 12 1/2 amps, not the common lamp-type extension cord. Avoid using an electric heater on the same circuit with another high-wattage appliance, such as a refrigerator or a toaster oven. Regularly inspect the wall outlet and heater plug and cord for excessive heat. Heater cords naturally get warm, so don't cover or leave the cord coiled or knotted. If the heater has a compartment for cord storage, withdraw the cord fully during use. Unplug the heater when it's not in use.

Ratings of portable electric heaters

(As published in an October 1985 report.)

Listed by types; within types, listed in order of estimated overall quality. Dimensions and weights are rounded. Prices are mfr. suggested retail; + indicates shipping is extra.

	Excellent	Very good	Good	Fair	Poor
	◉	◒	○	◓	●

Brand and model	Price	Dimensions (HxWxD), in.	Weight, lb.	No. wattage settings	Temperature distrib. [1]	Temperature swing	Spot-heating	Control convenience	Portability	Safety	Advantages	Disadvantages	Comments
Convection models													
TPI 6LBB1	$185	11x72x5	23	1	○	◉	●	◓	●	◉	F,G,I,K,T	n,o,s	A,F
EDISON 324029G [5]	65	20x11x8	9	2	○	◒	◉	◓	○	◉	B,E,M,S	g,v	D
PATTON FL15A	80	10x51x7	12	3	◓	◉	◒	◉	◒	◉	B,C,F,K,T	g,k	D,F
INTERTHERM NP1500	139	10x74x6	24	1	○	◉	●	◒	●	◓	C,F,K,T	g,n,s	B,F
PRESTO 07860	77	19x12x8	11	2	○	◒	●	◓	○	◉	B,K,N,Q,S	g,k,v	D
TITAN T800	70	19x12x9	9	2	○	◉	●	◉	○	◉	A,C,K,P,S	s,t	—
MARKEL 484T	79	8x49x6	11	2	◓	○	●	◉	◓	◓	B,C,F,K,M,T	g	E,F
SHETLAND OF8A	90	27x18x9	37	3	◉	◉	●	◒	●	○	C,F,M,T	g,y	A,F
SLANT/FIN AQ1500	157	10x73x7	23	1	○	○	●	◉	●	◉	B,C,F,G,K,S,T	k,n,q	B,F

Model	Price	Size (in.)			Ratings							Features	Advantages/disadvantages	E
MARKEL 57TN	65	15x11x8	9	2	◐	○	○	○	○	○	◐	B,C,I,K	r,x	—
SEARS Cat. No. 7256 [5]	60+	17x13x9	10	2	○	●	⊙	○	○	●	○	C,I,K	s,u,y	—
ARVIN 29H92	45	15x7x15	8	2	◐	○	[2]	◐	○	◐	○	I	b,f,r,s	—
EMBASSY RAHX1500	99	10x49x7	14	1	◐	◐	●	◐	◐	●	◐	B,C,F,K,T	g,n	A,F
DELONGHI 5108	45	26x19x8	33	3	○	◐	●	●	●	●	◐	C,F,R,S,T	g,k	A,F
DIMPLEX 8115	168	27x20x9	38	1	○	⊙	●	◐	◐	●	◐	C,F,S,T	g,k	A,F
THERMAL 695A	50	26x19x12	34	3	◐	●	●	●	●	●	◐	C,F,R,S,T	g	A,F
WELBILT 2001	50	19x23x10	12	1	●	●	◐	◐	◐	◐	◐	F,K,M	i,o,r	F
DELONGHI DF-15	69	17x23x7	11	2	[4]	◐	○	◐	◐	○	◐	C,D,J,K,Q	a,h,k,r,v	—
DIMPLEX 9215	241	24x58x9	48	1	○	○	◐	◐	◐	◐	◐	A,F,S,T	k,p,s,t,u	A,F

Radiant and radiant/convection models

Model	Price	Size (in.)			Ratings							Features	Advantages/disadvantages	E
TITAN BB42A	70	9x39x6	10	2	○	○	○	◐	○	○	○	I,K,M,T	c,y	—
PRESTO 07875 [3]	98	31x15x16	9	2	◐	●	⊙	◐	◐	◐	◐	G,L,O,Q,S	d,k,o,s,w	C,F
ARVIN 49H2001	59	10x41x5	11	2	◐	○	[2]	○	◐	◐	◐	I,K	c,f,r,s,w	—
SEARS Cat. No. 7206 [5]	50+	10x41x5	11	2	◐	○	[2]	○	◐	◐	◐	I,K	c,f,r,s,w	—
PRESTO 07840	74	16x21x7	10	2	◐	○	●	◐	○	◐	◐	I,S	c,s,x	—
LAKEWOOD 415	35	13x19x10	8	2	○	○	○	◐	○	○	○	G	c,s,x	—
MARKEL 198TE	62	16x12x12	7	3	◐	○	○	◐	◐	◐	○	F,G,K	d,o	—
EDISON 324053C [5]	42	13x18x10	8	1	◐	○	◐	◐	◐	◐	○	G,S	d,s	—
ARVIN 30H33	55	13x19x10	7	3	◐	◐	[2]	◐	◐	◐	◐	—	d,f,r,s,w,x	—
SEARS Cat. No. 7139 [5]	45+	13x19x6	7	3	◐	○	[2]	◐	◐	◐	◐	I	d,f,m,r,s,w,x	—
TITAN RT26B1	47	13x17x12	8	1	○	○	○	◐	○	◐	◐	—	d,q,r,s,w	—
TITAN RT40A1	70	14x21x12	9	4	○	○	○	◐	○	◐	●	—	d,r,s	—
LASKO 969	40	12x20x6	8	3	◐	○	◐	○	◐	◐	◐	H,L	b,d,l,s,y	—

Not acceptable [6]

Brand and model	Price	Dimensions (HxWxD), in.	Weight, lb.	No. wattage settings	Temperature distrib. [1]	Temperature swing	Spot-heating	Control convenience	Portability	Safety	Advantages	Disadvantages	Comments
ARVIN 30H1101	39	12x19x6	7	1	○ [2]	◐	○	○	●	—		e,f,r,s,w,x	—
SEARS Cat. No. 7136 [5]	38+	12x19x6	7	2	◐ [2]	◐	○	○	●	I		e,f,r,s,w,x	—
TOASTMASTER 2477	37	6x30x7	5	2	●	○	⊙	◐	●	F,G,M,S		e,j,x	F

[1] At highest wattage setting.

[2] Judgment is average of several tests (see Disadvantages).

[3] Radiant-only quartz model.

[4] With fan off, judged ●.

[5] According to mfr., a later designation of Edison 324029G is Toastmaster 2529; Sears Cat. No. 7256, 36223; Edison 324053C, Toastmaster 2453; Sears Cat. No. 7206, 36400; Sears Cat. No. 7139, 36003; Sears Cat. No. 7136, 36002.

[6] Listed alphabetically.

Specifications and Features

All have: ● Adjustable thermostat.
Except as noted, all: ● Are rated 1500 watts at 120 volts. ● Have tipover switch that turns off power if heater is tipped forward or backward. ● Produced little or no scorching in CU's 30-min. draped-fabric test. ● Have on/off control separate from thermostat. ● Have a fan.

Key to Advantages

A – Signal light on when heater is plugged in.
B – Signal light on when heater is switched on.
C – Signal light cycles on with heating element.
D – Signal light also on with fan-only operation.
E – Grille judged safer from probing than others.
F – Quieter than most.
G – Multidirectional tipover switch.
H – Has tipover switch as well as manually resettable overheat sensor.
I – Has tipover switch and overheat sensor.
J – Has fan-only setting.
K – Front grille very safe from probing.
L – Thermostat has positive off position.
M – Thermostat operated well over a wide variation in room temperature.
N – Both fan and wattage range changed as heater shifted automatically.
O – Heater can oscillate back and forth.
P – Very easy to see on/off light.
Q – Top-mounted controls, easy to read/ use.
R – Control layout easy to use.
S – Thermostat labeled with numbers.
T – More stable than most.

Key to Disadvantages

a – No tipover switch; internal overheat fuse blew when heater was operated on its side.

b – Front grille judged less safe from probing than most.
c – Scorched fabric lightly during 30-min. draped-fabric test.
d – Produced moderate to heavy scorching in 30-min. draped-fabric test.
e – Produced smoke and flame in 30-min. draped-fabric test.
f – Thermostat showed wide test variations.
g – No tipover switch; heater operates face down but did not damage vinyl tile in 90-min. test.
h – Can operate with fan off; may soften vinyl tile floor.
i – No tipover switch; lifted vinyl tile and scorched carpet in 90-min. tipover test.
j – Softened vinyl floor tile and scorched carpet when tipped over face down for 5 min.
k – Narrower thermostat range than most.
l – On/off switch hard to operate.
m – Less stable than any other.
n – Sharp edges beneath heater.
o – Handle area marginally hot.
p – Control markings may be difficult to see.
q – Thermostat knob has no pointer.
r – Thermostat lacks intermediate markings.
s – On/off control is part of thermostat.
t – No tipover switch but overheat sensor prevented damage to vinyl tile.
u – No on/off markings on power switch.
v – Fan noisier than most.
w – Makes annoying noises during cycling.
x – Made floor hotter than most.
y – Few markings to aid thermostat setting.

Key to Comments

A – Manufacturer claims oil filled.
B – Manufacturer claims antifreeze filled.
C – Utilizes quartz glass tubes.
D – Wattage-shift type heater.
E – Wattage-setback type thermostat.
F – No fan.

Electricity vs. kerosene

Kerosene heaters cost less to operate than electric heaters—typically only half as much. But along with a kerosene heater's savings comes a risk. The dangers of kerosene heaters were widely publicized a few years ago. A Consumers Union October 1982 report on test models notes that they present three major hazards: they give off potentially hazardous fumes, require you to store flammable fuel, and, in two cases, pose a distinct fire hazard if tipped over.

As a result of improved safety features on some newer kerosene heaters, their overall safety has improved.

Nonetheless, even these new kerosene heaters can still produce dangerous levels of indoor pollution, particularly in small rooms that lack adequate ventilation. An electric heater is still the better choice.

Floor polishers

An electric floor polisher is one of those appliances a lot of households can do nicely without, even if carpets do not cover most of the floor. There are, after all, workable alternatives. As a once-or-twice-a-year proposition, floors can be polished with a rented machine, or a service company can be called in. For floors that can take a water-based polish, "self-polishing" floor wax may meet some people's standards all year round.

That's not to say an electric floor polisher is of no interest, particularly if you take pride in near-perfect floors and want to keep them buffed to a mirror sheen.

If you want a machine of your own, you should have little trouble buying one that works well enough at polishing bare floors or hard-surface floor coverings. Differences will more likely be in convenience features than in performance.

Shampooing

Manufacturers of floor polishers usually offer devices for wet-shampooing rugs. These attachments can work quite well, but their use may entail some risks of damage to the rug from the brushes' abrasive action. Therefore, you should always try shampooing a small inconspicuous area of rug or carpet first to find out whether the rug can withstand the machine. Better still, rent one for a trial run. Besides checking for damage, see whether you're satisfied with the shampooing. You may find the shampooing technique difficult—and the results may not satisfy you.

Converting a machine from polisher to shampooer should be easy and quick.

In shampooing, it's important to work a good thick foam into the pile of the rug to loosen the dirt. Machines that dispense the shampoo as a liquid sometimes wash too much dirt to the bottom of the pile, where it will tend to escape the vacuum cleaner later on. Moisture can also promote mildew or rotting if it soaks into the carpet and underlay. Most shampooing machines have some means of agitating the shampoo into foam before it hits the rug.

If shampooing sounds like too much trouble, remember that rugs may not require shampooing very often, and when they do, may be sent out for cleaning, if you can take them up easily enough, or shampooed in place by a professional firm.

A floor-polishing machine, with or without special attachments, can be used for wet scrubbing on hard-surfaced floors. It can be a real boon on extremely dirty floors—much better than hand scrubbing or wet mopping. But for tidying up a slightly dirty floor, damp mopping is easier and quicker.

A machine with vacuuming action offers a special advantage: it can suck up the dirty water, eliminating tedious sponge squeezing. But don't be surprised if the holes in the water-pickup entrance become blocked by particles of dirt. You can minimize that by sweeping or vacuuming the floor before you scrub. And dirty water may continue to drip from the machine even after you have emptied it. Even a little water on a polishing pad or brush can smear a newly waxed surface, so be sure the machine has dried completely before you use it for polishing.

Waxing

Two-brush models, the most common type, tend to leave a narrow strip of less-well-polished floor in the space between the brushes. To get reasonably even polishing, you have to push the machine through overlapping strokes.

Most machines have dispensers for such liquids as wax and sudsy water. A dispenser is likely to be much more useful for shampooing rugs or washing floors than for waxing. For one thing, a machine tends to spread wax unevenly. For another, a solvent-based liquid wax can clog a machine's dispenser and perhaps damage it. So if you do put wax in the dispenser, use a water-based emulsified liquid type. All in all, it's better to spread wax—liquid or paste—with an applicator and use your machine only for polishing and buffing.

Other factors

In general, the faster the brushes rotate, the more you can expect them to spatter wax, shampoo, or water. It would seem, then, that you'd want a machine with a fairly low speed for shampooing and scrubbing and a high speed to switch to for polishing after your wax is spread. Apparently with some such idea in mind, the manufacturers of many models provide two speeds. But such a model isn't likely to perform noticeably better than a one-speed model.

A floor polisher often seems to have a mind of its own. If you aren't careful, it can spin out of your hand and go careening across the floor. That's most likely to happen when the handle is vertical. So keep the handle at an angle.

It helps if there's an indication whether the motor switch is on or off before you plug the machine in. With a push-button switch there may be no way to tell.

Recommendations

Price is not a guide to effectiveness: even a cheap polisher will do the job. As you move up the price ladder you get more accessories and more features. Those things are best left to independent decision, based on needs and preferences.

Flue-heat recovery devices

If your furnace hasn't been cleaned and serviced for a year or so, you should have it done. A thorough tune-up can improve the furnace's performance so it will extract as much heat as possible from the fuel it consumes.

However, even a properly tuned furnace releases some heat, along with the products of combustion, up the chimney. You wouldn't want those noxious vapors around, but you may want to consider trying to recover some of that excess heat using a device especially intended for the purpose.

Such a device won't necessarily lower your gas or oil bills. Unless you can move the heated air to an area near the thermostat, the heating system will burn nearly as much fuel as it always does; you'll get a somewhat greater amount of usable heat from that fuel. In practice, a flue heat recovery device won't recover a large amount of heat (no more than 6 percent of the energy in the fuel burned in Consumers Union tests). And it's not practical to move that amount of heat more than about 20 feet away from the furnace. However, if there is an unheated room near the furnace (or a nearby room that's now warmed with a portable electric heater), then a flue-heat recovery device can be useful.

These units serve, in effect, as secondary heat exchangers: a metal surface, heated by flue gases, transfers its heat to room air. Some are simply clamped over the flue pipe,

drawing heat from the pipe itself. Others, which replace a section of flue pipe, circulate the heated air.

A flue-heat recovery device (also called a "chimney robber") works best on a very inefficient furnace—one with very high temperatures in the stack, or flue; the amount of heat to be reclaimed drops as the stack temperature falls. Most are heavy, bulky, and can be difficult to clean. Note, too, that these devices may not conform to the building code in your area. Before you buy one, make sure it can be installed legally.

It's not possible to say exactly how much heat one of these devices will save in your home, but there are some guidelines. The absolute amount of heat recoverable depends on the stack temperature and the size of the furnace. A furnace rated at 140,000 Btu per hour will provide more reclaimable heat than a 100,000 Btu-per-hour model, assuming that both furnaces have the same stack temperature. And on two furnaces of the same capacity, the one with the higher stack temperature will send a greater amount of heat up the flue. It's unlikely that any flue-heat recovery device will provide a significant saving if the temperature of the flue gases entering the device is less than about 450°F. This consideration more or less limits these devices to use on oil-fired furnaces or older gas-fired models, which have fairly high flue temperatures.

How usable is the heat?

An electric space heater (see pages 100–102) will supply heat until the room reaches a preset temperature; a

flue-heat recovery device, on the other hand, will stop supplying heat when the furnace shuts down. That may leave the room too cool for your liking. Or—if the room is very small—the recovery device might supply *too much* heat.

A flue-heat recovery system can also create problems if you have a tankless coil in your boiler to provide hot water; production of hot water in the summer would then be accompanied by unwanted space heating. To avoid this problem on a fan-equipped model, simply switch the fan off.

Fan-operated models have a stub of pipe to which stovepipe or ducting can be connected to move heated air away from the furnace. Most use a 4- or 6-inch-diameter pipe. You'll do best using as little ducting as possible, since air velocities drop off sharply as the length of the duct increases.

Installation and maintenance

These gadgets are not likely to be difficult to install, but some may be too bulky to fit on your flue. You'll need between 10 and 24 inches of straight pipe between the furnace and the draft diverter to accommodate one of those devices. You'll also need ample space around the pipe to be able to install, service, and clean.

Before you install a flue-heat recovery device, have a service technician adjust the burner and check the flue

temperature. Have the temperature checked after installation, too. In some cases, a "chimney robber" could drop stack temperatures below 300°, which could bring on excessive condensation in the flue.

Any flue device should be cleaned periodically; a monthly inspection can't hurt. Cleaning can be a messy chore, involving brushing and vacuuming hard-to-reach sooty surfaces.

Recommendations

Recovering heat from the furnace flue may sound like a good way to save energy. In fact, not too much heat can be reclaimed, and it can't be moved too far away from the furnace. You should consider installing a flue-heat recovery device only if your situation meets four conditions: if your local building code permits it; if your furnace is incurably

inefficient; if you need to heat an area within about 20 feet of the furnace; and if your annual heating costs are high enough to justify an investment of $100 or more to get no more than 5 to 6 percent more heat from the fuel you buy. In short, a flue-heat recovery device won't be effective on every furnace, but it can be useful in certain situations.

Food fixers

This section will tell you about the pluses and minuses of the five major types of motorized kitchen food fixers.

Blender. This is the appliance of choice for pureeing foods, mixing exotic drinks, and preparing mayonnaise. A

At a glance: Which does what well?

Consumers Union's test program for the following series of articles on food fixers includes some 29 food-preparation tasks. This table shows how the typical blender, portable mixer, stand mixer, and food processor should perform in appropriate tasks. Where a particular chore is inappropriate for the appliance, there's no judgment in the table. The judgments are generalizations, based on how easily the task can be done and on the quality of the food produced.

	Excellent ⊙	Very good ◓	Good ○	Fair ◒	Poor ●

Task	Blender	Portable mixer	Stand mixer	Food processor
Whipping cream	Very good	Excellent	Excellent	Fair
Puréeing for soup	Excellent	—	—	Good
Blending mayonnaise	Excellent	Fair	Fair	Excellent
Beurre manie	—	—	—	Good
Mixing pudding	Fair	Fair	Fair	Fair
Mashing potatoes	—	Fair	Good	—
Mixing cake batter	Fair	Excellent	Excellent	Good
Mixing cookie dough	—	Good	Fair	Fair
Blending pie dough	—	Good	Good	Fair
Making peanut butter	Good	—	—	Excellent
Kneading bread dough	—	—	Fair	Good
Crumbing crackers	Good	—	—	Fair
Grating Parmesan	Fair	—	—	Good
Chopping parsley	Fair	—	—	Fair
Chopping mushrooms	—	—	—	Fair
Mincing onions	Fair	—	—	Fair
Chopping prosciutto	Fair	—	—	Fair
Grinding beef	—	—	—	Good
Shredding cabbage	Fair	—	—	Fair
Shredding cheddar	—	—	—	Fair
Shredding zucchini	—	—	—	Fair
Liquefying solids	Fair	—	—	Good
Mixing frozen drinks	Excellent	—	—	—

Task	Blender	Port-able mixer	Stand mixer	Food processor
Slicing carrots	—	—	—	◑
Slicing cucumbers	—	—	—	◉
Slicing mozzarella	—	—	—	◒
Slicing mushrooms	—	—	—	○
Slicing pepperoni	—	—	—	◓
Slicing onions	—	—	—	◒

blender falls short for grating, shredding, and chopping, and is poor at whipping cream. It doesn't do well with cake batters either.

Portable mixer. It makes excellent whipped cream and can mix cake batter with ease. A model with a hefty motor can even make its way through heavy butter cookie batter or small batches of bread dough.

Stand mixer. This mixer takes over where the portable leaves off—especially the heavy-duty model. It can handle big batches of dough as well as a multitude of other chores.

Food processor. It's advertised as a wonder worker that can chop, grind, and slice, as well as blend, mix, and knead. Not all brands and models do all chores equally well.

Multipurpose food fixer. This may not be the answer to the cook's prayer. Nevertheless, the idea of having one appliance for fixing foods that can do the work of three is appealing, especially on a crowded countertop.

Blenders

It's hard to tell from all the buttons on a blender's control panel, but the appliance is really less versatile than it appears at first glance. The table on pages 108–109 shows you what you can and cannot expect a blender to do well, compared with the other food-fixing appliances in this section.

CU tested blenders from the three major manufacturers: Hamilton Beach, Oster, and Waring. Hamilton Beach manufactures the **Sears** models tested; Oster and Waring supply the **Penneys** models.

The blenders tested range in suggested retail price from $22 to $64.

All the blenders perform about the same in their assigned food-preparation tasks. The primary differences are related to convenience.

Blending. The blender's claim to fame rests with its ability to mix perfect piña coladas and to puree smooth baby food. What else can it do? The tests tell the tale:

Whipping cream. Volume is the key to good whipped cream. The air introduced into the cream should double its volume, and the resulting whipped cream should fluff up into peaks. Good cook's rules for making whipped cream call for chilling the container and blades ahead of time. CU testers filled each container to the eight-ounce mark and pressed the "whip" button.

But the design of the blender defeats the ability to introduce air. As soon as the cream begins to thicken, it packs around the sides of the container. You have to stop the blades and use a rubber spatula to scrape the cream back into the center of the container. With all the blenders, the result is about 10 ounces of a creamy substance that forms soft peaks. When tasted, the substance feels coarse on the tongue.

Pureeing for soup. Each blender was used to puree a carrot, a scallion, some chopped parsley, and canned sliced potatoes in a chicken bouillon. All the blenders can produce lump-free puree with an excellent creamy texture. They do the job neatly and tidily because their ample containers don't leak or spatter liquid (a problem with some food processors).

Blending mayonnaise. The blenders can make 1½ cups of thick, creamy mayonnaise in about 40 seconds.

Making peanut butter. All the blenders are able to turn salted cocktail peanuts into a fairly thin, smooth peanut butter. Only the **Waring** family of blenders (which includes the **J.C. Penney 3420**) leaves a few chunks in the peanut butter. But the process is slow going. As the peanuts transform into goo you have to stop the machine and prod the peanut butter with a rubber spatula to keep the substance circulating.

Making graham-cracker crumbs. After breaking graham crackers into pieces and placing them into the blender's container, the instruction booklets recommend using an on/off pulsing technique, which is particularly convenient with models that have a special pulse provision (they're noted in the Ratings).

None of the blenders is likely to do a perfect job, leaving a few cracker chunks to be reprocessed. That usually happens when a piece of cracker lodges in the narrow neck of the container, keeping other pieces away from the blades.

Grating Parmesan cheese. Hard cheese poses a bit of a challenge to person or appliance. A blender tackles the task with considerable noise and vibration. But in CU's tests, the blenders, try as they would, could not turn four ounces of cheese, broken into chunks, into a uniform, finely grated powder. Instead, they produced a pebbly, coarse substance that contained a number of good-sized ungrated pieces.

Chopping parsley. Freshly chopped parsley is decidedly more flavorful in soups and casseroles than the dried variety is. But some people find it difficult to produce an evenly chopped batch. Those people might want to try chopping parsley in a blender.

Most of the instructions recommend covering solid foods with water before blending to aid circulation within the container. So the testers covered the parsley with water. All the blenders can chop a half-cup of parsley evenly and very fine. But draining and drying the parsley is a tedious chore (if you immediately add the parsley to soup or stew, of course, you don't have to worry about drying it thoroughly).

Mincing onions. Using a blender to mince onions will not spare you many tears. You have to peel the onions, trim the ends, and cut them into pieces small enough to fit in the area down by the blades. You'll be so teary by then that you might as well have chopped the onions by hand—an opinion that will be reinforced when you see how the blender minces the onion. A blender tends to produce onion slush. If you try mincing onions without the water, using the "chop" setting (or the manufacturer's recommended setting) in two, five-second pulses, the results won't be much better—onion slush with a few large pieces mixed in.

Shredding cabbage. Coleslaw calls for long, stringy shreds of cabbage. That takes quite a while if you're chopping by hand. But a blender isn't the answer: if you try to shred the cabbage in a blender, you get chips rather than shreds.

CU testers cut heads of cabbage into eight sections and put four sections into the container of each blender. They covered the cabbage sections with water. Since few of the machines have a setting labeled "shred," they used the setting suggested in the directions. As a rule, the cabbage comes out in small, jagged pieces that are fine for soup but not for slaw.

Liquefying solids. A juice extractor is about the only appliance that can turn raw carrots into carrot juice. But a blender can make a pretty good carrot mash from four ounces of short, thin carrot strips. The test was intended to measure the machines' ability to start up with a hard solid in the way of the blades, to chop, to maintain speed, and to circulate the contents with a minimum of spattering. If you cover the carrots with water and use the recommended speed (or the highest speed), after about three minutes you'll have a carrot mixture of fine consistency with nearly all particles small enough to pass through a sieve. It won't be juice, but it will be puree.

Making ice slush. Most blenders made for the consumer market aren't meant to reduce large chunks of ice to smaller chunks. That task is intended for an ice crusher. But a blender can turn ice cubes covered with water into the slush that's basic to a variety of exotic summer drinks. The **Waring** models are the fastest.

Convenience and safety

Controls. All the tested blenders have push-button speed controls. On most, the motor starts when you press one of the speed controls. Two **Oster** models have a separate, rotary on/off control; that's a lot less convenient.

Speeds. All those push buttons on a blender may look impressive. But you really don't need many speeds to do the chores blenders do well. The difference from one speed setting to the next on a blender is often minuscule. With one 14-speed **Waring** blender, for example, measurements show that the three highest speeds are almost identical.

Pulsing. The pulsing feature that some blenders have is operated by push button, slide control, or rotary knob; the feature doesn't always operate on all speeds—the Ratings give specifics. Pulsing is a convenience. But it's no great burden to do the pulse operation manually by alternately pushing the speed-control and off buttons on most models.

Leakage. Most blenders can be filled to their marked five-cup capacity without overflowing or splashing during the operation. On all the tested blenders, you can remove the center part of the lid (which acts as a small measuring cup) to add ingredients while the blender is working, and still not make a mess of yourself or the countertop.

The **Waring** blenders with plastic containers have rather ill-fitting lids that occasionally allow leakage. And when the center part of those lids is removed to add ingredients, all the **Warings** slosh over, even when only half filled.

The **Waring BL132** has a funnel that can be inserted into the lid opening to aid in adding ingredients. The testers used it while making mayonnaise. None of the mixture spatters out around the funnel, but they nonetheless didn't think much of the feature—it just slowed them down.

Two **Sears** models, noted in the Ratings, tend to leak a bit around the lid when they are filled to capacity.

Stability. Little rubber feet generally prevent a blender from moving about. The relatively hard rubber feet on the **Waring** models don't grip as well as the softer feet on other models.

On two **Oster** models, some push buttons are mounted vertically on the face of the base. Pushing the buttons may move the blenders, although the machines stand still when operating.

With most models, the container fits securely onto the base, so it's not likely to topple if jostled. The **Waring** containers are more topple-prone than others because they fit loosely into the base.

Noise. Blenders are not quiet, but they're used in such short bursts that they aren't likely to cause a great disturbance.

Cleaning. None of the blender bases are easy to clean. The push buttons and grooves in the housings trap dirt. The chrome-base models look slick when all polished up, but the slightest fingerprint or smear shows all too quickly.

Most of the blenders have a glass container, which stays better-looking longer than plastic because it resists scratching and staining. Glass containers should safely go into the dishwasher. Plastic containers can't since they might soften or melt if they come too close to the dryer element. But plastic containers are lighter than glass ones, and less likely to break if dropped.

For hand-washing, you will like the convenience of the wide-mouth glass containers on some **Hamilton Beach** models.

Cord length. Consider cord length and the location of outlets in your kitchen when choosing a small appliance. The cords on the blenders range from two to three feet long (see Ratings).

Safety. Modern blenders are designed so that you'd actually have to reach down into the container to be slashed by the blades. Manufacturers have taken measures to prevent containers from loosening and exposing the blades during operation. The blades on the **Hamilton Beach** models, for example, turn clockwise so they tighten the container as they turn. The **Waring** and **Oster** blenders rely on projections in the neck of the base that mate with projections on the container.

Recommendations

A blender probably won't find much to do in your kitchen unless you plan to puree lots of foods, make exotic drinks, and whip up your own mayonnaise and peanut butter.

If you enjoy baking, you'll find more use for a portable mixer (page 128) or even a stand mixer (page 133).

Chopping, mincing and other chores involving solid foods are best left to a food processor (page 114).

Theoretically, the multitalented and expensive kitchen centers can replace all of these appliances. To find out if they do, see page 120.

Blenders are best suited for blending, mixing, pureeing, or doing almost anything that involves combining liquids.

Ratings of blenders

(As published in an August 1984 report.)

Listed by groups in order of estimated overall quality. Within groups, listed in order of increasing price. Differences between groups were judged small. Prices are list; + indicates shipping is extra.

Brand and model	Price	Container (plastic or glass)	Number of speeds	Advantages	Disadvantages	Comments
SEARS COUNTER CRAFT Cat. No. 82918, *A Best Buy*	$22+	P	7	B,H	—	A
SEARS COUNTER CRAFT Cat. No. 82902, *A Best Buy*	27+	G	7	B,H	b	—
SEARS 82938	30+	G	14	B,E,H	b	E
HAMILTON BEACH 610	34	P	14	B	—	A,B
HAMILTON BEACH 600	36	P	7	B	—	A
HAMILTON BEACH 582	38	G	7	B,G	—	—
HAMILTON BEACH 626	39	P	7	B	—	A
HAMILTON BEACH 653	39	G	14	B,F,G,I	—	—
HAMILTON BEACH 651	47	G	14	B,F,G,H	—	B
HAMILTON BEACH 632	50	G	14	B,F,G	k	—
HAMILTON BEACH 640	50	G	14	B,E,F,G	—	B
OSTER 869	40	P	10	D,H,I	j	A
OSTER 890	40	P	10	D,H,I	j	A
J.C. PENNEY Cat. No. 4525	40+	G	10	D,F,H	j,k	A
OSTER 643	47	G	10	D,F,H	j	A
OSTER 862	47	G	12	C,H,I	j	A
OSTER 855	53	G	14	C,H	e,f,j	A
OSTER 848	55	G	14	C,F,H	j,k	A
OSTER 861	64	G	16	C,F,H	e,f,j,k	A,C
WARING BL208	37	P	7	A	a,b,c,g,h,i	—
WARING BL590	41	P	7	A	a,b,c,g,h,i	—
WARING BL132	41	P	14	A,I	a,b,c,d,g,h,i	D
WARING L14	43	P	14	A,H	a,b,c,d,g,h	—
WARING BL130	52	G	14	A	a,c,d,g,h	B

SPECIFICATIONS AND FEATURES
All ● Stand 13¾ to 15¼ in. tall with container on base. ● Have a removable cap in the container lid. ● Have a removable cutter assembly. ● Are designed to prevent container from loosening and exposing blades in operation. *Except as noted, all:* ● Have push buttons labeled with functions. ● Lack pulse feature. ● Have 5-cup (40 oz.) container marked in 1-cup intervals. ● Have maximum liquid capacity when operating of more than 32 oz. ● Have 2-ft. cord.

KEY TO ADVANTAGES
A – Turned ice cubes to slush faster than most.
B – Container is marked at ½-cup intervals.
C – Has pulse provision; operates on all speeds.
D – Has pulse provision; operates on 3 speeds.
E – Has pulse provision; operates on 1 speed.
F – Container more stable on base than most.
G – Has wide-mouthed container, judged fairly easy to empty and clean.
H – Has fairly long cord, about 3 ft.
I – Has cord-storage feature in base.

KEY TO DISADVANTAGES
a – Left chunks when making peanut butter.
b – Container lid allowed some leakage.
c – With lid insert removed, liquid splashed out if container held more than 20 oz.
d – Buttons are numbered, not labeled.
e – Rotary on/off control judged less convenient than push buttons.
f – Activating some push-button controls jostled blender.
g – Feet less skid-resistant than most.
h – Container less stable on base than most.
i – Noisier than most.
j – Requires care in drying after washing to prevent corrosion of metal blade assembly.
k – Chrome-like base finish shows dirt more readily than plain plastic bases.

KEY TO COMMENTS
A – Has 5½-cup marked capacity.
B – Comes with extra containers.
C – Comes with extra ½-pt. container.
D – Comes with sauce and salad-dressing attachment, judged of little value.
E – According to Sears, it has been replaced by Cat. No. 82968 ($37+), identical except that it includes extra containers.

All the models tested do those chores well enough. The **Hamilton Beach** models have the edge in convenience. And two **Hamilton Beach** blenders made for Sears—**Counter Craft 82918** and **82902**—at $22 and $27 (plus shipping) have the edge in price, too. Both are best buys.

Do not bypass an **Oster** model if the price is right. The **Osters** perform as well as the **Hamilton Beach** blenders, though some are a little less convenient to use.

Food processors

Food processors have become the pulse of the American kitchen. A novelty little more than a decade ago, they're now standard operating equipment. Food processors provide hurry-up haute cuisine, onions without tears, even baby food. They blend, crumb, emulsify, flake, grate, grind, mince, puree, shred, slice. They chop just about everything but wood. Some even knead bread dough.

About ten years ago, only one company, Cuisinarts Inc., marketed a food processor in the United States. The machine was made in France by Robot Coupe. Now Cuisinarts has dozens of competitors—including Robot Coupe itself.

Speed is the food processor's forte. It can mix cookie batter in minutes, slice a cucumber in seconds. All the tested models are lightning fast.

Power is also important. Grating carrots or Parmesan cheese by hand is a demanding chore, but it's a snap for a good food processor. Kneading bread dough is a time-consuming, tough job that a powerful processor can do quickly and easily. The high-rated models all take heavy-duty tasks in stride.

Capacity is important, too. The food processors at the top of the ratings hold a good deal more processed food than the low-rated models.

Processing

Consumers Union challenged the processors with chores demanded of mixers (blending, mixing, kneading) and of blenders (some mixing and some chopping), as well as with a host of cutting, slicing, and shredding jobs.

Conclusions concerning overall quality are based primarily on how well and how easily each machine does a vast assortment of typical food-fixing tasks.

Blending and mixing. Chores that fall into this category range from easy ones such as combining pudding mix with milk to difficult ones such as kneading bread dough.

Processors can combine pudding mix and milk with ease. Whipping cream would seem to be another easy chore, but none of the food processors do it as well as the mixers. The processors cannot introduce enough air to produce the volume and peaking one wants of whipped cream. Incorporation of air is important for many cake batters, too. Since food processors don't do this very well, a finished cake is apt to be rather compact instead of light (which is fine for recipes such as carrot cake).

The processors can blend mayonnaise easily, but they aren't quite as fast as the blenders.

Combining butter and flour for a little *beurre manié* to thicken sauces and stews won't tax the machines, even if you use refrigerator-hard butter. To blend pie dough CU testers use cold hard butter, too, for a nice flaky piecrust.

A cookie-dough recipe calls for creaming butter with sugar, then adding flour and other ingredients to form a thick, sticky dough. Soft butter will make the job easier, but any food processor should be able to do the job—even with cold butter—in about 90 seconds.

Kneading bread dough is the ultimate challenge. As the Ratings show, half the tested models did very well with small batches of brioche dough. (The large-capacity models performed equally well with much larger batches of bread dough.) Processors noted in the Ratings as good or fair at mixing bread dough slow down considerably during the kneading.

Grating and grinding. To evaluate grinding chores the testers turned graham crackers into crumbs and peanuts into peanut butter. They grated Parmesan cheese and they ground up beef. Many of the food processors did a very good job of crumbing crackers in just 10 on/off pulses. All the machines made a pound of very good or excellent creamy peanut butter in about 3½ minutes. And all turned out coarse but uniformly grated Parmesan cheese. Grinding the beef was the toughest chore, but the processors did an adequate job. The meat didn't look like conventional ground beef, but it made perfectly good hamburgers.

Chopping. Parsley is easy to chop, and all the tested food processors can produce a uniform batch in just two on/off pulses. Soft mushrooms are also easy to chop—perhaps too easy: all the processors do a very good job, but the mushroom pieces they produce may be too small for anything but gravy.

Most of the processors are very good or excellent at chopping onions, without tears.

Carrots are a more challenging chore. With four ounces of carrots in four ounces of water as a test of each machine's ability to start up with a hard object in the blade's path, maintain chopping speed, and keep spattering inside the container to a minimum, we find none of the machines are able to reduce the carrots to a smooth consistency in three minutes, as the blenders can.

Although prosciutto is a tough, hard food, most of the processors chop inch-sized cubes with little difficulty.

Slicing. The texture of food affects the way a processor slices. Crisp, firm cucumbers are easy. Nearly all the machines do an excellent job, producing cleanly cut, uniform slices. Some machines are judged only good because they do not slice the cukes as cleanly as the others, or because they produce extraneous small pieces, or both.

Mozzarella cheese is gummy and rubbery and a difficult slicing task for a food processor. Most of the machines are judged poor or fair.

Most of the processors slice carrots quickly and easily. A few models, judged only fair, produce more irregular pieces than the others.

Most of the processors do a very good or excellent job of slicing onions.

Hand-slicing mushrooms is easy and, perhaps, the best way to do the job. Using a processor, you must take time to stack the mushrooms on their sides in the feed tube to obtain attractive cross sections.

Shredding. Whether with a shredding or a slicing disk, most of the processors make a disappointing coleslaw: they chip rather than shred the cabbage.

Zucchini works better with the shredding disks. Most of the models do a very good or excellent job.

Shredded cheddar is a success, too. Most of the machines give excellent or very good results.

Design variations

Capacity. Most of the containers can hold more dry (or semidry) ingredients than liquid, because liquid capacity is limited by the height of the tube in the center of the container. The chopping or mixing blade fits over the tube, but it usually doesn't form a tight seal. Liquid can ooze under the hub of the blade, through the tube, and onto the base.

Continuous feed. The four models with this handy feature have unlimited capacity for shredding or slicing food, since the food goes out the chute and into a container. The food, however, occasionally misses the container and lands on counter or floor. And bits of food sometimes drop into the processor's container, so you have to clean two bowls when you're finished. Food also tends to clog the chute occasionally, and you have to stop the machine to clear the blockage.

Blades and disks. Most food processors come with an S-shaped metal chopping blade, an S-shaped plastic mixing blade, a slicing disk, and a shredding disk. Some models come with additional tools. Many models also have optional blades and disks.

Speeds. A number of the tested processors offer a selection of speeds. Most of the machines—including the best-performing models—have only one. That single speed, operated continuously or pulsed, is adequate for most chores.

Controls. Differences in control design—push buttons, rocker switches, slide controls are usually insignificant.

Most models have a pulse setting built into the on/off switch. As a rule that arrangement works well.

Noise. Small food-fixing appliances most often are noisy, but the noise level depends upon the speed at which the machine is operated and the food being processed.

Cleaning. The simpler the design of a processor, the easier it is to clean.

On the body of the food processor, gaps around the switches and trim can trap food and dirt.

Storage. Many parts can mean storage problems. Disks and blades often have to be stored apart from the machine. It's not a good idea to put them in a drawer with other kitchen tools; someone reaching blindly into the drawer could easily be cut.

Most of the machines are about 15 inches tall when assembled for use. Several are taller. But most can be stored with the feed tube inverted, bringing them down to size. A few models weigh nine pounds or less, but many are quite a bit heavier.

Safety and durability

Food-processor blades have to be sharp to do their job. Some of the manufacturers try to keep your contact with the blades to a minimum by designing disks with finger grips or holes. Some have shredding and slicing disks on a separate stem with a knob at top; that arrangement lets you handle the knob instead of the blade.

All processors have some kind of interlock mechanism to prevent operation of the machine when its lid is off.

It's important that the processor's blades stop turning before the lid can be removed, or you might get cut reaching in. In measuring stopping time we find that most of the processors coast to a halt in less than $3^{1}/_{2}$ seconds.

There are complaints about the plastic parts on some food processors, especially the **Moulinex** models. Apparently, the stems on the shredding and chopping disks and the plastic hub of the chopping blade sometimes crack. Our experience with hubs on a few samples in the CU lab confirms that. It's not possible to say whether the problem affects all **Moulinex** models, since not all samples have the same blade design.

Recommendations

Active cooks—particularly those with an advanced repertoire and those who cook in quantity—appreciate a food processor.

But no food processor can do all tasks equally well. You can use the performance judgments in the Ratings as a guide to determine the best models for the chores you do most often.

A food processor isn't the appliance of choice for mixing cake batter, mashing potatoes, or whipping cream. A portable mixer is. So the well-equipped kitchen needs both.

Ratings of food processors

(As published in a November 1984 report.)

Listed in order of estimated overall quality, based mainly on ability to perform heavy tasks and on capacity. Bracketed models were approximately equal in quality and are listed alphabetically.

Prices are suggested retail; + indicates shipping is extra.

Legend: Excellent ● · | Very good ◐ | Good ○ | Fair ◓ | Poor ●

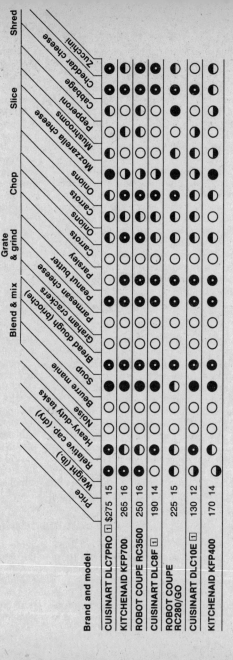

Brand and model	Price	Weight (lb.)
CUISINART DLC7PRO [1]	$275	15
KITCHENAID KFP700	265	16
ROBOT COUPE RC3500	250	16
CUISINART DLC8F [1]	190	14
ROBOT COUPE RC280/GO	225	15
CUISINART DLC10E [1]	130	12
KITCHENAID KFP400	170	14

Rating categories (columns): Relative cap. (dry), Heavy-duty tasks, Noise, Beurre manie, Soup, Bread dough (brioche), Graham crackers, Parmesan cheese, Peanut butter, Parsley (Grate & grind), Carrots (Grate & grind), Onions (Grate & grind), Carrots (Chop), Onions (Chop), Mozzarella cheese, Mushrooms, Pepperoni, Cabbage, Cheddar cheese, Zucchini (Slice), Cheddar cheese, Zucchini (Shred)

ROBOT COUPE RC2000	100	13
SUNBEAM LE CHEF [1] 14021	132	14
PANASONIC KITCHEN WIZARD MK5070	70	7
SUNBEAM 14056	70	8
TOSHIBA TFP2200	160	10
J.C. PENNEY Cat. No. 3246	60+	8
MOULINEX REGAL LA MACHINE LMII	96	8
MOULINEX REGAL LA MACHINE LMIII	129	8
HAMILTON BEACH 702	75	7
HAMILTON BEACH 712	95	7
HAMILTON BEACH 737	113	7
J.C. PENNEY Cat. No. 3337	80+	14
SEARS Cat. No. 82288	50+	7

[1] *Model discontinued at the time the article was originally published in* **Consumer Reports**. *The test information has been retained here, however, for its use as a guide to buying.*

SPECIFICATIONS AND FEATURES

All have: ● On/off switch and pulse provision. ● Transparent plastic bowl and lid. ● Plastic food pusher. ● Effective safety-interlock. ● Multipurpose S-shaped steel blades.

All were judged: ◑ at making pie dough. ○ at grinding beef.

Multipurpose food fixers

Oster calls it the **Kitchen Center.** Sunbeam and Sears each have a **Food Preparation Center.** RonMatic's multipurpose food fixer is the **Ronson Foodmatic Preparation Center.** Nutone's is the **Food Center.** Given such names, you might expect the appliances to do everything but set the table.

Consumers Union tests include five of these fancy food fixers, along with a simpler machine, the **General Electric FP3.** All have a motor that drives a variety of attachments for blending, mixing, and other chores.

Two models—the **Nutone Food Center** and the **Ronson Foodmatic Preparation Center**—are built-ins. You must have a hole cut in the kitchen counter so the motor can be installed underneath. A plate over the opening in the counter provides a fairly flat surface when the machine is not in use.

The **Nutone** and the **Ronson** can create a hole in your wallet as well as in your counter. The **Nutone,** with food processor, blender, and mixer components, is $529; the **Ronson,** $888. The four other machines are freestanding appliances about the size of stand mixer; they range in price from about $100 to $260.

Both built-ins are full-function machines, since they can do the work of a blender, a mixer, and a food processor. The **Sunbeam** is the only full-function freestanding model. The **Oster** doesn't include a true food processor. It has a salad maker for slicing and shredding, but it lacks a chopping blade. The **Sears** claims to include a mixer function, but has only a paddle attachment for its food processor. And the **GE** has no special mixer component. Since the machines vary so much in design they are not strictly comparable, and the Ratings list them in alphabetical order.

Preparing food

Overall, the blender attachments perform about as well as typical blenders. The mixer and food-processor components are the problem: they don't usually do as well as regular mixers and processors.

Blending and mixing. Combining butter and flour into a *beurre manié* to thicken sauces is a job for a food processor. The processor attachments do the job fairly easily, though the results are better in some cases than in others. Since the **Oster** lacks a processor, you would use its blender, which unfortunately does a poor job.

But blenders are ideal for soup processing—pureeing vegetables in liquid.

All the machines can do an excellent job with their blender attachment.

Making mayonnaise or combining pudding mix with milk is simple for any blender or food processor. These machines are no exception.

Mixers are ideal for whipping cream and mixing cake batter. The four machines with a mixer attachment do those chores very well. Since the **GE** lacks a mixer, you have to use its food-processor attachment. The results: only fair for whipped cream, good for cake batter. The **Sears** paddle attachment for the food processor does a good enough job with large batches of whipped cream, but its cream-whipping falls a bit flat in average-sized batches of about eight ounces. It does fairly well at mixing cake batter.

Mashing potatoes is another typical mixer chore. The **Nutone** will do a very good job, in about a minute. The **Ronson** makes good mashed potatoes, but needs time.

All the appliances can make very good or excellent cookie dough.

Food-processor chopping blades easily cut cold butter into flour to make flaky piecrust. Results with the processor-equipped machines are very good. The **Oster**'s mixer does a good job.

Power is the key to kneading bread dough. In CU's initial small-batch test, the testers used the food-processor attachments. The **Sears**, the **Sunbeam**, and the **Nutone** completed the task easily in less than a minute. The **GE** makes very good dough, but works hard doing it. The **Ronson**'s food processor stalled, shut off, and would not finish the job. Using the **Ronson**'s mixer and dough hooks, kneading took longer than with the other machines (about six minutes), but results were good. The **Oster** mixer could do a comparable job in comparable time.

Design limitations complicated making large batches of bread. Only the **Sears**' food-processor bowl and the **Sunbeam**'s mixer are large enough to accommodate recipes calling for about seven cups of flour.

Chopping. Most chopping chores are best done with a food processor. All the blender attachments operate smoothly and cleanly when chopping carrots but only four reduce the carrots to a smooth consistency.

The processor attachments do a good job chopping onions and mushrooms.

The toughest chopping test is prosciutto. All the processor attachments—and the **Oster** blender—do well.

Grating and grinding. All performed adequately in reducing graham crackers to crumbs and grating hard Parmesan cheese.

All the models with a food processor yield a pound of very good or excellent peanut butter. The **Oster**'s blender makes very good peanut butter, but only half a pound at a time.

Shredding. Many food processors (see pages 118–19) produce cabbage chips instead of shreds for cole slaw. The multipurpose machines have the same problem. The **Ronson**'s slicing disk does the best job.

Cheddar cheese shreds well, although a little will be lost in all the machines, lodged between the disk and the container lid.

Slicing. Crisp, firm cucumbers are easy to slice, and all the machines are good or better at the job.

The machines generally do a good job slicing carrots, onions, and mushrooms. But rubbery mozzarella cheese is a challenge.

Design

Blenders tend to look alike. Mixers look like other mixers. Most food processors are similar in design. Not so these multipurpose food fixers.

Containers and capacity. Most blenders have a working liquid capacity of a quart. The blender attachments do, too.

The **GE,** the **Nutone,** and the **Ronson** containers are plastic, which is lighter than glass and won't break if dropped. But the glass containers of the other machines resist scratches better and are easier to clean.

On most, the bottom of the blender, which houses the blades, is removable for easy cleaning. But the **Nutone** and the **Ronson** don't offer that convenience. The **Ronson** has a special tool that allows removal of the blade for thorough cleaning.

The mixer components of all but the **Nutone** include a large and a small bowl. The **Sunbeam**'s bowls are stainless steel—light, unbreakable, and easy to clean. The others have glass bowls. The **Nutone**'s large single bowl is opaque and easily scratched, so it's likely to be harder to clean after repeated use.

Food processors generally can hold more dry ingredients than liquid ingredients, because the latter can easily leak out. The **GE** and **Ronson** processors have a blenderlike design that reduces the likelihood of leakage. But either way, the capacity of most of the food-processor attachments is a skimpy one quart or so. In contrast, the **Sears** holds more than most regular food processors—2½ quarts wet, 3½ quarts dry.

The **Oster**, the **Sears**, and the **Sunbeam** have a continuous-feed design that lets you slice or shred unlimited quantities of food. A "slinger plate" diverts the sliced food from the processor bowl to a separate bowl placed alongside the machine. Unfortunately, it sometimes may sling food to the floor, too.

Feed tubes and pushers. The feed tube on most models can fit foods no larger than about 1½ inches in diameter. The tube on the **Sears** can take whole onions or tomatoes up to 2½ inches in diameter. But you pay for that convenience in added complexity: There are two pushers, one with a pair of hollow inserts for feeding skinny foods, the other for feeding larger ones. There's also a hinged plate, part of the feed tube, that you must swing out of

the way to load large food items. It's a difficult puzzle to figure out, but it works.

A number of the food pushers are open, for use as a cup. But only the **GE** has a one-cup measuring mark.

Blades, disks, and mixing equipment. The food-processor attachments consist of a chopping blade (used for chopping and mixing), a slicing disk, and a shredding disk. Some models have extras, noted in the Ratings. The **Oster**, which does not include a true food processor, has disks for slicing and shredding.

Most of the mixer attachments have a pair of beaters and a pair of dough hooks. The **Nutone** has a single beater. Unfortunately, it lacks a dough hook.

Speeds and controls. Individual appliances usually don't require a vast assortment of speeds to do their jobs well. Multipurpose machines are different. Since a single motor drives an assortment of appliances, a selection of speeds can be useful. Half the models have an adequate selection of speeds; the others have a continuously variable speed control.

Most of the controls are clearly marked and convenient to use. But the dial on the **Ronson** is inconvenient to use, since you must turn through all its speeds to get to High. And the control on the **Nutone** is somewhat hard to grasp because it's recessed.

Noise. The attachment, the speed, and the food itself all affect noise level. None of the machines can be called quiet, but the **Nutone** and the **GE** are the noisiest. The **Ronson**'s blender, mixer, and processor are very quiet at low speed but the processor is very noisy at high speed. Fortunately, the **Ronson**'s processor is usually run at lower speeds.

Convenience and safety

Differences in design make for differences in convenience—sometimes big differences.

Changing functions. Party-food preparations that call for switching from food processor to mixer and then to blender can drive a cook to distraction.

The **GE** is the easiest to convert from processor to blender. Both attachments simply twist into place on the motor housing.

The **Oster** has one mechanism to run the blender and the salad maker, both of which are fairly easy to attach. A turntable for the mixer is off to one side. Use of the cumbersome mixer arm and grinder adapter, which attach to the motor housing, gets easier with practice.

The **Sears** has separate mechanisms to run its blender and processor. To operate one, you must cover the other. It's a complicated procedure—and it requires care, since incorrectly positioning the plate over the blender mechanism can break parts of the plate. The **Sunbeam** has similar interlocking requirements, plus the added complexity of a removable mixer arm and a mixer turntable that must be carefully positioned.

The **Nutone**'s blender assembles easily. But there are a series of steps to set up the food processor or mixer, though it's not complicated or difficult.

The **Ronson** has an array of latches, releases, and drive spindles for assembling its appliances. Once you figure out how each functions, the machine isn't hard to set up or take down.

Cleaning. They're all a nuisance. Those with more parts give you more work, of course.

Storage. The compact **GE** is the easiest model to store. Its chopping blade and slicing and shredding disks fit into the processor bowl, which can sit on the motor housing for storage. The only extraneous piece is the blender.

The other freestanding models require separate storage of their myriad parts. The built-ins pose the same difficulty. Each manufacturer markets a storage rack or drawer, which would probably be a wise investment.

Safety. The cutting blades of all the models are designed so you have minimum contact with sharp edges. All have an effective interlock mechanism, so the food processor can't operate when the lid is removed. And the blades on all models coast to a stop within $3\frac{1}{2}$ seconds of shutting off the motor.

The **Ronson** ascends and descends through the hole in the counter with hydraulic assistance. With the press of a button on the counter panel, the motor rises abruptly. Pressing the same button lowers it partway, with some risk of it painfully pinching your hand.

Recommendations

Hybrid appliances seldom do all things well, and a multipurpose food fixer is no exception. Though the machines are pretty good as blenders, and though some aren't bad as mixers, they're mostly mediocre as food processors.

Most cooks would probably be happier with individual appliances that do their respective jobs well. Individual appliances don't require the back-and-forth conversions that these machines do.

Nonetheless, you may want the space

saving these fixers provide. If so, think about the **Oster** first. This $260 machine does not a have a true food processor, so it can't chop—a serious drawback. But it can slice and shred. Its unique grinding attachment can do a better job with ground meat than a food processor. Its mixer and blender perform well.

The **Sears** is one of two models that can handle large batches of bread dough. Its blender works well. It lacks a mixer. Still, it's a good value at $160 plus shipping. You can always pick up a good portable mixer to go with it.

The $240 **Sunbeam** has a mixer

(which can produce large batches of bread dough), a food processor, and a blender.

The built-in **Nutone**'s greatest failing is its lack of a dough hook. Without it, and with its small food processor, the machine is unable to knead large batches of bread.

When you pay $900 for a small appliance—more than the price of many refrigerators and ranges—you expect an exceptional machine. The **Ronson** is not. Its mixer and blender work well, but the performance of its food processor is mediocre.

Ratings of multipurpose food fixers

(As published in a January 1985 report.)

Listed in groups; within groups, listed alphabetically. Food-preparation judgments are based on the best results obtained and on convenience, no matter which component was used to do the job. Price is suggested retail, as stated by manufacturer, for model with blender, mixer, and processor included or with those components as options, when available; + means shipping is extra.

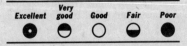

Food processors

Free-standing models

	Price	Blender	Mixer	Food processor	Food-processor capacity	Heavy-duty ability	Conversion ease	Noise	Beurre manie	Soup	Mayonnaise	Pudding	Whipping cream	Cake batter	Mashed potatoes	Cookie dough	Pie dough	Bread dough (small amount)
GENERAL ELECTRIC FP3	$104	✓	—	✓	◐	◑	◑	●	◉	◑	◐	◑	○	●	●	●	◑	◐
OSTER 98816	260	✓	—	✓[1]	—	—	○	○	●	●	◐	◉	◉	○	◉	○	○	○
SEARS Cat. No. 82408	160+	✓	✓[2]	✓	◐	●	◐	◐	●	●	◐	◐	●	●	◉	◐	◐	◑
SUNBEAM 83036	240	✓	✓	✓	○	○	○	◐	●	●	◐	◐	○	○	●	◐	◐	◐

Built-in models

	Price	Blender	Mixer	Food processor	Food-processor capacity	Heavy-duty ability	Conversion ease	Noise	Beurre manie	Soup	Mayonnaise	Pudding	Whipping cream	Cake batter	Mashed potatoes	Cookie dough	Pie dough	Bread dough (small amount)
NUTONE 251	529	✓	✓	✓	◐	○	○	●	●	●	◐	●	●	◐	●	●	◐	◐
RONSON 68120	888	✓	✓	✓	◑	○	[3]	◉	●	●	◐	●	●	◐	●	◐	◐	○

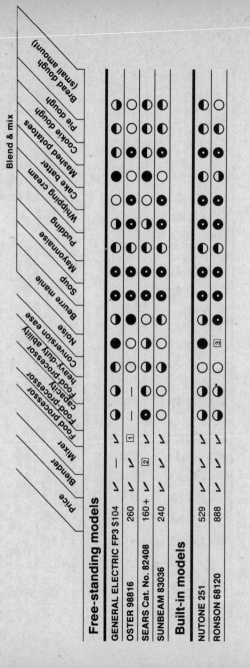

Free-standing models

	Chop					Grate & grind					Shred						Slice		Advantages	Disadvantages	Comments
	Parsley	Onions	Carrots	Mushrooms	Prosciutto	Graham crackers	Parmesan cheese	Peanut butter	Beef	Cabbage	Cheddar cheese	Zucchini	Cucumbers	Carrots	Onions	Mushrooms	Pepperoni	Mozzarella cheese			
GE FP3	●	○	○	◐	◐	◐	○	◐	◐	◐	◐	◐	◐	◐	◐	◐	◐	◐	D,J,K	g	A,E
OSTER 98816	◐	○	○	◐	○	◐	○	◐	◐	◐	●	◐	◐	◐	●	●	◐	◐	A,B,G	f	D,G,H,I,F
SEARS 82408	●	●	●	●	●	●	●	○	◐	◐	●	●	●	●	●	●	◐	●	C,F,H,I,M	c,h,i	—
SUNBEAM 83036	◐	◐	◐	●	●	◐	●	○	◐	◐	●	◐	●	●	○	○	●	●	E,F,M	—	E,J

Built-in models

	Chop					Grate & grind					Shred						Slice		Advantages	Disadvantages	Comments
NUTONE 251	○	◐	◐	◐	◐	◐	●	○	◐	◐	◐	◐	◐	◐	○	○	○	○	D	b,d,e,j	A,B,C,E,F,G,K,L
RONSON 68120	◐	◐	◐	●	◐	●	●	○	◐	◐	●	◐	◐	●	◐	◐	◐	◐	C,L	a,b,j	A,B,E,F,K

1 Has salad maker instead of food processor. Can slice and shred, but not chop.

2 Has paddle attachment for food processor, which then functions as mixer.

3 Relative noise for mixer, blender, and processor at low speed, ◐; for processor at high speed, ●.

SPECIFICATIONS AND FEATURES

All: ● Come with 1 yr. warranty. ● Built-ins are installed in counter, with motor underneath; when not in use, they are almost flush with counter.

Except as noted, all have: ● Many parts, which are not easily stored without optional racks. ● Convenient controls.

Mixer attachments. *Except as noted, all have:* ● 2 conventional beaters, 2 dough hooks. ● 1 large, 1 small glass bowl.

KEY TO DISADVANTAGES

a – Mechanism to raise and lower motor judged somewhat hazardous because hand or fingers can be pinched between counter and motor.

b – Blender base is not removable (though cutter blade for **Ronson** can be removed for cleaning with special tool); container cannot be washed in dishwasher.

KEY TO COMMENTS

A – Has plastic blender container, judged less breakable than glass but susceptible to scratching and fogging.

B – Power drive assists mixer-bowl rotation.

C – Mixer has only 1 conventional beater.

D – Has 2 slicing disks, 1 thick and 1 thin.

E – Food processor lacks plastic mixing blade.

Blender attachments. *Except as noted, all have:* ● Removable base to prevent container from loosening and exposing blades in use. ● Glass container, marked in 1-cup intervals.

Food-processor attachments. *All:* Come with slicing and shredding disks designed for handling with minimum contact with blade. ● Have safety interlock to prevent operation without lid. ● Have plastic bowl that can be washed in dishwasher.
Except as noted, all: ● Have multipurpose S-shaped steel chopping blade, plastic S-shaped mixing blade, and 2 separate disks, mounted on separate stems, for slicing and shredding. ● Lack measuring graduations on food pusher.

KEY TO ADVANTAGES

A – Touch-panel controls easily cleaned.
B – Lights show switch is on, speed selected.
C – Blender marked in $1/2$-cup intervals.
D – Blender marked in $1/4$-cup intervals.
E – Has stainless-steel mixer bowls.
F – Food processor operates in conventional and continuous-feed modes.
G – Shredder, slicer, and grinder operate in continuous-feed mode.
H – Large feed tube allows slicing and shredding $2^1/_2$-in.-wide foods (but see Disadvantages).
I – Food-processor bowl holds $2^1/_2$ qt. liquid.
J – Food pusher has 1-cup mark.
K – Food-processor blades, disk store in bowl; blender jar is only extra piece to store.
L – Has built-in timer.
M – Handled large quantities of bread dough better than most.

c – Blender interlock plate must be removed with care to avoid damage.
d – Mixer has only one 4-qt. plastic bowl, judged inconvenient for mixing small quantities and easily scratched.
e – No dough hooks available for mixer; food-processor attachment handles only small batches of bread dough.
f – Salad maker does not chop; blender and food grinder used for some chopping tests.
g – Food processor spattered when mixing liquids.
h – Food-processor disks and blade held in place by easily misplaced plastic nut (but manufacturer provides 2 extra nuts).
i – Food processor has 2 pushers, one with separate inserts for feeding narrow foods. Feed tube has pivoted section that must be moved to load large items. Pushers and feed tube judged complex and difficult to clean.
j – Controls judged inconvenient.

F – With some tasks, best performance achieved at speeds or in ways other than those recommended by manufacturer.
G – Comes with french-fry disk (not tested).
H – Comes with juicer (not tested).
I – According to the company, this model was discontinued and replaced by the $290 **988.26**, identical except for stainless-steel bowls.
J – Model discontinued at the time the article was originally published in *Consumer Reports*. The test information has been retained here, however, for its use as a guide to buying.
K – Model no. is for power unit only; the other components have different model numbers.
L – Tested blender was model **272**; model **232** blender attachment (not tested) is available, needs adapter.

Portable mixers

Portable mixers have changed little over the years. There's really no reason for them to change: their basic design—lightweight and fairly compact—is well suited for the chores they do.

The primary differences among the various models in the stores are in the power of the motor, the number of speeds offered, and the configuration of the beaters. Price differences seem large: from $13 all the way up to $44. But those are approximate retail prices; steep discounts, especially on the higher-priced models, should narrow the range.

Motor power is the most significant feature affecting performance. A strong motor can keep beaters turning through the stickiest dough, where a weak one may stall. A motor with a governor will help to maintain the desired speed setting when you're doing a demanding task with a mixer. A few have a special power switch intended to provide the boost needed to keep the beaters going through gooey mixes.

A wide range of speeds isn't necessary. Although many models offer from 5 up to 12 speeds or a continuously variable speed, only three speeds—slow, medium, and fast—are really necessary. The slower the slow speed, the better; it keeps spattering to a minimum when starting to mix liquids. Mixers with a governor control offer the slowest speeds, although you sometimes have to set the control below its lowest marked setting to achieve the slowest speed.

The shape of the beaters varies from model to model, but shape doesn't seem to affect performance. The **Moulinex** and the **Waring HM9518,** however, have unusual wire-whisk beaters that to seem to have an edge in some chores. The **Moulinex** has dough hooks in addition to the whisks, so it can tackle small bread-making chores the other mixers won't be able to handle.

Mixing

Portable mixers vary a lot in their mixing ability.

Whipping cream. Consumers Union started its tests with chilled bowls, chilled beaters, and eight ounces of whipping cream for each machine. When the manufacturers provided instructions, the testers followed them; otherwise, they used their best judgment. Either way, the portable mixers produced billows of whipped cream, more than double the volume of the liquid. The mixers succeed where the blenders fail because they are able to introduce air into the cream continually while beating, and it is air that provides volume. The mixers do the job quickly, too. The **Moulinex** and the **Waring HM9518,** with their unusual whisks, whip through the task in less than 2½ minutes. Other models take up to about 4 minutes.

Blending mayonnaise. A portable mixer is much slower than a blender at making mayonnaise—taking more than 5 minutes—primarily because the oil has to be dripped into the mixture at a tedious rate.

Mixing pudding. Most mixers can combine the pudding mix with milk uniformly, easily, and quickly, with just a little spattering.

Mashing potatoes. For each portable mixer, CU's testers cut up and boiled two pounds of potatoes. They added some milk and butter and, following instructions for speed settings, put the mixers to work. After mashing the potatoes for 1½ minutes, they ran them through a sieve to see what lumps remained. The Ratings note models that made the creamiest potatoes.

The whisklike beaters of the **Moulinex** mash the lumps, too, but only when the instructions, which call for a low-speed setting, are disregarded. The high-speed setting gives much better results, though with quite a bit of messy potato spattering.

Mashing potatoes isn't a particularly strenuous chore for a mixer, but a number of **Hamilton Beach** and **Sears** models have to work hard, at first, to break up the chunks. The burst-of-speed feature on the **Hamilton Beach 113** and **116** helps.

Making cake batter. Most cake batters need to be beaten to incorporate a certain amount of air. Otherwise the cake may not rise properly, and its texture is likely to be leaden instead of light. Mixers are the masters of cake batter, much better than a blender because a blender does not introduce enough air into the mixture. A cake made using a blender is denser than a cake made using a mixer.

Mixing cookie dough. Cookie dough will separate a powerful portable from a pretender. CU's butter-cookie recipe has two steps that are particularly challenging to mixer motors: creaming a half-pound of butter at the beginning, and adding two cups of flour to the butter-and-sugar mixture at the end.

Many models, including most of the **Sunbeams** and two **Hamilton Beach** models, handle both heavy tasks without pause. The whisk-equipped **Moulinex** and **Waring** models do a good job, too. The **Hamilton Beach** and **Sears** mixers without a governor control slow and occasionally stall when fighting the thick cookie dough. The burst-of-speed switch helps the **Hamilton Beach 113** and **116** through the tough times, but those mixers still don't perform as well as the best models.

Thick dough often creeps up the beater shafts during mixing, so you have to shut off the mixer and scrape the beaters clean. With three **Waring** models, the dough tends to creep up. The dough doesn't creep at all with the whisk-fitted **Moulinex** and **Waring** mixers.

Kneading bread. Most portables don't have the tools or the punch for kneading bread dough. The **Moulinex,** with its dough hooks, is exceptional. CU made a brioche dough using 2½ cups of flour, and the testers were impressed with the results. The **Moulinex** did the job in about 8 minutes. But the mixer wouldn't easily handle batches much larger than that, and it would be tiring to use.

Convenience and safety

Performance isn't everything. Some mixers are easier to use than others.

Controls. On all models, the speed switch is located toward the front of the handle so the machine can be held and set to different speeds (or shut off) with one hand. On most models, the control is a switch that moves forward from "off" to increasingly higher speeds. A few models have a thumbwheel control that turns in one direction from "off" to "low" through high speeds. Both types are convenient.

Most **Waring** models have a switch with "off" in the center location and low speeds on one side, high on the other. It's too easy to bypass "off" with this arrangement. The **Waring HM8** has a selector dial with twelve speed settings. But the mixer has only six speeds. The overly elaborate control is confusing.

Handle comfort. The men who participated in the tests found all the mixers comfortable to hold while doing any mixing chore. A woman tester, however, said that some handles were quite uncomfortable when a firm grip was required for fairly heavy-duty chores such as mashing potatoes. The seven models she criticized (see Ratings) have a hollowed-out channel on the top of the handle.

Beater insertion and ejection. With all but one of the mixers, the beaters are interchangeable: Either beater fits either socket in the mixer housing. The **Moulinex** is the exception. But you can easily determine which beater goes where on that model, since the sockets are different sizes to match the large and small diameters of the beater shafts.

All the mixers have a push-button or lever arrangement for ejecting the beaters. On many, it's conveniently positioned in front of the speed control, so the beaters can be ejected with the same hand that holds the mixer. On some models, the ejection button is on the side of the mixer. That may provide convenient, single-handed operation when you're using the mixer with the right hand, but not when you're using it with the left hand. A few models have the ejection control on the bottom of the mixer. With that design, removing the beaters is always a two-handed effort.

Stability. A mixer with its beaters in place should stand on end on a countertop, without toppling over. That can be a lot to expect, however, when the beaters are loaded down with cookie dough. The high-rated **Sunbeams** and the **Penney**'s **5195**, made by Sunbeam, maintain their balance even under such difficult circumstances. In less severe situations, most of the other models manage to stay upright, but the **Moulinex** topples over very easily.

Noise. Noise level varies with speeds used and the foods being mixed. But none of the tested mixers are exceptionally quiet or exceptionally noisy.

Cleaning. Removing sticky foods from the beaters can be a chore. The **Moulinex** and **Waring HM9518** whisks are easiest to clean.

Portable mixers, like many food-fixing appliances, have crevices and grooves that seem designed to trap food and dirt. The plastic housings hide dirt better than the chrome-finished ones, which show every fingerprint and smear.

Cord length. A long cord can be an asset if your kitchen hasn't many electrical outlets. All the mixers have a generous cord, from about $4^1/_2$ to 6 feet long. (Left-handed users might be annoyed by a cord that affixes to the right side of the housing, and most do.)

Storage. A portable mixer doesn't require much storage space: it can share space in a drawer or cupboard with other kitchen gadgets. Most models have a keyhole-shaped opening on the mixer body so they can be hung on a wall hook. A few come with a wall storage rack. The Ratings note models that have clips on the side of the housing for storing the beaters. That's a nice feature, particularly for wall hanging. Most portables have a detachable cord to simplify storage.

Safety. There's always the possibility

of injury from rotating beaters. Fortunately, the beaters aren't sharp edged and they tend to repel objects—including fingers—in their path.

Recommendations

A portable mixer is an ideal appliance for making cake batter, mashing potatoes, and whipping cream. It's lightweight and fairly compact for convenient storage.

If a portable mixer is the *only* mixer in your kitchen, choose from among the top-rated models. Most of the **Sunbeam** and **Penney**'s models and the governor-controlled **Hamilton Beach** model can handle heavy-duty chores, such as mixing cookie dough, without difficulty. Several can be operated at an extra-slow speed, which is convenient in the early stages of mixing liquids.

The **Moulinex** is good with heavy-duty loads, too. And it can whip cream faster than most other models. If you occasionally like to make small batches of bread, the **Moulinex,** with its dough hooks, is the only portable that can handle the job. The **Moulinex** has some drawbacks. It may spatter when mashing potatoes, and it tends to topple when its beaters are loaded. But its advantages outweigh its disadvantages.

The **Penney**'s **5190**, at $13, may be the best value among the high-rated mixers. But bear in mind that discounts may lower the prices of other models to a competitive level.

The ability to do heavy chores isn't important if a portable mixer is to be an adjunct to a stand mixer that you already own, or if you rarely use a mixer. In both cases, then, let price be your buying guide.

A portable mixer holds the cook captive. You have to hold it and maneuver it. That may not be a problem if you occasionally spend a few minutes mashing a pot of potatoes or whipping a batch of cream. But if you use a lot of recipes that call for beating a batter for five minutes, or if you tend to do things in large quantities, you'll be better off with a stand mixer, which demands a bit less of you.

Ratings of portable mixers

(As published in a September 1984 report.)

Listed in order of estimated overall quality. Bracketed models were judged approximately equal in quality and are listed alphabetically. Prices are approximate retail; + indicates shipping is extra.

Brand and model	Price	No. of speeds	Advantages	Disadvantages	Comments
SUNBEAM 03196	$39	1	A,B,E,F,J	—	A
MOULINEX V394	44	3	A,B,C,D,G,I,K	b,f,k,m	—
J.C. PENNEY 5195	28	5	A,B,F,J	—	—
SUNBEAM 03076	35	5	A,B,F,J	—	—
HAMILTON BEACH 0912T	40	1	A,E	—	A,B
HAMILTON BEACH 103	33	1	A,E,	—	A
J.C. PENNEY 5190, *A Best Buy*	13	5	A	m	—
SUNBEAM 03056	27	5	A	m	—
WARING HM9518	29	12	A,C,D,G	a,l,m	—
GENERAL ELECTRIC M22	26	5	H	—	—
GENERAL ELECTRIC M24	19	3	H	j	—
BETTY G HM26	18	5	—	j,m	—
RIVAL 433	20	3	B	j,m	—
RIVAL 439	21	5	—	m	—
SUNBEAM 81286	28	7	—	l,m	—
WARING HM8	27	6	—	c,i,l,m	—
WARING HM12	28	12	—	c,l,m	—
WARING HM110	22	3	B	c,g,h.l,m	—
HAMILTON BEACH 97	20	3	—	d,g,j,l,m	—
HAMILTON BEACH 107	24	3	—	d,g,j,l	—
HAMILTON BEACH 108	26	5	—	d,g,l	—
HAMILTON BEACH 113	30	3	F	e,g,l	—
HAMILTON BEACH 116	32	5	F	d,e,g	—
HAMILTON BEACH 121	20	3	H	d,g,j,l	—
SEARS Cat. No. 82878	16+	3	—	d,g,j,m	—

1 *Speeds are continuously variable.*

SPECIFICATIONS AND FEATURES

All: ● Come with power cord about 4½ to 6 ft. long ● Weigh 1¾ to 2½ lb.

Except as noted, all: ● Have beaters that fit in either left or right socket. ● Have convenient beater-ejector mechanism. ● Have easy-to-read control markings. ● Have good stability when placed on heel rest with beaters in place. ● Lack built-in provision for beater storage. ● Have keyhole slot for hanging mixer on wall hook. ● Have detachable power cord.

KEY TO ADVANTAGES

A – Handled heavy loads better than most.
B – Excellent at mashing potatoes.
C – Whipped cream fastest of all.
D – Less likely than most to collect heavy mixtures on beater shafts.
E – Can run at lower speed than most.
F – Has extra-speed or power-burst feature.
G – Has whisks instead of conventional beaters; judged easier to clean and more corrosion-resistant.
H – Clips on motor housing store beaters.
I – Only model that comes with dough hooks. Capable of mixing small quantities of bread dough.
J – Very stable on heel rest, even with dough on beaters.
K – Motor housing easier to clean than most.

KEY TO DISADVANTAGES

a – Spattered greatly when mixing pudding.
b – Spattered greatly when mashing potatoes on high speed.
c – Heavy batter tended to collect on beater shafts more often than most.
d – Labored more than most under heavy loads.
e – Labored somewhat more than most under heavy loads, even when using power-burst feature.
f – Beaters are not interchangeable.
g – Beater ejection less convenient than most.
h – Hard-to-read speed markings.
i – Rotary speed selector shows 12 settings, but mixer has only 6 speeds. Judged confusing and inconvenient.
j – Handle somewhat less comfortable than most.
k – Less stable on heel rest than other models.
l – Lacks provision for wall mounting.
m – Power cord is not detachable.

KEY TO COMMENTS

A – Has governor speed control; helped maintain constant speed under load.
B – Comes with plastic wall rack or cabinet for storing mixer, beaters, and cord.

Stand mixers

Although relatively few of them are on the market, stand mixers vary considerably in performance. In particular, differences in power have a great effect on performance. But other differences—in the design of the beaters, the size of the bowl, and bowl rotation—also affect performance, convenience, or both.

With portable mixers, the cook is in control, moving the whirring beaters around the inside of the mixing bowl. With most stand mixers, the pair of beaters is stationary; the bowl rotates on a turntable in the base of the stand.

The **KitchenAid** and **Waring** mixers are different. The expensive, heavy-duty **KitchenAid** uses a single mixing tool—either a large, whisklike beater, a flat beater, or a dough hook—that automatically revolves around the inside of the bowl. With the inexpensive, light-duty **Waring** mixers, beaters and bowl remain stationary; you must manually rotate the bowl to circulate the food properly.

Stainless-steel bowls are an asset. They're lightweight and won't break when dropped.

Some stand mixers with glass bowls are Not Acceptable because the glass occasionally chipped during Consumers Union's dough-kneading tests.

Working

CU used a number of representative food-fixing chores to evaluate the performance of stand mixers. They range from the light-duty task of whipping cream to the heavy work of kneading bread dough.

Here's how the stand mixers performed in the tests:

Whipping cream. Most of the stand mixers, like the portables on page 132, produce perfect whipped cream that is more than double the volume of the liquid cream. Mixers excel at chores such as whipping cream or beating egg whites because they're able to introduce air into the food continuously while beating, and it's air that provides volume.

The **KitchenAid,** the **Sunbeam 01096,** and the Sunbeam-made **Wards 45754** prove to be the fastest models. They can whip cream in a little more than 2 minutes. Most of the others take 6 minutes to whip cream. But, except for the **Waring** models, they all can produce excellent cream. The whipped cream made in the **Warings** doesn't have as much volume and doesn't peak as well as the cream made with the other stand mixers or with the portables.

Blending mayonnaise. The stand mixers, like the portables, will take four to seven minutes to make mayonnaise—a job a blender can do in 40 seconds. Extra time is needed to drip the oil ever so slowly into the mixture.

Because mayonnaise is rather thin, the beaters on some models don't get enough friction to keep the bowl rotating.

Any such inconveniences aside, the stand mixers produce a very good mayonnaise.

Mixing pudding. Mixers have no difficulty combining pudding mix with milk. The **Sunbeam 01096** and the **Wards 45754** are noteworthy for their smooth operation and good circulation within the mixing bowl during this light-duty task.

Mashing potatoes. In each stand mixer CU's testers mashed two pounds of potatoes—cut up, boiled, and drained—with some milk and butter. The chunks of potato were broken up and then mashed, using the manufacturers' recommended speed. After the potatoes were mashed for a minute, they ran them through a sieve to see what lumps remained. The **Hamilton Beach 60** and the **Waring HS158** and **HS3218** left more lumps than the others.

Given a choice, a portable mixer is better for mashing potatoes. Its light weight and compact size makes it easy to use right in the potato pot, and by maneuvering the mixer manually, you can chase down lumps. Although many stand mixers can be used as hand mixers, their weight and size make them awkward to handle.

Making cake batter. Like portable mixers, stand models are masters of cake batter. That's because they incorporate air as they work, so the finished cake has a lighter texture.

Mixing cookie dough. Heavy and thick dough poses a significant challenge to some portable mixers, but requires only modest effort from most stand mixers. CU's butter-cookie recipe includes two fairly difficult chores: creaming a half-pound of butter at the beginning, and mixing two cups of flour into the butter mixture at the end. Only the light-duty **Waring**s seem to work hard at those tasks. (The butter used was fairly hard. Had it been softened a bit, as some manufacturers recommend, the chore would have been less difficult.) The heavy-duty **KitchenAid,** the **Sunbeam 01266,** and the similar **Sears**

82188 can plow effortlessly through the cool butter and incorporate the flour without hesitation.

Blending pie dough. To achieve flaky piecrust or pastry, you cut cold butter or shortening into the flour. It's not a chore mixers do very well. The friction of the beaters softens the butter or shortening, so it acts more like oil. Oil can be used, of course, but the resulting pastry is usually less flaky. The **Kitchen-Aid** makes a very good piecrust using the flat beater.

Kneading dough. All the mixers except the light-duty **Sunbeam 02066** and the two **Warings** come with dough hooks for kneading bread. The testers made a small batch of sticky brioche dough that used 2½ cups of flour. The **KitchenAid,** the **Sunbeam**

01266, and the Sunbeam-made **Sears 82188** mixers seemed to handle the job with the least effort, but none of the machines was overtaxed by the job. All were able to have the brioche dough ready for baking in seven or eight minutes.

The **KitchenAid** instructions say it can work with recipes that call for as much as 8 cups of flour. In the tests, it did so with ease. The baked bread was excellent. The heavy-duty **Sunbeam 01266** and **Sears 82188** can manage up to 7 cups of flour. Most other models set the limit at 6½ cups. The instructions with some models say the maximum quantity that the machines can handle is 3½ cups or so, yet the instructions include bread recipes that call for nearly twice as much flour.

A look at convenience

Speeds. The stand mixers have far more settings than most cooks can use. It does help, however, to have an exceptionally slow speed to keep spattering to a minimum when mixing liquids. Most of the mixers have a governor-controlled motor, so a very slow speed is available if the control is set at or below the lowest speed setting.

Controls. The controls are well designed on the heavy-duty **Sunbeam 01266** and **01096** and the similar models made for **Sears** and **Wards.** The entire rear section of the motor housing turns to set the speed. And the markings are quite clear.

The controls on the **Hamilton Beach 60** are less convenient. The small metal thumb slide control operates stiffly and is difficult to move from its "off" setting. But the thumbwheel controls on the light-duty **Warings** frequently slip their connection with the speed-controlling mechanism. As a result, you

can't rely on the thumbwheel's labeling to reflect the actual speed of operation.

Handling the beaters. With most stand mixers, the beaters are separate and interchangeable: Beaters fit either socket in the motor housing. The beaters on the **Sunbeam** family of stand mixers are an exception. The conical beater has to fit in one socket, the squarish one in the other. Markings on the motor housing show which one goes where, and the operating manual gives clear instructions, so there's not likely to be a mixup with the beaters. The **Sunbeam**'s dough hooks, however, can easily be inserted incorrectly. When CU's testers intentionally did so, the bowl rotated backward and dough crept up the beater shafts and into the sockets. The manufacturer warns that incorrectly inserting the beaters or dough hooks could damage the mixer.

The beaters of the **Hamilton Beach 60**

mixer are joined together by a frame. There's no problem inserting them into the sockets, but once inserted, they have to be tightened with a thumbscrew. That's an annoying extra step. To use the dough hooks, you have to clamp them in place and then clamp the stainless-steel bowl to prevent it from moving.

The same bowl clamping is required with the **Sears 82768** and the **Hamilton Beach 43**, both of which have glass bowls. And with those models, the problem is more than just annoying. When the metal clamp is adjusted "finger tight," as the manufacturer recommends, it occasionally chips the glass. The testers were left with a hazardous ragged edge on the bowl. They threw out the dough because it was likely to have fragments of glass in it, and they rated the three models Not Acceptable.

To eject the beaters on most of the mixers, you turn part of the motor housing or push a button. Some of the buttons are positioned so they're more easily used by right-handed cooks than left-handed, but that's a minor problem. The **KitchenAid**'s mixing tools come out with a single turn and pull. Loosening the thumbscrew to remove the beaters from the **Hamilton Beach 60** is a messy and annoying chore.

Noise. None are particularly quiet. And the most powerful—the high-rated **KitchenAid, Sears,** and **Sunbeam** models—are quite noisy at high speed.

Cleaning. Rinsing a liquid mixture from all the beaters is easy. But sticky substances can be annoying. There are twelve wires on the **KitchenAid** whisk, all of which may need to be wiped. Cleaning the joined beaters of the **Hamilton Beach 60** requires even more patience because it's hard to get in between the beaters and their frame.

Cleaning the machine itself is easiest

with the **KitchenAid.** Its rounded shape and lack of decoration leave few areas to trap food or dirt. The **Hamilton Beach 60,** on the other hand, has many crevices and grooves.

Portable potential. If a stand mixer is the only mixer in your kitchen, you might appreciate the convenience of being able to use it as a portable on occasion. The motor housing on the high-rated heavy-duty **KitchenAid, Sunbeam,** and **Sears** models doesn't detach from the stand for hand-held use. Even if it did, it'd be too heavy and bulky for comfort. All the other mixers are detachable, but only the light-duty models are really suitable for hand-held use. There are various arrangements that release a mixer from its stand. The **Waring**'s is the simplest: you rock the mixer loose. But that arrangement allows the mixer to release when you may not expect it to. The lock on the **Hamilton Beach 60** is not easy to figure out. Directions are on the stand itself, a good idea. However, you must *remove* the mixer to see the instructions.

Safety. The stand mixers pose no unusual dangers other than those associated with the glass bowls on the models judged Not Acceptable. There is some risk of injury from the rotating beaters. But the beaters aren't sharp edged and they tend to repel objects—including fingers—in their path.

Stability. The heavy-duty stand mixers are heavy enough to stay put on a counter even while doing tough chores.

The motor housing of the high-rated models can be locked on the stand so that if you lift the entire unit by the housing, the stand and bowl stay in place. The housing on the **Sunbeam 01266** and **Sears 82188** also locks in place when it's tilted back, away from the bowl. It won't crash down into the bowl when the beaters are laden with dough.

Cord length. Appliance cords should be long enough for comfortable use, but not so long that excess wire drapes over kitchen counters. The cords on the stand mixers range from 3½ to 5 feet long—all generous enough. The cord on some models is detachable. There's little advantage to that feature, but some cooks might find it helps to eliminate a tangle of cord for storage.

Storage. The larger, more powerful stand mixers are taller and slightly bulkier than the small, light-duty models, so they require more storage space. But stowing them away in a cupboard may turn out to be more trouble than it's worth, because they can be heavy. The twenty-two-pound **KitchenAid** is the heaviest. Weights are given in the Ratings.

Recommendations

Unless you do a lot of baking—particularly of the kind that calls for heavy work such as kneading bread—you really don't require a stand mixer. A good portable (page 132) would serve you well. Those models can handle sticky cookie dough without pause. And the **Moulinex,** which comes with dough hooks, can even make small batches of bread dough. Both sell for less than $45.

If you really need a stand mixer, you'll need a heavy-duty model. That means the **KitchenAid.** It handles large batches of heavy dough with remarkable efficiency. Its $270 price tag is significantly higher than any of the other models tested, but you're likely to find substantial discounts if you shop around.

In the next Ratings group, the **Sunbeam 01266** and the **Sears 82188** offer a little less performance for a little less

money—around $200 suggested retail. Both are convenient to use.

As you move down the Ratings, the price of the machines decreases. But so do performance and convenience. The last of the heavy-duty models, the **Hamilton Beach 60,** offers more disadvantages than advantages. Unless it's absolutely bargain-priced, we see no point in buying it. And there's no apparent reason to buy a light-duty stand mixer.

A number of stand mixers offer optional attachments to expand their versatility. But a large-capacity food processor can handle some of the heavy-duty mixing chores—cookie dough and bread dough—that are often assigned to a stand mixer. And it can chop, grate, and slice, too. For the serious cook, a food processor may be a better investment than a stand mixer (see pages 118–19).

Ratings of stand mixers

(As published in an October 1984 report.)

Listed in order of estimated overall quality. Bracketed models were judged approximately equal in quality and are listed in order of increasing price. Weight includes beaters, cord, and large bowl. Prices are suggested retail; + indicates shipping is extra.

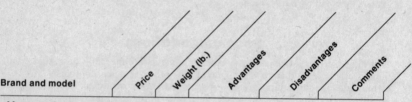

Brand and model	Price	Weight (lb.)	Advantages	Disadvantages	Comments
Heavy-duty models					
✪ KITCHENAID K45SS	$280	22	A,B,D,E,F,H, I,K,L,M	d,l,q,s	E,G,M,N
SEARS Cat. No. 82188	200+	12	A,I,J,K,M,N,O	e,q	J,K,N
SUNBEAM 01266	227	11	A,I,J,K,M,N,O	e,g,q	K,N
WARDS Cat. No. 45754	90+	8	C,D,K,M,O	g,j	L,N,O
SUNBEAM 01096	131	13.5	C,D,K,O	g,j	L,N,O
HAMILTON BEACH 60	154	9.5	M	b,c,i,k,m,r	C,I,O

Not Acceptable

■ *The following models were judged not acceptable because clamp for dough-hook adapter can chip their glass bowl, leaving jagged edges on bowl and glass particles in food.*

SEARS Cat. No. 82768	50+	8	—	h,i	F,O
HAMILTON BEACH 43	74	8	—	h,i	F,O

Light-duty models

SUNBEAM 02066	72	4.5	G,M	—	A,O
WARING HS158	46	5.5	G	a,b,f,j,l,n,o,p	A,C,D,H
WARING HS3218	50	5.5	G	a,b,f,j,n,o,p	A,B,C,D,H

SPECIFICATIONS AND FEATURES

All: ● Heavy-duty models come with dough hooks as well as beaters. ● Light-duty models are intended for relatively light mixing tasks and do not come with dough hooks. ● Have power cord 3½ to 5 ft. long. *Except as noted, all:* ● Come with two beaters that can be inserted into either socket in motor housing. ● Come with large, approx. 4-qt. glass bowl and 1 to 1½-qt. glass bowl. ● Have governor-controlled motor with continuously variable speed control and 12 to 16 marked settings. ● Have bowl that automatically revolves on turntable.

KEY TO ADVANTAGES

A – Excellent (**KitchenAid**) or very good (**Sears** and **Sunbeam**) for heavy dough and batter.

B – Suitable for mixing larger amounts of dough than any other model.

C – Excellent at mixing pudding.

D – Whipped cream faster and with less effort than most.

E – Spattered liquids less than others.

F – Required less assistance than any other model; only model to make very good pie crust dough using cool butter.

G – Detachable mixer works well as portable because of light weight and maneuverability.

H – Easy-to-clean base and motor housing.

I – Lock on stand allows entire unit to be picked up and carried by motor housing.

J – Lock on stand keeps mixer from dropping down when tilted away from bowl.

K – Beaters easier to clean than most.

L – Single whisk, flat beater, and dough hook easier to remove and replace than beaters and dough hooks of other models.

M – Comes with stainless-steel bowls.

N – Small bowl allowed ample clearance to add ingredients while mixing.

O – Easy-to-set speeds.

KEY TO DISADVANTAGES

a – Did not whip cream as well as most.

b – Tended to leave lumps in mashed potatoes.

c – Base and motor head difficult to clean.

d – Weight and size make mixer more difficult to store than most.

e – Lock on stand must be released each time head is tilted up or down.

f – Stand lock relatively insecure.

g – Beaters and dough hooks should not be inserted into either socket.

h – Inconvenient beater-ejector button location.

i – Using dough hook involves clamping dough-hook adapter to mixing bowl, inconvenient.

j – When tilted back with beaters loaded with batter, mixer tended to fall back into bowl.

k – One-piece beater assembly difficult to clean, insert, and eject.

l – Does not come with small bowl. Mixed small quantities well but has less clearance than most for adding ingredients during use.

m – Awkward slide control made setting speeds more difficult than with most.

n – Speed settings shifted during use.

o – Motor lacks governor control and cannot be run at very low speeds.

p – Less suitable than others for frequent, long mixing, chores.

q – Noisier than most at high speeds.

r – Required more manual assistance than most heavy-duty models.

s – Hard-to-clean whisk-type beater.

KEY TO COMMENTS

A – Not tested for dough mixing since manufacturer does not provide dough hooks or claim mixer can be used for that function.

B – Comes with plastic cup and drink-mixer paddle instead of small bowl.

C – Large bowl has only 2½- to 3-qt. capacity.

D – Lacks drive button on beater to rotate bowl.

E – Bowl is stationary; beater rotates.

F – Has 9 marked speeds.

G – Has 6 marked speeds.

H – Speed control is not continuously variable.

I – Instructions for removing mixer from base are visible only when mixer is removed.

J – Has built-in timer with alarm; timer doesn't turn mixer off or on.

K – Has built-in light in mixer body.

L – Cord is detachable.

M – Cord has three-prong plug.

N – Has power take-off for running attachments (not tested).

O – Detachable mixer, but not suitable for portable use because of size and weight.

Immersion gadgets

You may have seen advertisements for a relative newcomer to the food-fixing field: the hand-held immersion device. It's often marketed as a blender, sometimes as a mixer, as a food processor, or as all those things.

This food fixer sounds too good to be true. But in fact, it's not very good at all.

Consumers Union bought four models for evaluation: The **Braun MR6 Minipimer Vario** (approximate retail price, $70), the **Braun MR30 Minipimer 2** ($35), the **Maxim M11 Maxi Mix** ($60), and the **Taurus SR2C** ($55). Size, shape, power, and speeds differ somewhat, but the four devices are used in a similar way.

All of the devices can whip cream the way a blender does—not very well. The **Braun MR6 Minipimer** can do a better job than the others because of the shape of its whisk attachment; still, the whipped cream will be nothing to write home about.

All of these models are worse than a blender at chopping cabbage. A blender chips the cabbage; these gadgets barely cut it. They are poor at chopping parsley, although the **Braun MR6 Minipimer** and the **Maxim** are no worse than a blender. Other chopping and grinding chores—crumbing crackers, making peanut butter, grinding Parmesan cheese, chopping onions, and grinding carrots—are beyond the capabilities of the devices.

There is one task that the immersion appliances can do well—making mayonnaise—but that's not a very challenging chore.

Generally, an immersion device is best suited for mixing liquids. But they spit and spatter furiously while doing such jobs. They can't do chores that involve reducing solids from large-cut pieces, because the guard around the cutter admits only small particles, and circulation of ingredients is poor. They would probably work better without the guard, but then they'd be unreasonably hazardous. (As it is, the guards offer only marginal protection.)

You can derive far greater value and versatility if you buy one of the high-rated mixers and a good, inexpensive blender. Together, the mixer and the blender would probably cost less than the price of one of these immersion gadgets.

Freezers

A well-stocked freezer can provide raspberries in January and winter squash in July, stews and sauces, casseroles and codfish, mail-order filet from the stockyard, and take-out pizza—nearly everything, in short, to nourish a carload of unexpected guests or to let you stay indoors when the wind-chill factor is high. But a freezer probably won't save you much money.

Even if you use your freezer to stock up on sale items or store the bounty from a backyard garden beyond what you can store in your refrigerator's freezer compartment, your economies will have to exceed $75 to $100 a year before you will start to save.

If you amortize the purchase price over the fifteen years you can expect a freezer to last and then consider the $50 or $60 worth of electricity a full-sized freezer is apt to use in a year, one of these appliances is more likely to be a convenience than an economy.

A freezer can actually cause you to waste money if you don't manage what you freeze prudently. Thrown-away steak is no bargain, even if you bought it at a bargain price. Without proper wrapping, frozen food can suffer "freezer burn," or dehydration. And without proper rotation of your stocks, you can ruin a lot of food. Frozen food is ultimately perishable, and too long a stay in the freezer can make the tastiest of foods inedible.

Freezer management, however, requires only a little time and organization; the box on page 143–44 gives some tips. Most freezers entail one other chore: periodic defrosting. (Self-defrosting models exist, but they're relatively expensive to buy and to run.)

There are two types of freezers: chests and uprights. Each has its advantages and disadvantages, noted in the Recommendations on page 144.

Freezers also come in several sizes. Large-capacity models—rated by manufacturers at 15 to 16 cubic feet—may be too big for a lot of families. Compact freezers, with a capacity of less than 10 cubic feet, may be a more practical choice.

Freezing

A good freezer should keep everything inside it at a temperature of 0°F, give or take a couple of degrees, even on the hottest of summer days.

Theoretically, a chest freezer has a big advantage over an upright freezer in maintaining uniform cold temperatures. In a chest freezer, the freezer coils are within the walls, surrounding the food. The door is on top; cold air, which tends to sink below warm air, naturally wants to stay in the chest. Most of the chest freezers keep cold enough when faced with an outside temperature of 70°. The best are a family of six models that appear to be made by the same manufacturer—the **Wards, Frigidaire, Kelvinator, White-Westinghouse, General Electric,** and **Gibson.** Only the top of their storage compartment is a few degrees above zero; the rest is closer to zero.

The design of an upright freezer, while convenient, is at odds with physics. Cold air within that tall, narrow cabinet tends to sink to the bottom, letting warmer air collect at the top. Each time you open the door, cold air spills out from the bottom while warm air comes in at the top. The freezer coils, instead of surrounding the food, are built into the shelves of the main compartment.

Still, some upright models do nearly as well in the tests as the chest freezers do.

The door shelves of an upright freezer, relatively far from the freezer coils, are likely to be far warmer than the main compartment. Food on those shelves is unlikely to remain at the ideal 0°.

Extreme conditions. Reserve capacity gives a measure of a freezer's performance on really torrid days. It's also an indication of how well a freezer will endure through the years.

Resistance to condensation in humid weather used to be more of a problem with freezers than it is now. Keeping the outside walls warm deters droplets from forming and dripping to the floor. Freezers commonly use one of two methods to warm the outside of the freezer. Most are designed so the "hot tubes," which carry the heat extracted by the refrigeration process, run though the walls. A few have electric heating elements instead; you can switch the elements on to discourage condensation—or off to save energy.

The cost of freezing

Because of their more efficient design, chest freezers as a rule use less electricity than upright freezers. The chest models in CU's tests will cost about $55 to $60 to run per year, compared with $60 to $67 for most of the uprights. Because that range is so narrow, you generally need not place great emphasis on energy consumption as a criterion for choosing one freezer over another.

Defrosting and cleaning

You will probably have to defrost an upright freezer twice a year, the frequency depending on the humidity and on how often you open the door. With a chest freezer, however, you may be able to go for 12 to 18 months between defrostings, especially if you occasionally scrape off the ice that builds up around the rim.

Most of the models have a drain to draw off the water coming from the melted ice.

As a further aid to defrosting, most uprights have a hose attached to the drain. Chests generally are designed so a standard, half-inch hose adapter can be attached.

Chest freezers are relatively easy to defrost and clean. They have a smooth interior and removable wire baskets or dividers instead of shelves. You can use an ice scraper to hasten defrosting and then swab down the walls.

You have to be more patient when defrosting an upright, waiting for the ice to melt around the coils in the shelves. Using tools to speed the process could lead to a very expensive puncture of the coils. Since the shelves are stationary, cleaning is a bit difficult.

Most uprights have wire shelves. So, if something spills on the top shelf, it's likely to drip all the way to the bottom.

Freezers with a plastic or porcelain-on-steel interior resist scratching, chipping, and rusting better than those with painted-steel insides.

Convenience

Freezers have little in the way of controls—merely a temperature control dial. On chest freezers, the dial is outside, generally low and on the left. On uprights, it's inside, generally on the right. For most effective use of that dial, you should use a thermometer to set your freezer's temperature to 0°F. See the section on page 143 for more information.

Most of the models in our test have an interior light; the bulb on some is more vulnerable to breakage than that on others.

Some models have an outside light to indicate that the freezer is receiving electrical power, to guard against accidental unplugging or an overlooked blown fuse.

Space organizers are minimal in chest freezers—a wire basket and maybe a divider. Most uprights have three fixed shelves in the main compartment.

A typical upright has five solid shelves in the door. The metal retainers that keep cans and packages on the shelves are generally fixed.

Freezer management

It takes only a little extra time and organization to manage your freezer wisely:

Use a freezer thermometer (see page 269). When you're first setting up the freezer, leave the thermometer in the center of the empty compartment for at least a day to be sure the freezer is working properly. Depending on the reading, adjust the temperature control. Aim for a reading of 0°F.

Learn which foods freeze best. Check the freezer's manual, or send $3.50 for the U.S. Department of Agriculture booklet "Home Freezing of Fruits and Vegetables" (it's item 142M; the address is: Consumer Information Center (P), P.O. Box 100, Pueblo, Colo. 81002). Nutritionally, meat, fish, poultry, and eggs are the same frozen or fresh. Fruits and vegetables can lose vitamins if they're not handled the right way before freezing.

Wrap the food correctly. The wrapping needs to be vapor-proof. In tests of freezer wraps, Consumers Union finds **Saran Wrap** to be the best. You can also use aluminum foil or plastic containers with snap-on lids. Don't rely on waxed paper, butcher's paper, regular polyethylene plastic wraps, or even cardboard ice cream cartons. Rewrap all supermarket-packaged meat. Try to expel as much air as possible from a package before you seal it. Freezer tape, rubber bands, twist ties, or even string can seal the wrapping.

Label the packages with their contents, serving size, and date of freezing. The date is especially important, so you can use up food before it's been in the freezer too long. Try to balance the flow of food into the freezer with the flow out—that's one way to ensure that no foods stay frozen overlong. Try to move the oldest packages to top or front, as a supermarket should.

Make sure the food freezes quickly. When using the freezer to put by food made from scratch, don't stack the packages; leave an inch or so of room between them so cold air can circulate and remove heat. Don't put too much warm food into the freezer at one time.

Thaw food in the refrigerator. If you must speed the process, run lukewarm water over the package. Never refreeze thawed-out food.

Defrost when the frozen-food supply is low. Transfer the remaining food to your refrigerator's freezer or to the refrigerator itself. Or wrap it in layers of newspaper while you defrost. Use pans full of hot water and a fan to speed the process. Never use knives, ice picks, or other sharp objects to loosen the ice.

Recommendations

First, decide between a chest freezer and an upright freezer. Chests cost less to buy, cost less to run, need less frequent defrosting—and their design helps keep food uniformly cold. However, they take up a lot of floor space and it's hard to keep track of food items. Upright freezers are easy to use and are about the same size as a refrigerator. But uprights cost more to buy, cost more to run, and their design encourages warm spots. Overall, a chest freezer probably has the edge over an upright model—if a rather bulky box of an appliance fits into your basement or kitchen arrangement.

When figuring how well a freezer will fit in your home, add a few inches at the back and sides to let heat dispel from the condenser. Note on which side a chest's controls are placed—you'll need to leave some extra room to reach them.

Every model offers a one-year warranty covering parts and labor for the unit as a whole, plus a refrigeration-system warranty for five years. Freezer companies also warrant your food in the event of thaw due to freezer breakdown; compensation varies from one company to another.

In the event of a power failure, your local utility company may accept some responsibility for food spoilage if the power outage lasts beyond a particular time period. But to be on the safe side it's a good idea to know about a nearby source for dry ice. A few pounds can help tide you over a fairly long shutdown.

Ratings of freezers

(As published in a February 1985 report.)

◉ ◓ ○ ◒ ●

Better ⟷ Worse

Listed by types; within types, listed in order of estimated overall quality. Differences in quality between closely ranked models were slight. Prices are approximate retail.

Brand and model	Price	Claimed capacity (cu. ft.)	Measured capacity (cu. ft.)	Temperature judgments: Cabinet uniformity	Door uniformity	Reserve capacity	Compensation for change	Energy cost per yr. [2]	Dimensions [1]: HxWxD, door closed (in.)	H (lid open) or D (door open) [3] (in.)	Food-loss warranty [4]	Advantages	Disadvantages	Comments
Chest models														
WARDS 8523	$299	15.7	15.1	◓	—	◒	◒	$58	35x44³/₄x30¹/₄	61	$100(1+4)	B,H	h	*D*
FRIGIDAIRE CF16J	570	15.6	15.1	◓	—	◒	◒	58	35x44³/₄x30¹/₄	61	150(1+2)	—	h	*D*
KELVINATOR HFS156SM	659	15.6	15.1	◓	—	◒	◒	58	35x44³/₄x30¹/₄	61	200(1+2)	H	a,h	—
WHITE-WESTINGHOUSE FC164D	520	15.6	15.1	◓	—	◒	◒	58	35x44³/₄x30¹/₄	61	150(1+2)	H	a,h	—
GENERAL ELECTRIC CB15DF	419	15.6	15.1	◓	—	◒	◒	58	35x44³/₄x29³/₄	60³/₄	150(1+4)	B	a,h	—
GIBSON FH16M2WM	360	15.7	15.1	◓	—	◒	◒	58	35x44³/₄x30¹/₄	61¹/₂	150(1+2)	—	a,h	*C*
AMANA C15B1	450	15.0	14.8	○	—	○	●	54	36³/₄x41³/₄x31	61³/₄	125(1+4)	A,C,H	h,i,m	*D*
MAGIC CHEF C15D	369	15.2	14.8	◒	—	◉	◓	59	35³/₄x43³/₄x30	61¹/₂	200(1+2)	—	c,h,m	*A*
SEARS 14152	335	15.1	14.5	◒	—	○	○	53	36x43¹/₄x28¹/₄	62¹/₄	100(1+4)	A	a,h,m	*A,L*
WHIRLPOOL EH150CXL	400	15.2	14.5	◒	—	○	○	53	36x43¹/₄x28¹/₄	62¹/₄	125(1+4)	A	a,h,m	*A*

Upright models

Brand and model	Price	Claimed capacity (cu. ft.)	Measured capacity (cu. ft.)	Cabinet uniformity	Door uniformity	Reserve capacity	Compensation for change	Energy cost per yr. [2]	Food-loss warranty [4]	H×W×D, door closed (in.)	H (lid open) or D (door open) (in.) [3]	Advantages	Disadvantages	Comments
AMANA ESU15C	540	15.0	13.7	◐	◐	●	●	58	125(1+4)	63x28¼x30¼	55¾ 13½	B,E,G,I	b,f,l	E
ADMIRAL DF15	540	15.2	12.9	●	●	●	◐	67	200(1+2)	60¼x30x28½	55¾	G	f,k,l	K
MAGIC CHEF DF15	399	15.2	12.9	●	●	●	◐	67	200(1+2)	60¼x30x28½	55¾	G	f,k,l	K
SEARS 24152	335	15.1	13.2	○	◐	◐	◐	58	100(1+4)	63½x28x26½	51¾	A,G	a,d,e,h,j,k,l	B,G,K,L
GIBSON FV16M5WN	430	16.0	14.3	○	◐	◐	○	66	150(1+2)	59¼x28x30	55½	H	c,k	H
GENERAL ELECTRIC CA16DF	469	16.0	14.3	◐	◐	◐	○	66	150(1+4)	59¼x28x30	55½	B	k	I
WHITE-WESTINGHOUSE FU166E	580	16.0	14.3	○	○	○	○	66	150(1+2)	59¼x28x31½	56¾ (20¾)	—	k	F,I
KELVINATOR UFS161SM	669	16.0	14.3	○	◐	○	○	66	150(1+2)	59¼x28x30	55½	B	a,k	I
FRIGIDAIRE UF16	590	16.0	14.3	○	◐	◐	○	66	150(1+2)	59¼x28x30½	55½	—	c,k	I
WARDS 4633	339	16.0	14.3	○	◐	○	○	66	100(1+4)	59¼x28x30	55½	—	f,g,k	H
WHIRLPOOL EV160FXK	480	15.9	13.0	○	◐	●	○	84	125(1+4)	65¾x29¾x31¾	58¼	D,F,H	j,k,l	B,G,H,J

[1] Dimensions are rounded to next higher ¼ in. Uprights' height is with levelling legs fully retracted (chests have no legs). Up to ½ in. may be needed to level freezer.

[2] Cost figured at 7.63¢ per kilowatt-hour.

[3] Numbers in parentheses give extra width needed to open door wide enough to remove basket or shelf.

[4] Maximum coverage and duration of warranty (in years) in event of freezer malfunction. First number = length of coverage for loss due to malfunction. Second number = subsequent years of coverage for loss due to malfunction in sealed refrigeration system.

Specifications and Features

All: • Have 1-yr. warranty on parts and labor for entire unit, 5-yr. on refrigeration system. • Need periodic defrosting. • Lock with key.
All chests have: • Temperature control outside freezer. • 1 suspended storage basket.
All uprights have: • Temperature control inside freezer. • Nonreversible door.
Except as noted, all have: • Interior light but no power-on light. • Painted steel interior. • Drain and hose or pan to dispose of defrost water.
Except as noted, all chests have: • Temperature control on lower left side.
Except as noted, all uprights have: • 3 fixed, refrigerated shelves in main compartment and 5 shelves on door. • Door stop to keep door from opening 180° or more. • Gate across bottom of storage compartment.

Key to Advantages

A – Temp. control much more legible than most.
B – Temp. control more legible than most.
C – Temp. control on front; may be more readily accessible than that of other chests.
D – Interior lighting better than most.

E – Shelves are solid and door shelves have swing-up retainers, easier to clean than other uprights (but see Disadvantages).
F – Interior is porcelain on steel.
G – Interior is plastic.
H – Has power-on light.
I – Has sliding basket instead of gate across bottom of main compartment.

Key to Disadvantages

a – Lacks interior light.
b – Has solid shelves, making interior lighting dim on lower shelves.
c – Temperature control not as legible as most.
d – Temperature control on rear wall; may be less accessible than on other uprights.
e – Recessed grip on side of door less convenient than handle of other uprights.
f – Interior light not well protected.
g – Has no defrost drain.
h – Has defrost drain, but no hose or pan.
i – Defrost drain lacks standard hose adapter and will not take standard hose adapter, making defrosting less convenient than with most.
j – Somewhat more condensation on exterior walls in humid conditions than most.

k – Shelves are wire and contain refrigerant tubes; cleaning relatively difficult.
l – Has no stop to prevent door from opening 180° or more.
m – Lid not balanced as well as that of other chest models.

Key to Comments

A – Temperature control is on lower right side.
B – Has switch to reduce condensation; with switch on, energy use increased slightly.
C – Has 1 adjustable, removable divider.
D – Has 1 single position, removable divider.
E – Has 4 fixed shelves.
F – Has 1 3-position, removable shelf.
G – Has trivet instead of gate across bottom of main compartment.
H – Has 2 juice-can shelves on door.
I – Has 1 juice-can shelf on door.
J – Has 6 shelves on door.
K – Has 4 shelves on door.
L – Model discontinued at the time the article was originally published in Consumer Reports. The test information has been retained here, however, for its use as a guide to buying.

Garbage disposers

A food waste disposer is, simply, a motor-driven grinder installed under the sink. The drain that opens into the grinder is part of the disposer and replaces your present sink opening. When the disposer is on, a turntable spins rapidly, forcing the wastes outward against the shredder. Hammers or vanes on the turntable help push the wastes against the grinding surface. While the spinning and grinding is going on, running cold water flushes the pulverized products through small holes in the turntable or grinding ring and down the drain.

Some local governments *forbid* using a disposer because of the additional load on the sewage system, while some towns *require* the use of disposers. Thus, before you even consider installing one, check with city hall to find out which side your town is on.

Even if your town allows disposers, your existing plumbing may still make it difficult or impractical to install one. The first question is whether a disposer will fit into your space and meet your plumbing connections. As for fit, the key dimensions are the overall height of the disposer and the distance from its top to the center of its drain outlet. The second question is whether your household drain and waste system can handle the extra burden. A plumber will help you answer these questions and suggest any modifications that might be needed.

If you have your own septic tank instead of being connected to the town's sewer system, there is a further question. Ask the septic-system service company if the tank can handle the additional wastes. A disposer probably won't be a problem if the tank and leaching fields are large enough, though the tank may have to be cleaned out more often than before. A disposer is probably not a good idea if you have a cesspool; the soil-clogging properties of undecomposed food may eventually cause the cesspool's failure.

If disposers are legal and your waste system is adequate, you should know something about the care and use of disposers before you spend your money.

Disposers come in two basic types: batch-feed and continuous-feed. Batch-feed models are activated by positioning the stopper in the drain so it triggers a built-in switch. Since the stopper blocks the drain opening, you can't feed the machine while it's running. You must alternately feed and grind a batch at a time.

A wall-mounted switch near the sink starts the continuous-feed models. Since no stopper is required, the drain opening isn't blocked and you can feed the disposer continuously, whether it's running or not. The continuous-feed models have a splash guard—rubber fingers that push open for feeding but provide partial closure of the grinding chamber the rest of the time.

Pulverizing

A food waste disposer isn't omnivorous, but it should be able to consume a varied diet. It should accept and digest banana peels, celery stalks, potatoes, corn cobs, walnut shells, and apricot and olive pits, bones (a hard load), corn husks (tough and fibrous), and grapefruit rinds (bulky and leathery). A good disposer will grind up any type of food waste you can put into it. And it should grind the waste into particles fine enough to pass easily through the household drain system—quickly, safely, and without jamming. Three of the most difficult-to-grind wastes are bones, corn husks, and grapefruit rinds.

A typical disposer will reduce 90 percent of the bone in the first two or three minutes and take a while longer to get rid of the remaining 10 percent. The last few bone fragments tend to rattle around before they become trapped by the hammers or vanes on the turntable and are ground.

Some manufacturers suggest slow feeding of tough, fibrous material like corn husks. Others recommend combining such waste with less taxing material. Still others advise against putting this type of waste into their machines at all. Because corn husks may end up as long strands that can easily become tangled clumps, clogging the drain or the disposer itself, it's probably best to toss them into a garbage pail.

A disposer should be able to reduce the halved rinds of a grapefruit to a soft, finely ground pulp in a minute or two. With some units that have fairly small grind chambers, you have to cram the rinds into the drain opening. The tight fit sometimes results in labored operation of the motor, and in some cases, the rind will just sit on top of the turntable without being ground. Cutting the rind into smaller pieces should solve the problem.

Jamming

Jamming was sometimes a problem when we put bones into machines with fixed hammers or vanes. Occasionally a bone fragment or some other piece of waste is too hard to be crushed and gets stuck between the rotating and stationary parts of a disposer.

When such a jam occurs, the disposer should be shut off as quickly as possible. If you don't shut it off, the built-in circuit breaker will shut it off for you. Doing it yourself reduces overheating of the motor. It may save some time, too; when a disposer heats up enough for the circuit breaker to go off you may have to wait as long as twenty minutes before the machine cools enough to restart. A machine should have a manual "reset" button to restart it.

Manual resetting is essential for safety; automatic restarting would be a serious hazard, since the user might well be trying to clean out the grinding chamber by hand. (Actually, it's not a good idea to put your hand into a disposer. A pair of ice tongs may do the trick. If they don't, and you have to reach in with your hand, shut off the electricity in that circuit before doing so.)

Some models have special features designed to help free jams. Those features can help free some jams, but keep a broomstick handy to apply leverage and reverse the turntable, freeing the machine. Some models come with a special wrench to do just that.

Safety

The greatest safety concern is the potential for starting the machine accidentally. Continuous-feed models require a separate on/off switch. It should be installed within reach of the sink, but not where the disposer is apt to be turned on inadvertently. And it should be high enough so a small child can't reach it.

Batch-feed models don't have an external switch. Putting the stopper in place closes a built-in switch, turning the unit on. Since the disposer supposedly won't run without the stopper in place, there should be minimum risk of its going on if you absentmindedly reach into the machine.

Continuous-feed models have splash guards in the drain opening. They're flexible enough to allow wastes to be pushed in. And, ideally, they're closed enough to prevent water from splashing and particles from flying out, but they aren't always effective. But that's not a real hazard, just a nuisance.

Other factors

Beyond their basic grinding job, disposers should be convenient to use, self-cleaning, and reasonably quiet.

Convenience. Continuous-feed models as a type are generally more convenient to use than the batch-feed models. Disposing can be done in one operation, and there is less chance of overloading the machine just to get the job done (as is possible with the batch-feed models).

Cleaning. Disposers are essentially self-cleaning except for the problems noted with corn husks, noted earlier. Most manufacturers suggest letting the disposers run—or at least letting the water run—for thirty to sixty seconds after grinding is finished. Some also suggest purging the disposer (filling the sink halfway with water and then removing the drain stopper) from time to time. This could be particularly important with disposers that require a comparatively low water flow.

Noise. The noise produced by a disposer is in about direct proportion to the hardness of the waste it's grinding.

Any machine will be noisy when grinding bones and fairly quiet when grinding soft wastes. Although most models have muffling jackets, the actual noise level can be greatly influenced by the cabinets in which they're installed. One cabinet may help to muffle sound while another may act as an echo chamber.

Operating costs. A typical 500-watt disposer used for five minutes a day would use only $1\frac{1}{4}$ kilowatt-hours per month. At 7.75 cents per kilowatt-hour, that's less than a dime.

Water consumption is a consideration in some areas. With a slower machine, water consumption can be reduced by simply using a lower faucet setting. A rate of about $1\frac{1}{2}$ gallons per minute might prove adequate if the drain is well flushed once a day (if you wash dishes by hand in a sink filled with water, draining the dishwater will accomplish this automatically). Of course, in areas where water has been cut back radically—either voluntarily or by law—a disposer is a luxury you can probably do without.

Recommendations

Though a batch-feed model has a slight edge in safety, the continuous-feed variety is more convenient to use, and, generally, can be expected to do a bit better in performance.

Heat-recovery ventilators

Older homes, even those that have been extensively weatherized, still let a lot of fresh air leak in. Typically, such homes are leaky enough to undergo about one air change per hour. Smoke and smells may linger a bit in such a house, but a ventilation rate of even 0.5 air changes per hour is usually sufficient to prevent excess moisture from building up and to keep the level of indoor pollutants reasonably low.

New houses are another story. Modern construction techniques such as lining the entire house with a plastic vapor barrier can create a house with a ventilation rate of 0.2 air changes per hour or even lower. In a house that tight, fumes from heating, cooking, solvents, cleaning agents, and smoking, as well as moisture from breathing and bathing, can make the air not only unpleasant but unhealthful. That's the type of house that needs deliberate ventilation, and

for which a heat-recovery ventilator is worth considering.

A heat-recovery ventilator—or air-to-air heat exchanger, as the device is often called—typically has a pair of fans to blow stale air out of the house and draw in fresh air. As the fresh air comes in, it picks up heat from the stale air being exhausted. By the time the fresh air enters the house, it can be quite a bit warmer than the outdoor air.

Most heat-recovery ventilators are intended to ventilate the whole house. They're usually installed in the basement or attic, and they require their own ductwork. Cost: anywhere from $500 to more than $1,000, not including installation.

There are a few small models meant to ventilate only a room or two; they mount in the window or wall like a room air conditioner, so they're fairly easy to install in an existing structure. Cost: about $400.

How they work

There are three basic ways for these devices to recover heat.

One design—the so-called *fixed-plate* type—involves no moving parts other than fans. Cold air and warm air run in narrow, thin-walled passages routed alongside one another; heat from the warm, stale air transfers through the walls to the incoming fresh, cold air. The passages can run the length of the ventilator (a "counter-flow" model) or at right angles, in a sandwich of corrugated channels (a "crossflow" model).

A second design, the *rotary* type, uses a wheel that rotates across both air passages to transfer heat. The wheel is por-

ous enough for the air to pass through; the warm outgoing air gives up some heat to the wheel, which then warms the incoming air as it rotates across that passage.

The third type, the *heat-pipe*, has a design similar in concept to a heat pump. An array of sealed pipes containing a refrigerant lies across the air passages. One end of each pipe is in the warm airstream, and the other end is in the cold airstream. The refrigerant evaporates at the warm end and condenses at the cold end, repeating in a cycle that transfers heat from one airstream to the other.

These devices can also help ventilate

an air-conditioned house in the summer, cooling and dehumidifying incoming fresh air instead of warming it. However, the temperature difference between indoor and outdoor air is likely to be much narrower in summer than in winter, so the potential for energy saving is smaller.

Performance

Manufacturers often cite effectiveness percentages in their literature as a measure of how much the incoming air is warmed. Say the temperature outside is 20° and the temperature in the house is 70°. Say, too, that the ventilator warms the incoming air to 50°. The ventilator is said to have an effectiveness of 60 percent, since it has made up 60 percent of the difference between the outdoor and indoor temperature.

A stream of 50° air is still uncomfortably chilly. That's why ducts for heat-recovery ventilators are often arranged so the cool, fresh air enters high up in a room, where the air is already warm, maybe even too warm.

There are two other ways to make the incoming air a comfortable temperature. If you have a forced-air heating system, you can direct the ventilator's fresh air to the furnace (but the job has to be done carefully to keep the heating system's airflow balanced). Or you can install auxiliary electric heaters in the ducts (but that adds to energy costs, of course).

While the effectiveness percentage gives you an idea of how warm the incoming air will be, it's not an ideal measure of how efficiently a ventilator works. Most ventilators allow some cross-leakage of air between incoming and outgoing passages; simple calculations of effectiveness don't take that leakage into account. The more warm, stale air that leaks into the incoming air passages, the warmer the air entering the house—and the more effective the heat-recovery would seem. But obviously, the greater the cross-leakage, the less fresh air you get.

"Heat recovery efficiency" is a measure of how well a ventilator actually warms the incoming air—a figure corrected for cross-leakage, for heat generated by the fans in the ventilator, and for differences in airflow between incoming and outgoing passages.

But even that measure is less than perfect, since it doesn't take into account different fan sizes. A model with a powerful fan will move more air than a model with a weak fan. So Consumers Union engineers make a further calculation, resulting in what they call the "warm-air ventilating equivalent." That number, arrived at by multiplying the heat-recovery efficiency by the airflow in cubic feet per minute, gives an overall measure of the fresh, warm air that a ventilator can provide.

Humidity

Any house that is tight enough to need a heat-recovery ventilator will end up trapping a lot of moisture during cold weather. Many heat-recovery ventilators expel all the humidity contained in indoor air. As a result, moisture condenses inside the device whenever the moist, stale air is cooled below its dew point. Such ventilators, therefore, need to be hooked up to a drain.

Normally, condensation is no problem. But in freezing weather, ice can sometimes build up and block the air-flow on the exhaust side. With some models, it is important to remember to check for ice in cold weather and turn the device off if it needs defrosting.

A heat-recovery ventilator may be designed to defrost automatically by closing off the incoming air if there's a buildup of ice. That appears to be a smart design—but there's a problem. In a tight house, shutting off the intake of fresh air while still expelling stale air can throw the airflow out of balance. That can be a serious problem if it limits the draft available to a furnace or wood stove. A more recent design solves the problem by recirculating the indoor air through the ventilator until the ice melts.

The rotary models and one fixed-plate model recover some of the moisture from the stale indoor air, sending one-quarter to one-third of it back into the house with the fresh air. Depending on the house, that can make it feel either more comfortable or too damp.

There are two other effects of recovering moisture:

By recycling some water vapor, these models tend to minimize condensation within the unit. Drainage may well be unnecessary.

Models that recover some water vapor are also conserving extra heat—heat that was absorbed by the water when it was vaporized in the first place.

Opening a window

Even if you do have a tight house badly in need of ventilation, opening a window or two may well be more cost effective than a heat-recovery ventilator.

Consider a small house (1,250 square feet with 8-foot ceilings) and the following conditions:

- Indoor temperature, 70°.
- Outdoor temperature, 20°.
- Oil heat (80 percent efficient), and oil at $1.11 a gallon.
- Windows that are open partway to admit 50 cubic feet of air per minute (the equivalent of 0.3 air changes per hour in a house that size) or an exhaust fan of comparable flow.

We calculate the cost of warming that incoming cold air to be about 2.7 cents an hour. That's 64 cents for 24 hours, or $19 per month.

Now consider the same house, under the same conditions, but with a heat-recovery ventilator whose heat-recovery efficiency is typical of the high-rated models (55 percent) and whose airflow rate is also 50 cubic feet per minute. Assuming that electricity costs about 8 cents per kilowatt-hour, it would cost about 17 cents a day to run a 90-watt ventilator. Add another 29 cents a day to replace the heat that's still lost to the outside. Total: 46 cents a day, or about $14 a month.

Clearly, operating a heat-recovery ventilator won't save much on a month-to-month basis. With such a small monthly saving, recovering the cost of the device and its installation could conceivably take twenty to forty years or even longer.

Other assumptions, of course, lead to different payback periods. (These calculations are based on heating with oil. Gas or wood heat makes the heat-recovery ventilator look even less economical. Electric heat makes it look better.) But in most reasonable scenarios, opening a window or turning on an ordinary exhaust fan when fresh air is needed can provide cheaper, simpler ventilation than running a heat-recovery ventilator for twenty-four hours a day, as manufacturers recommend.

In this country, at least, the best argument for a heat-recovery ventilator is that it provides continuous, controlled ventilation, as opposed to the periodic, uneven ventilation of open windows.

Recommendations

A heat-recovery ventilator makes sense mainly where supplemental ventilation is truly needed, as in extremely tight houses in extremely cold climates or where unusual pollution problems are present.

A window/wall model is fairly easy to install. In an existing house, it's probably cheaper to install several of them than to install one whole-house model.

Like any central system, a whole-house ventilator is best installed at the time the house is being built, when it's much easier to put in the ductwork.

Ratings of heat-recovery ventilators

(As published in an October 1985 report.)

Listed by groups; within groups, listed in order of overall quality based mainly on warm-air ventilating equivalent at 5° F. All measurements given are for balanced airflow at device's lowest speed, which is speed mfr. recommends for routine use. Window/wall models are rated on same scale as whole-house models; for their size and ease of installation, they function well.

Except as noted, all have: • Sheet-metal case. • Two-speed centrifugal blower fans. • Adequate access for cleaning. • No defrost provision. • Adequate installation and operating instructions. Unless noted, all whole-house models can be installed in basement or attic and have 6-in. duct connections.

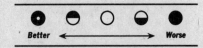

Better ⟵   ⟶ Worse

Whole-house models

Q-Dot SAE-150 (Q-Dot Corp., 701 N. First St., Garland, Tex. 75040), $599. 13x29x22 in. Heat-pipe design.

Performance at 5° ●
Performance at 45° ○

Specifications: Warm-air ventilating equivalent, 68 cfm at 5°; 48 cfm at 45°. Heat-recovery efficiency, 55% at 5°; 38% at 45°. Airflow capacity, 123 cfm at 5°; 127 cfm at 45°. Cross-leakage, 2%. Wattage, 100 at low speed; 195 at high speed.

Comments: In case of ice blockage, fresh-air fan shuts down. Has condensation outlet requiring drain connection.

Aldes VMPH 3-5 (American Aldes Ventilation Corp., Northgate Center Industrial Park, 4539 Northgate Ct., Sarasota, Fla. 33580), $1161. 11x40x20 in.; separate fan module, 15x16x12 in. Fixed-plate design.

Performance at 5° ◖
Performance at 45° ◑

Specifications: Warm-air ventilating equivalent, 55 cfm at 5°; 59 cfm at 45°. Heat-recovery efficiency, 55% at 5°; 46% at 45°. Airflow capacity, 99 cfm at 5°; 127 cfm at 45°. Cross-leakage, 0%. Wattage, 65 at low speed; 145 at high speed.
Comments: Comes with substantially more installation hardware and instructions than do other models tested. But because of two-part construction, more ducting may be needed. Has condensation outlet requiring drain connection; drainage was incomplete when ventilator was installed at recommended slant. Cleaning of core requires removal of duct connections. Lack of cross-leakage indicates good design and workmanship.

Nutone AE-200 (Scovill Housing Group, Madison and Red Bank Rds., Cincinnati 45227), $540. 10x17x29 in. Rotary design.

Performance at 5° ○
Performance at 45° ●

Specifications: Warm-air ventilating equivalent, 42 cfm at 5°; 89 cfm at 45°. Heat-recovery efficiency, 62% at 5°; 71% at 45°. Airflow capacity, 67 cfm at 5°; 126 cfm at 45°. Cross-leakage, 8%. Wattage, 90 at low speed; 125 at high speed.
Comments: Can be installed either side up, as convenient. Instructions recommend 7-in. ducts. Optional adapter ($67) required if stale-air intake is ducted so it draws air from several rooms. Frost formed on outside of case and on wheel during 5° operation, a serious disadvantage; case may require insulation for low-temperature use.

Air Changer DR-150 (Air Changer Co. Ltd., 334 King St. E., Suite 505, Toronto M5A 1K8), $950. 22x14x51 in. Fixed-plate design.

Performance at 5° ◐
Performance at 45° ●

Specifications: Warm-air ventilating equivalent, 17 cfm at 5°; 9 cfm at 45°. Heat-recovery efficiency, 62% at 5°; 32% at 45°. Airflow capacity, 28 cfm at 5°; 27 cfm at 45°. Cross-leakage, 10%. Wattage, 45 at low speed; 60 at high speed.
Comments: Basement installation recommended by mfr. Has variable-speed axial fans, weakest of those tested. Built-in humidistat increases fan speed when stale-air humidity rises above adjustable set point. In case of ice blockage, fresh-air fan shuts down. Has condensation outlet requiring drain connection, but model generated little condensation in our tests. Cleaning of core requires removal of duct connections.

Vent-X-Changer WR-25 (P.M. Wright Ltd., 1300 Jules-Poitras, Montreal, H4N 1X8), $594. 13x16x13 in. Rotary design.

Performance at 5° ●
Performance at 45° ◐

Specifications: Warm-air ventilating equivalent, 10 cfm at 5°; 9 cfm at 45°. Heat-recovery efficiency, 19% at 5°; 15% at 45°. Airflow capacity, 55 cfm at 5°; 59 cfm at 45°. Cross-leakage, 40%. Wattage, 45 at low speed; 120 at high speed.
Comments: This model is sold as the *Fan-X-Changer* in Canada. Mfr. recommends basement installation. Sturdy, compact construction. Has variable-speed fan. Heating elements may be required in fresh-air ducts during cold weather to heat air to comfortable level. Frost formed on outside of case during 5° operation; case may require insulation in low-temperature use. Because of high cross-leakage, model recirculates a substantial amount of stale air.

Window/wall-mounted models

Lossnay VL-1500C (Mitsubishi Electric Sales America Inc., 5757 Plaza Dr. P.O. Box 607, Cypress, Calif. 90630), $420. 10x19x15 in. Fixed-plate design.

Performance at 5° ◐
Performance at 45° ◐

Specifications: Warm-air ventilating equivalent, 21 cfm at 5°; 22 cfm at 45°. Heat-recovery efficiency, 49% at 5°; 49% at 45°. Airflow capacity, 43 cfm at 5°; 45 cfm at 45°. Cross-leakage, 8%. Wattage, 20 at low speed; 50 at high speed.
Comments: Looks like air conditioner with plastic front. Fans have 3 speeds, with pull cord for controlling speed.

Berner Economini EM-120 (Berner International Corp., P.O. Box 5205, New Castle, Pa. 16105), $390. 12x22x9 in. Rotary design.

Performance at 5° ◐
Performance at 45° ◐

Specifications: Warm-air ventilating equivalent, 21 cfm at 5°; 17 cfm at 45°. Heat-recovery efficiency, 45% at 5°; 40% at 45°. Airflow capacity, 48 cfm at 5°; 43 cfm at 45°. Cross-leakage, 15%. Wattage, 35 at low speed; 45 at high speed.
Comments: Looks like air conditioner. Has plastic outer case. Fans have 2 speeds, with pull cord for controlling speed. With wall installation, connects to outside weather cap with 3-in. ducts (all supplied). Condensation formed on case during 5° operation. Only sketchy operating instructions provided.

Hot plates

As a stopgap cooking appliance, the hot plate is invaluable to travelers, office workers, students, and others who don't have standard kitchen facilities. And, in its fancier role as "buffet range," it can be useful to the host or hostess who wants to keep party food warm.

A sealed-rod hot plate has burners essentially similar to those on electric ranges; the insulated heating element is enclosed in a metal tube, resembling a flattened spiral, on which the cooking utensil rests. That sealed-rod design makes for electrical safety. Hot plates of the open-coil type are potentially hazardous. Their heating coils are simply laid in a groove in the surface of a ceramic plate. When the hot plate is in use it is too easy to touch those coils accidentally, with a metal spoon or fork, and all too easy for a metal-bottomed pot or pan to touch coils that might work out of the grooves. If you touch the fork or the pan and a ground at the same time, you might get a lethal electric shock. You should discard an old open-coil hot plate and replace it with one of the sealed-rod type.

Heating

Don't count on a hot plate for speedy cooking, as you might expect from their wattage. It may take about 15 minutes to bring two quarts of tap water to boil in a three-quart covered saucepan, starting with the burners turned off. That's longer, as a rule, than it would take to boil twice that quantity on the eight-inch burners of an electric range.

Hot plates should have a very low heat setting to keep cooked food warm—between servings during a meal served buffet style, for example. If you get a unit that's reluctant to maintain a low temperature, try setting its heat controls a bit below that lowest indicated setting.

Overall dimensions vary, but the space between centers of the high and low burners is usually about eight or nine inches—so a hot plate may not take two large pans at the same time.

A unit should have legs of smooth metal or of such relatively soft materials as wood or plastic that would be unlikely to scratch a tabletop. If food spills, it should be possible to raise a unit's non-removable heating elements to clean the drip pans beneath them. If food spills or boils over, you may also need to clean beneath the drip pan. That's easy if the drip pans are removable, not so easy if you need tools to get to that area. Unplugging should always precede cleaning. And a hot plate should never be immersed for cleaning.

With an appliance such as this, which may draw nearly 15 amps, try to operate it without an extension cord. If you can't, keep the cord as short as possible. And be sure any extension cord is in the heavy-duty class; the usual 18-gauge lamp cord may overheat.

Even if there's a warning light to let you know that the burners are on, a light obviously can't tell you about burners that are still hot after you turn the unit off. You can get a nasty burn from burners that don't seem to be on. In operation, too, some units get quite hot in places you might not expect—on the sides, for example, or around the controls. Because of the unpredictable hot spots, a hot plate shouldn't be moved until it cools.

Recommendations

No hot plate is likely to be exceptionally good or bad in performance or convenience. If you eliminate from consideration units with obvious hazards (sharp edges, open-coil heating elements, wobbly feet), you won't go far wrong if you buy the least expensive model available that suits your aesthetic eye.

Large humidifiers

You can blame the dryness of the air in heated houses during the winter in part on a basic law of nature, in part on central heating systems. Warm air can hold more moisture than cold air. So, when the furnace heats your house, it also reduces the "relative humidity," air's moisture content expressed as a percentage of its moisture-holding capacity. Consider, for example, outdoor air at 20°F and 80 percent relative humidity. In heating that air to 70°, a furnace may reduce its humidity to a parched 12 percent, merely because the added heat also makes a gigantic increase in the air's ability to hold moisture. That drastic change in the relative humidity—even though the actual amount of moisture in the air may not have changed—encourages the air to remove moisture from household furnishings and occupants.

The moisture that laundering, cooking, and bathing add to the air aren't likely to offset the resulting uncomfortable dryness, especially in northern parts of the country. The most convenient and often the most effective solution—but only if you have forced-air heat—is a humidifier built into your heating system. A built-in will distribute moisture throughout the house along with the heated air. You don't have to bother keeping the unit filled with water; it's connected to the plumbing.

If you have hot-water, steam, or electric-resistance heating, you can't use a built-in humidifier. But you can use a plug-in, portable console model. These humidifiers are capable of evaporating a lot more water than a tabletop ultrasonic humidifier can (see pages 162–63) and are therefore suitable for humidifying large areas of your home.

Usefulness

Some houses shouldn't be humidified at all. Don't think of using a humidifier unless the outside wall of your house has a vapor barrier. If your house is insulated, that barrier will probably be a layer of plastic film, metal foil, or asphalt-coated paper on the room side of the insulation. If the house has insulation but no barrier (not an unusual situation in houses built before 1950), moisture from inside the house will enter the insulation, where it may condense on cold days. At best, that water will reduce the insulation's effectiveness; at worst, it may cause wood within the walls to swell or rot, or paint to peel from the siding.

A humidifier won't moisturize the air adequately in an uninsulated house that has no vapor barrier; instead, moisture will escape directly to the outdoors. And if an uninsulated house does have a vapor barrier (a coat of special interior paint or a vinyl wall covering), a humidifier may not provide much benefit in colder climates; the humidifier may cause condensation to form on walls.

Assuming that your house is properly insulated and has a vapor barrier, a humidifier should enhance your comfort during the winter.

A typical console consists of a rather bulky cabinet, styled to look like a piece of furniture, that conceals a water reservoir. On most models, an absorbent pad turns slowly through the water reservoir. The majority of units have the pad mounted on a drum. On others, the pad is a belt mounted on a pair of rollers. With either type, a fan draws in room air and forces it through the pad to pick up moisture. The air then flows out through vents in the top of the appliance and back into the room. Rarely, there is a stationary pad, kept moist by a pump that throws water onto it.

A more up-to-date design would operate through the use of one or more ultrasonic transducers, much like the small ultrasonic humidifiers on pages 170–71.

There is a standardized system for specifying moisture output. Moisture capacity is certified under a voluntary program overseen by the Association of Home Appliance Manufacturers (AHAM). Those ratings are stated for performance in a 70° room at 30 percent relative humidity, reasonably typical home conditions.

Using a humidifier

There are differences from one model to another that affect ease of use.

Effective capacity. In a cold snap, you may have to refill a humidifier as often as three times a day. When a typical humidifier is thirsty, its fan shuts off automatically and a signal light shines. Those automatic shutoffs may work prematurely, turning the unit off when there is still about 1 to 1½ gallons of water in the tank. Even a model without a shutoff would fail to make efficient use of the last gallon of water. Therefore, the water actually available for use may be less than that claimed by the manufacturer.

Moving and filling. Rolling an empty humidifier to the sink is easy if it has casters. But, even with casters, a humidifier is much harder to push when filled, especially if you have to pilot it over doorsills or carpets. And you have to be very careful when moving most humidifiers to avoid sloshing water.

Bring water to the appliance, then? That will involve several trips with a pail or watering can. The smaller a unit's effective capacity, the more often you'll have to refill the unit.

The most convenient models to fill have a large fill opening, interior baffling that keeps water from slopping through openings at the back or over the edges of the reservoir, and an easily seen "full" marking on the reservoir. Somewhat less convenient are units with an external fill gauge, often near other controls on top of the unit. You must look away from your pouring to read the gauge. When filling one of those models for the first time, you should pay close attention to the water level to be sure that the gauge is accurate and doesn't permit overfilling.

Access to the water reservoir is handiest if the unit has a tilt-up lid, but only eight models do. With the rest, you remove either the entire lid or several louvered panels.

Some models have a fill chute to direct the water and minimize drips and spills. On a few, the chute is just an opening in the surface of the drum; you may have to turn the drum to point the chute's opening upward.

For filling directly from a sink faucet, you need a hose.

Cleaning

Unless you clean a humidifier periodically, minerals from the water will clog working parts and slime will tend to build up on the surfaces of the reservoir. The manufacturers recommend cleaning anywhere from twice a season to every two weeks, depending on the condition of your water. But CU's medical consultants believe weekly cleaning is best to minimize the spread of molds and bacteria.

The evaporation pad should also be cleaned regularly and may need to be replaced annually. Pads on drum models are generally easiest to remove for cleaning or replacement. With the belt units, you have to remove a pair of rollers before you can disengage the pad, and it may be hard to reassemble the functional interior parts after cleaning the reservoir.

Disposable plastic reservoir liners, which cost about fifty cents each, are sometimes available. They save you from having to stoop to reach into the reservoirs for cleaning.

A number of models use a float in the reservoir to operate their fill gauge, signal light, and automatic low-level shutoff. The floats are delicate, so a gentle touch is in order. Drum models with a slide-out reservoir generally have some setup that lifts the float out of harm's way when the reservoir is removed.

Never store your humidifier with water in it; remove all water at the end of every season. And empty the humidifier if it won't be used for a week or so during the winter, to prevent odors and the growth of bacteria and fungi.

Emptying is fairly straightforward with a removable reservoir, which a majority of the units have. With some, you merely lift the reservoir out of the console. With others, you lift the console off the reservoir. But emptying can be quite a bother with a fixed reservoir, even if the humidifier has a drain. The drains are so low that you really can't get anything under them but a shallow pan—hardly an ideal receptacle for moving $1\frac{1}{2}$ to $2\frac{1}{2}$ gallons of water. With fixed-reservoir models in particular, you'll save yourself trouble if you run them until most of the water evaporates, then bail out the rest.

Controls, noise, safety

All humidifiers have a humidistat, a device that will sense the relative humidity indoors and cycle the machine on and off to keep conditions fairly steady. None of the controls, however, are marked with actual humidity percentages; instead, they have arbitrary calibrations. Numbered settings can be more precisely reset than dials with only a few reference marks or a Spartan "high, low."

It's also preferable that any on/off switch be separate from the humidistat. That way, you can shut the appliance off without disturbing the humidistat setting. Some models, all with a multispeed fan, provide that amenity. Other multispeed models and other single-speed appliances lack a separate on/off switch; dialing the humidistat to its lowest setting (or to "off") controls the fan.

Noise and drafts. At their noisiest, console humidifiers can be as loud as a small air conditioner. You can quiet the multispeed units by turning down the fan, although that will reduce moisture output as well as noise.

A few models expel cool, humid air at an angle, so that a person standing four or five feet away can feel the draft. That shouldn't be a major problem, however. You aren't apt to be close enough to one of these noisy machines to be bothered by drafts.

Safety. If there are small children about, check to be sure the rear grille openings are small enough to keep small fingers out of the fan. Though all the fans have plastic blades, they still might cause injury.

You should unplug a humidifier before refilling or cleaning it, as many of the manufacturers recommend. If you don't, the machine's humidistat could start the fan while you're reaching inside.

Many units come with separately packaged tablets or liquid that you can add to the reservoir's water. Those additives may help prevent mineral buildup in the humidifier, but they won't disinfect water.

The labels on the additives warn that all may be hazardous if swallowed and that all are to be kept out of children's reach.

The best location

Moisture will spread outward from the humidifier until stopped by a vapor barrier. Interior walls will slow the spread, but won't stop it. Positioning a humidifier in the center of the living space or near a stairway will help distribute moisture more evenly. Or, if your home has forced-air heat, try putting the unit near a return register. That, too, will help spread moisture.

The kitchen or a bathroom isn't a good spot for a humidifier; those rooms get plenty of moisture in the normal course of activities. Don't put a humidifier near an outside wall; that may decrease the unit's output or promote condensation within the wall. Leave at least a few inches of space between the humidifier and interior walls. Finally, be sure that cool, humidified air isn't directed at your heating system's thermostat.

Recommendations

Before you settle on a specific brand and model of humidifier, figure the maximum moisture output you're apt to need for your house. The box on page 161 can help with the calculations.

The more expensive models usually offer extra convenience features as well as a higher output than cheaper models. But not always.

How much humidity do you need?

There's an easy way to figure humidifier capacity: merely calculate the volume of your house. Multiply the total floor area (in square feet) by the ceiling height (in feet). Include closet space in your calculations. Also include the basement and attic unless vapor barriers separate them from the rest of the house. The result will be in cubic feet. For every 10,000 cubic feet, allow 6½ gallons per day of humidifying capacity.

For example, consider a two-story house that has 980 square feet on the second floor, 8-foot-high ceilings, and a vapor barrier in the outside walls, the attic, and the basement. That house will contain 15,200 cubic feet: (980 × 8)+(920 × 8). That house will need a humidifier with a 10-gallon-per-day capacity (10 gallons is about 1½ times 6½ gallons).

Our way of estimating a humidifier's capacity assumes a house of typical "tightness"—one with loose storm windows, say, so that enough air leaks in and out to change the air completely every hour. If you have a tight house, with snug storm windows and doors and dampered fireplaces, you can assume less than one air change an hour. You'll then need only about half the capacity our method indicates. In a loose house (no storm windows, storm doors, weather stripping, or dampers), you may have to assume 1½ air changes an hour. In that case, choose a humidifier with a capacity 50 percent larger than our method suggests.

You should also correct your calculation for moisture given off inside the home by cooking, bathing, laundering, and the like. A typical family of four can deduct 2 gallons per day from their calculated capacity needs.

You can, of course, buy a unit with a higher capacity than your calculations dictate. The humidistat will prevent overhumidification. But if you buy a unit that has a lower capacity, it may never deliver enough moisture.

When using your humidifier, adjust its humidistat so that the unit doesn't cause excessive condensation on your windows. As soon as droplets appear, shut off the humidifier or turn down its humidistat.

Ratings of console humidifiers

(As published in a September 1984 report.)

Listed, except as noted, in order of estimated overall quality. Closely ranked models differed little in overall quality. Dimensions are HxWxD. Weights are rounded to next higher lb., weights to nearest lb. Prices are approximate retail; + indicates shipping is extra.

Better ● ◐ ○ → Worse

Brand and model	Price	HxWxD, in.	Weight, lb. Empty	Weight, lb. Full [1]	Evaporative capacity, gal./24 hr. [1]	Fan speeds	Effective reservoir capacity, gal. [2]	Filling	Mobility	Emptying	Replacing belt or drum, cleaning [3]	Humidistat marking	Noise At high speed	Noise At low speed	Ease of use Advantages	Disadvantages	Comments
SEARS Cat. No. 7433	$130+	25x26x17	20	106	13	3	9	●	●	◐	○	●	●	●	A,B,C,E,H,J,K,L,N,P	h,j	A,B,D
SEARS Cat. No. 7437	160+	25x26x17	22	110	15	V	9	●	●	◐	○	●	●	●	A,B,C,E,F,H,J,K,L,N,P	h,j	A,B,D
■ *The following models had significantly lower effective reservoir capacity than those preceding and will require more frequent filling.*																	
HERRMIDIFIER 130	175	25x26x17	23	98	12	3	6½	◐	○	○	○	◐	●	●	A,C,H,J,K,L,N,P	c,h,j	A,B,D
EMERSON HD 133 [4]	154	25x26x17	21	94	12	3	6½	◐	○	○	○	◐	●	●	A,C,H,J,K,L,N,P	c,h,j	A,B
EMERSON HD 14V [4]	166	25x26x17	21	93	14	V	6½	◐	○	◐	○	◐	●	●	A,C,F,H,J,K,L,N,P	c,h,j	A,B
EDISON 534171A [4]	225	27x29x13	31	93	17	V	6	◐	◐	◐	●	◐	◐	◐	B,H,N	d	—
COMFORT-AIRE HCG15	155	26x24x13	34	83	15	V	4½	◐	◐	◑	●	◐	◐	◐	B,D,G,K	a,e,i	D

Brand & Model	Price[2]	Dimensions HxWxD (in.)						Ratings								Features	Comments	Rating
SEARS Cat. No. 7450	120+	28x17x17	16	61	10	2	5	◐	◐	○	○	○	●	○	A,C,H,P	d,e,g,k,l	C
COMFORT-AIRE HCG12D	124	26x24x14	33	83	12	3	4½	◐	●	◐	◐	◐	●	●	B,D,G,K	a,e,l	D
WARDS Cat. No. 96172 [5]	190	26x28x13	32	96	17	V	6	◐	○	●	●	●	◐	○	B,H,N	d,m	—
WARDS Cat. No. 96142 [5]	150	24x26x12	23	75	14	V	5	◐	◐	●	●	◐	◐	●	B,H,N	d,j	—
EDISON 534121	155	24x26x12	21	73	12	3	5	◐	◐	◐	◐	◐	◐	◐	B,H,P	d,m	E
J.C. PENNEY Cat. No. 2114 [5]	160	25x28x13	29	82	13	3	5	○	◐	●	●	●	●	●	B,D,H,M,O	f,i	—
WARDS Cat. No. 96082	80+	24x26x12	21	73	8	1	5½	●	◐	◐	●	◐	●	○	—	b,d	—
J.C. PENNEY Cat. No. 2110 [5]	95	26x27x13	26	80	9	2	5	◐	○	●	●	●	●	●	D,M	b,i	—
J.C. PENNEY Cat. No. 2111 [5]	140	25x28x13	29	86	11	3	5	○	◐	◐	◐	◐	●	○	B,D,H,M	f,i	—
EDISON 534081	106	23x26x12	20	72	8	1	5½	[6]	◐	◐	◐	◐	●	◐	—	b	E

■ The following models were judged very difficult to reassemble, which may be required frequently to maintain sanitation. Listed alphabetically.

Brand & Model	Price[2]	Dimensions HxWxD (in.)						Ratings								Features	Comments	Rating
COMFORT-AIRE HCG7	81	24x21x13	15	82	7	1	6½	●	◐	◐	◐	◐	●	○	—	A,I,K,L	h,j	B
EMERSON HD071	106	24x21x13	15	81	7	1	6½	●	●	◐	◐	●	◐	○	—	A,I,K,L	h,j	B
SEARS Cat. No. 7409	80+	24x21x13	15	82	7	1	6½	●	◐	●	◐	◐	●	○	—	A,I,K,L	h,j	B
SKUTTLE 411	158	24x21x13	15	82	7	1	6½	●	◐	◐	◐	●	◐	○	—	A,I,K,L	h,j	B

[1] When filled to internal "full" mark, if clearly visible; otherwise, when filled to about 1 in. from top of reservoir.

[2] Capacity as certified by Assn. of Home Appliance Mfrs. Capacity measured when unit is set to run continuously at maximum fan speed in chamber at 70° F and 30% relative humidity. Numbers are based on evaporation during 2-hr. period when unit is about half full.

[3] Excluding water left in reservoir when machine stops humidifying

[4] Discontinued but may still be available.

[5] Not listed in company's current catalog.

[6] Lacks casters; mobility is worse than with any other model.

[7] V = Variable.

SPECIFICATIONS AND FEATURES
All: ● Have adjustable humidistat. ● Lack switch that shuts unit off when top is opened.
Except as noted, all: ● Have casters. ● Are "drum" models. ● Have fixed louvers. ● Have removable reservoir, judged reasonably easy to clean. ● Tended to spill at least some water if moved when full. ● Have 5- to 6-ft. power cord.

KEY TO ADVANTAGES
A – Has easy-to-see fixed "full" mark.
B – Has gauge, regulated by float, that indicates water level.
C – Large fill opening and interior baffling make unit relatively easy and tidy to fill.
D – Has fill chute.
E – Comes with separate 7½-ft. fill hose, handy if unit is rolled to a faucet.
F – "On" light always indicates when unit is plugged in and turned on, even if humidistat cycles unit off.
G – Has low-water shutoff and refill light that stays on even if humidistat cycles off.
H – Has low-water shutoff and refill light that is on only while humidistat has cycled on.
I – Has low-water shutoff but no refill light.
J – Judged less likely to spill when moved than most.
K – Evaporative pad or belt less apt to slip while in motion than others.
L – 8-ft. power cord.
M – Power cord may be removed from humidifier when rolling unit to sink. Cord should not be left plugged in, especially in house with small children.
N – Has tilt-up lid, a convenience when filling.
O – Has built-in dispenser in top for water-treatment liquids.
P – On/off control is separate from humidistat.

KEY TO DISADVANTAGES
a – Water level in reservoir difficult to see, and lack of fixed internal "full" mark makes float gauge's accuracy hard to check.

b – Lacks low-water shutoff, refill light, water-level gauge, and visible "full" mark.
c – When full, unit may tend to bind when moved over doorsills and deep-pile rugs.
d – When full, unit judged relatively inconvenient to move over rug edges and doorsills; user must stoop to lift base of reservoir.
e – Judged somewhat less stable than others when moved on rugs or over obstructions.
f – Back of reservoir must be tilted up when being removed to avoid damaging float.
g – Pump judged apt to clog and require cleaning.
h – Fixed reservoir; judged relatively hard to clean.
i – Uncomfortably cool draft may be felt some 4 to 5 ft. from cabinet (but louver sections can be repositioned to redirect airflow).
j – Fan blades judged accessible to small fingers; model judged not suitable for use where small children have access to it.
k – Evaporative pad's mounting clips rusted heavily after relatively few uses of unit.
l – After cleaning or replacement of evaporative pad, unit requires care in reassembly to avoid damage to float and pump.
m – Casters swiveled sluggishly; unit judged relatively inconvenient to move even on hard-surface flooring.

KEY TO COMMENTS
A – Has drain, but drain is so low that there is clearance only for a shallow pan, impractical for removing of leftover water.
B – Belt model.
C – Pump model.
D – Optional kit (not tested) allows permanent connection of unit to house plumbing.
E – Disposable plastic liners, **999040**, approx. $6 per doz., make reservoir easier to clean.

Ultrasonic humidifiers

Ultrasonic humidifiers promise relief from the discomforts of dry indoor air, especially during the winter heating season. By comparison with these high-tech appliances, old-fashioned vaporizers and console humidifiers seem primitive. Ultrasonic humidifiers, intended for tabletop use, are compact (about the size of a bread box) and relatively lightweight (about twenty pounds, filled with water). Instead of a mechanical propeller or drum to turn water into mist, they use a small electronic component called a transducer, which transforms electrical energy into mechanical energy.

The part that actually vibrates is called the nebulizer—a disk about the size of a dime. The nebulizer, oscillating at about 1.7 million times per second, churns up the water into a very fine, cool mist. The vibration is so rapid that it's ultrasonic—you can't hear it.

A typical humidifier tank holds about a gallon of water. The water trickles down through a valve in the tank into a shallow reservoir, or "mist chamber," in which the vibrating nebulizer is immersed. When the water is used up, a float trips a switch that automatically shuts off the transducer. A small, built-in fan draws in air through an intake vent; the air blows the mist out through a nozzle or two. The nozzle can be swiveled horizontally to direct the mist in any direction. An infinitely variable volume control regulates the amount of mist blown into the air. You can adjust the control so that the appliance puts out 5 fluid ounces per hour or less. A humidistat, another common control, automatically cycles the humidifier on and off to maintain room humidity at the desired level.

Noise. Even when they are running full blast, all are very quiet, with only an occasional gurgle.

Output rate. The greater the maximum output of moisture, the faster a humidifier can raise the humidity of a room, and the more rooms it can humidify.

Just about any one can effectively humidify one or two large rooms—though not an entire house, as a few manufacturers claim. They also have two clear advantages over conventional humidifiers and vaporizers:

They're quiet.

They apparently kill molds and bacteria that might be in the water, as explained on page 173.

Coverage. An ultrasonic humidifier can easily humidify a very large living room or two large bedrooms. If your house is small, you might be able to humidify the whole house with a pair of these humidifiers.

Coverage also depends on how well sealed your house is. In winter, a house with loose storm windows and little weatherstripping loses not only heat but humidity. A tighter house will hold in more of the humidity generated by people breathing, cooking, and bathing, so it will need less additional humidifying.

Distribution of vapor. Some ultrasonic humidifiers disperse their vapor much better than others do. When set on the floor and run at their maximum settings, some models send all their vapor into the air, while some let a lot of it saturate the floor. The number and shape of the nozzles seems to be a determining factor. Multiple nozzles may be better than a single nozzle.

Poor distribution of vapor can mean a

ruined finish on your floor or tabletop or a soaked carpet. Humidifiers that leave a wet spot should be placed on the edge of a table; that placement will reduce the problem. The only surefire cure is to lower the setting of the volume control.

To play it safe, place your humidifier on the forward edge of a table so you don't ruin its finish. Better yet, use a table with a waterproof top or place a sheet of plastic under the humidifier. (Incidentally, keep the humidifier away from your heating thermostat. The cool mist could fool the thermostat into producing heat.)

White dust

Ultrasonic humidifiers have another problem—at least in areas that have hard water. They can deposit a coating of fine white dust onto furniture and other surfaces. Sometimes you can find the dust even in adjoining rooms.

The dust comes from minerals, primarily calcium carbonate, found in hard water. When ejected into the air by the humidifier, the minerals settle on surrounding surfaces. The dust is harmless to people but certainly a nuisance. And it can harm sensitive electronic equipment. Even relatively soft water may cause some dust, so we suggest that you keep your humidifier away from computers, VCRs, and the like.

If you live in an area with hard water, you can avoid the problem by filling your humidifier with distilled or demineralized water. But that can be quite expensive—more than one dollar per tankful.

Some manufacturers of ultrasonic humidifiers are beginning to address the problem of white dust by incorporating a special demineralization cartridge in the unit that treats the water before it flows into the mist chamber or by providing a separate water-treatment cartridge. Such cartridges significantly reduce the amount of white dust but may not eliminate it entirely. The amount of water a cartridge can treat before it needs replacement depends on the water's hardness.

Convenience

Ultrasonic humidifiers are simple devices, but some are easier to use than others.

Controls and indicators. Besides a volume control, most humidifiers have a humidistat control that automatically turns them on and off when the room humidity reaches a preset level. In a large room, you may prefer to leave the humidistat at its maximum setting and regulate humidification with the volume control alone. A humidistat may be a useful feature in a small room, however, to avoid oversaturating the air.

Turning the volume control all the way down shuts off most humidifiers. A few, however, have a separate power switch, so you can turn the humidifier on and off without disturbing the setting of the volume control.

Can a humidifier be hazardous to your health?

Manufacturers used to make medical claims for humidified air—for example, that it would protect against colds and nosebleeds, "help relieve discomfort of winter colds," or relieve allergy and asthma symptoms. Now manufacturers limit themselves to claims of increased comfort. And with good reason: there's no firm evidence that humidified air improves anyone's health.

Humidifiers can, in fact, cause health problems, particularly for people allergic to molds. Humidifiers can promote mold growth in two ways—by making a house *too* humid, or by accumulating molds in the machine's reservoir.

Houses with wintertime humidities much above 40 percent have an increased chance of developing a mold problem. A relatively inexpensive humidity gauge—available at many hardware stores—can help you monitor indoor humidity levels (though you shouldn't expect pinpoint accuracy).

The humidifier's reservoir can be a breeding ground for bacteria as well as for molds. These microbes, though usually harmless, can cause allergic reactions or (more rarely) respiratory infections. However, studies show that evaporative humidifiers, the kind we tested for the accompanying report, don't seem to propel the organisms into the air where they can be inhaled.

Nevertheless, it's important to clean a humidifier's reservoir regularly, to minimize the chance of infection and to prevent odor problems. CU's medical consultants recommend weekly cleaning.

Clean the reservoir with bleach, using about a tablespoon per pint of water or half a cup per gallon. Scrub or scrape debris off the reservoir walls, pour out the bleach, and rinse thoroughly with plain water. Don't use bleach on the evaporation pad; if you do, the odor of bleach may fill the room when the humidifier is turned on.

People with mold allergies should exercise particular care when cleaning their humidifiers. The scraping and scrubbing can stir up lots of mold.

Liquid or tablet additives (often provided by manufacturers with humidifiers, and also sold separately) may help prevent mineral buildup in reservoirs, but don't rely on them to disinfect. One study found that the additives had little or no effect on the microbe populations.

Evaporative humidifiers are intended to humidify a sizable area. But some people may want to humidify their living quarters with one of the newer ultrasonic humidifiers (pages 170–71).

Those machines seem to pose a much lower risk of spreading bacteria and molds. With allergic or seriously ill patients, it's better to use a steam vaporizer. In the steam units (page 321) any microbes entering the reservoir are killed by heat.

All have an indicator that lights when the power goes on. An extra indicator that goes on when the tank needs refilling is a useful addition.

Filling the tank. Filled, the tank of a small unit weighs about ten to fourteen pounds.

Some tanks have markings to gauge the water level. The markings aren't particularly useful, because you can see the level fairly easily through the tank.

Sometimes a humidifier will develop an air lock in the tank that prevents water from flowing into the mist chamber. Resetting the tank on its base usually corrects the problem.

Cleaning

Ultrasonic humidifiers, like other types of humidifiers, require periodic cleaning. Mineral deposits on the nebulizer—the vibrating disk that creates the mist—can reduce the output of mist. Most manufacturers recommend cleaning the nebulizer every week; if you have hard water, you may need to clean that component more frequently.

Cleaning the nebulizer is simple. Some humidifiers come with a small, stiff brush that does the job nicely, as would any small paintbrush. White vinegar or lime-deposit remover, which is available in hardware stores, loosens deposits and makes cleaning easier. Don't use solvents or scrape with anything hard; that could damage the nebulizer.

The float that turns off the humidifier when the water has been used up needs regular cleaning with a soft cloth or brush. The float on most models is easy to reach. On others you have to undo a screw and remove a shield to reach it.

Cleaning the mist chamber is best done with a soft cloth.

Various vinegar and bleach solutions recommended by the manufacturers are largely ineffective against deposits. You can get better results with *Clean Away*, a product available by mail from Bionaire, P.O. Box 582, Franklin Lakes, N.J. 07417. A 16-ounce bottle costs $4. You mix an ounce (phosphoric acid is an active ingredient, according to the label) with an ounce of water, shake it vigorously in the tank, and then let it sit for an hour. Thorough rinsing leaves the tank bright and clean. Used undiluted, *Clean Away* effectively removes scale from nebulizers—but it may also remove the paint from some.

Most models have fine-mesh air filters covering their intake ports. They require periodic cleaning as well.

Precautions

Most manufacturers warn against placing your hands in the mist chamber while the humidifier is running, so as to avoid burns or the possibility of electric shock.

A number of models have an interlock that requires you to remove the tank from the humidifier before you can lift the cover off the mist chamber. That's a good safety feature.

Most models have fast-blow fuses to protect you and the humidifier in case of a short circuit. Usually, the base has to be removed to replace a fuse; most manufacturers suggest you have a repair shop do that.

Sometimes there is a tip-over switch that shuts off the power if the humidifier is picked up or knocked over. Although a tip-over switch isn't a critical safety feature, as it is in a portable electric heater, it can save you considerable expense: most models that lack such a switch will require repairs if you accidentally tip them over while they are running.

A humidifier should always be level. Some are more sensitive to tilting than others and may not run until they are level.

Recommendations

Ultrasonic humidifiers have several advantages over conventional humidifiers and conventional motor-operated cool-mist machines. They're quieter. The moisture droplets are smaller and more easily dispersed. And they eject no live molds and significantly fewer live bacteria than cool-mist vaporizers.

New, large models, when introduced, would compete with the console humidifiers on pages 162–63.

If your house lacks a vapor barrier inside the exterior walls, be wary of using *any* humidifier. A vapor barrier such as plastic sheeting or foil on the warm side of the wall keeps heated,

moist indoor air from meeting cold surfaces inside the wall and causing condensation. Such condensation can make insulation less effective and can lead to swollen, rotting wood and peeling exterior paint.

Houses most likely to lack a vapor barrier are those built decades ago. While some oil-based interior paints and a special latex—Glidden's *Insul-Aid*—can create a vapor barrier, they're not nearly as effective as plastic in the walls. Be cautious about using a humidifier if your only vapor barrier is a coat of paint.

Ratings for ultrasonic humidifiers

(As published in a November 1985 report.)

Listed in order of estimated over-all quality; bracketed models are essentially similar and are listed alphabetically. Prices are mfr. suggested retail.

Better ● ◐ ○ → **Worse**

Brand and model	Price	Tank capacity, pt.	Maximum output rate, fl. oz./hr. [1]	Estimated coverage, continuous running, cu. ft. [1]	Estimated coverage, 2 refills a day, cu. ft. [2]	Running time between refills, hr. [2]	Vapor distribution	Ease of use	Ease of cleaning	Advantages	Disadvantages	Comments
TOSHIBA KA508DE1	$110	10¼	18	5200	4000	9¼	●	○	○	D,F,H,L,O	b,i,l	—
SUNBEAM 661	98	9	14 [3]	4000	3400	10¼	◐ [4]	●	◐	B,K,M,O	f	F,G
SUNMARK 665	80	9	14 [3]	4000	3400	10¼	◐ [4]	●	◐	B,K,M,O	f	F,G
WELBILT MW30	80	9	14 [3]	4000	3400	10¼	◐ [4]	●	◐	B,K,O	f	C,F,G
BIONAIRE BT200	130	7½	15	4300	2800	8	◐ [4]	●	◑	C,E,G,H,J	e,f,h,k	D
HOLMES AIR HM-200	79	11¾	11	3200	3200	17	●	◑	○	A,F,I,O	d,l	—
ROBESON 3002	140	10	15½	4500	3800	10¼	○	○	○	A,D,K	k	A
DOUGLAS 1001	110	9	12½	3600	3500	11½	◐	○	○	A,K,O	—	—
ROBESON 3004	129	11¾	11	3200	3200	17	◐	◑	◑	D,I,L,O	c,k,l	H,I
QUIET MIST AHE150U	100	9¼	17	4900	3500	8¾	◑	○	○	A,D,F,H,K,O	a,c	I

HANKSCRAFT 5930	99	10¾	15½ [3]	4500	4100	11	○ [4]	○	O	k,l	E
SANYO CFK H501	140	10¾	15½ [3]	4500	4100	11	○ [4]	○	O	k,l	—
CORONA UF40	135	8½	14½	4200	3200	9¼	◐ [4]	○	N,O	b,g	J
CORONA UF30	120	8½	14½	4200	3200	9¼	◐ [4]	○	N,O	a,b,g	K
SANYEI UH2	100	10	14½	4200	3900	11	◐ [4]	◐	—	k,l	K
WELBILT MW50H	90	10½	12½	3600	3600	13½	○ [4]	◐	D,F,J	b,j,k	B,C
SEA MIST SM200H	70	8	13½ [3]	3900	3100	9½	● [5]	◐	—	b,j,k	C
TATUNG TUH400H	70	8	13½ [3]	3900	3100	9½	○ [5]	◐	—	b,j,k	C

1 Average of 4 tests; 2 per sample, 2 samples per model.
2 Assumes tank is never allowed to become empty.
3 Results for bracketed models have been averaged together.
4 Performance varied widely. Results are average of 2 samples and, where applicable, of similar models.
5 Sea Mist and Tatung have different spray nozzles.

SPECIFICATIONS AND FEATURES

All: ● Have volume control to regulate amount of mist. ● Have automatic shut-off of transducer when water supply is depleted. ● Have water tank with top handle. ● Measure about 10-12 in. high, 13-17 in. wide, 6-8 in. deep. ● Consumed 41-49 watts (about $2.50 a month at average national rate of 7.75¢ per kilowatt hour).
Except as noted, all have: ● Humidistat to automatically regulate room humidity. ● A single nozzle that swivels 360 degrees. ● Cleaning brush with storage provision. ● Power indicator light.

KEY TO ADVANTAGES

A – Has tip-over switch.
B – Large filler allows easy filling, cleaning.
C – Has built-in demineralization cartridge.
D – Has additional handle on underside of tank.
E – Has convenient carrying handle.
F – Has convenient recess in base for carrying.
G – Fan shuts off when tank is empty.
H – Has empty-tank indicator light.
 I – Has light to indicate air is too dry.
J – Has separate power switch.
K – Has safety interlock that prevents access to mist chamber until tank is removed.
L – Has two nozzles.
M – Water level easier to see than in most.
N – Has external fuses, easy to replace.
O – Has intake air filter.

KEY TO DISADVANTAGES

a – Lacks humidistat.
b – Lacks brush for cleaning.
c – Lacks storage provision for brush.
d – Tank hard to remove and replace.
e – Tank harder to fill than most.
f – Tank hard to empty completely.
g – Nozzle hard to swivel.
h – Markings on controls hard to decipher.
 i – Float more difficult to clean than most.
j – Slightly noisier than most.
k – Intake air vent located on bottom, where it may pick up dust or be obstructed by rug.
 l – Volume and humidistat controls turn in opposite directions, a confusing design.

KEY TO COMMENTS

A – Has humidity gauge; it was inaccurate.
B – Has nozzle adapter that evaporates room freshener or "medicated vapor" into mist.
C – Tank has water-level markings.
D – Has three-prong grounded plug.
E – Has no tip-over switch, but neither sample suffered damage in tip-over test.
F – Can use **Sunbeam** demineralization filter.
G – **Samsung HU701A** seems essentially similar.
H – **Imarflex UH050.**
 I – Model discontinued at the time the article was originally published in *Consumer Reports.* The test information has been retained here, however, for its use as a guide to buying.
J – According to company, this model was replaced by **Corona UF400,** $110.
K – According to the company, this model was replaced by **Corona UF300,** $100.

Humidifier bacteria and molds

For patients with dry, inflamed nasal membranes caused by breathing air that's too dry, doctors commonly recommend a vaporizer or humidifier to boost humidity in the bedroom or even the whole house.

But there's a drawback. The reservoir in a humidifier provides a growing environment for molds and bacteria. The heat of a steam vaporizer kills those organisms. But conventional cool-mist vaporizers can spew bacteria and molds into the air along with the moisture. Those air-borne microbes can cause allergic reactions in sensitive people and may also cause respiratory infections.

Consumers Union ran tests to see whether ultrasonic humidifiers would also spread a microbe-laden mist. They compared the ultrasonic machines with convenient cool-mist vaporizers. First, they ran them almost continuously for three weeks, using tap water. Then they let them sit idle for a week, as might occur in the home. By the end of that week, molds were present in the water in all the test models.

Each machine was placed—molds and all—into a disinfected chamber, along with petri dishes filled with a medium designed to grow molds. The humidifiers were run for ten minutes, exposing the petri dishes to the mist from the outlets. The petri dishes exposed to the ultrasonic humidifiers showed virtually no mold growth. By contrast, those exposed to the cool-mist vaporizers grew significant numbers of molds.

The tests were repeated, using a medium that could grow bacteria. Growth on the petri dishes showed that the ultrasonic humidifiers expelled only a few bacteria, while the conventional vaporizers expelled a lot.

Why the difference? It's possible that the ultrasonic vibrations destroy microbes, perhaps by breaking them apart. But an ultrasonic humidifier may still spew bits and pieces of mold and bacteria. While those bits may not cause infections, they may still trigger allergic reactions in sensitive individuals.

To avoid such problems, it's especially important for allergic people to keep the humidifier clean. Empty out the old water before each day's use and rinse or clean the humidifier (preferably with a solution of one tablespoon of bleach per pint of water) before adding fresh water.

Electric insect killers

There's a good bit of lore about insects being attracted to light. Enough is known about that attraction to help people enjoy warm-weather evenings in their own backyards.

Although no single kind of light attracts all insects at all times, it's fairly well agreed that the mix of light most alluring to nocturnal insects contains a significant component in the near-ultraviolet spectrum—"black light," as it is commonly called. Although invisible to humans, black light exists plentifully in sunlight and is produced by special lamps.

The most common electric bug killers are for use outdoors. They lure flying insects with black light and kill with electricity. These appliances can pass casual inspection as illuminated birdhouses or old-fashioned streetlamps.

Some are cylindrical, some are roughly square-cut or lantern-shaped, some are spherical. Inside a typical unit is a U-shaped black-light fluorescent bulb, which glows with a bluish tint. Surrounding the bulb is an electrically charged grid, which itself is contained within an outside screen. A cover and base complete the unit.

In its effort to get to the light, an insect must first fly or otherwise make its way through the outside screen, then try to get through the grid. The grid may be made up of concentric screens or formed of narrowly separated parallel wires. A transformer in the top of the unit produces a high voltage between the elements of the electrical grid, something on the order of 4,000 volts, while a limiter in the circuit holds the current down to about nine milliamperes.

When the insect touches both ele-

ments, it completes the circuit, and a small surge of current arcs through it. At high voltages, sudden heat generated by the tiny blast is so intense that it vaporizes the bug, leaving only dried fragments.

The size of a unit is generally dictated by the size and wattage of its fluorescent bulb. Units with a 20-watt bulb or two 8-watt bulbs are about 20 inches high. Those with a 15-watt bulb are usually about a foot high. Ranging in weight from about 4 to 11 pounds, the units are deceptively top-heavy because of the concealed mass of the transformer. However, as these are devices meant to be hung from a hook throughout a long season, they should seldom have to be moved or carried. Once a unit is installed outside, it need be taken down only for cleaning, bulb changes, and winter storage.

Since your aim is to divert insects from the outdoor area that you want to occupy, a bug killer should be placed at least 20 to 30 feet from that area. At the same time, it should be positioned to intercept insects' flying invasion routes from nearby thickets or woods.

Most manufacturers recommend hanging their units at a height of six to eight feet above the ground, and out of the reach of children. One place to hang a unit is from the branch of a tree. But a tall post, fitted with a mounting bracket, will also serve as an adequate support. Such hardware is sold by most manufacturers.

Bug killers attract and destroy large numbers of insects. More important, though, fluorescent-bulb devices noticeably reduce the annoyance of bugs as you sit outdoors during the evening hours.

It isn't possible, on the basis of existing information, to confirm a unit's advertised lure range; the purpose of these bug killers is not, after all, to attract all the bugs in the neighborhood. But the fluorescent bulb models should prove effective in reducing the level of insect annoyance on average-sized suburban lots.

Models with incandescent bulbs can prove disappointing in that they attract far fewer insects than other units. The bulk of their light output is in the visible spectrum, while the peak output of other units is in the near-ultraviolet spectrum.

Most manufacturers recommend keeping a unit turned on 24 hours a day. A few advise full-time use, at least for the first two weeks of operation, to break up insect reproduction cycles. That's probably excessive. Dusk-to-dawn use should be enough because the daytime catch of bugs in a unit run all the time is virtually nil, which may not matter much, since the most serious insect annoyance comes with evening.

That's when the bug killers draw conventions of all kinds of moths. Other insects that fly in numbers to their own electrocution are beetles, gnats, flies, and some mosquitoes. In contradiction to some earlier studies by the U.S. Department of Agriculture, studies by several universities have cast doubts on black light as an effective control for mosquito infestation. While the debate continues, practical use tests suggest that mosquito problems are reduced by the more effective bug killers.

Convenience

A big attraction of electric bug killers is their potential for doing a rather dirty job cleanly and detachedly, enhancing human comfort with a minimum of human fuss and involvement. However, it's not always quite that easy. Here are factors touching on some cares of operation.

Noise. "Zap" pretty well describes the sharp, sizzling sound that announces bug electrocution on the grid of all models. But it doesn't signify the differences in sound intensity that our testers heard. Some of the crackling noises can be downright annoying to you or even your neighbors, and send shivers through the squeamish.

Grid clogging. The grids of some units clog up fast with insect remains. In times of high insect activity, they have to be cleaned every few days or so to retain their effectiveness. That's a nuisance chore at worst.

Buildup of insect remains. Thoroughly burned-out husks of insects usually present no disposal or cleaning problem. They simply slide from the grid and drop through the bottom of most units.

Disassembly for cleaning. It's rather futile to try cleaning a unit, particularly its grid, by poking through the outer screen or blowing through it with the exhaust of a vacuum cleaner. Usually it's much easier to take the unit apart, at least to the extent of taking off the sides so that the grid can be properly brushed off. But some units don't come apart easily.

Bulb durability. About half the bug-killer manufacturers recommend replacing fluorescent bulbs every season, while others suggest replacement every two seasons. Some say only that the bulbs will last about 7,000 hours (more than two long summers of continuous 24-hour use). At any rate, the manufacturers seem to agree, the bulbs deteriorate long before they burn out, and it's wise to replace them before their light output diminishes too far. After two months of continuous use, most fluorescent bulbs dim to between 50 and 60 percent of their original brightness. Light from the incandescent-bulb models dims more slowly, but their output isn't much of an insect attraction to begin with. Replacing a fluorescent bulb is expensive; they cost from $6 to $24. The incandescents cost $2 to $3 apiece.

Power consumption. Power to light the bulb and charge the killing grid normally ranged from 24 to 35 watts for the fluorescent units. You can run one every summer night for the price of one or two bottles of insect repellent.

Safety. Despite the lethal shocks they deal to insects, an electric insect killer shouldn't pose a serious hazard for people if used properly. The high voltage is largely mitigated by the tiny current flow. In normal circumstances, it would be hard to receive a hazardous shock from one of these units; fingers probing to touch the killing grid would almost certainly have to be in contact with the grounded outer screen as well, limiting the shock to the finger—perhaps painful, but not lethal.

Still, there's no point in hanging a unit where it can tempt the curiosity of children. And don't try to circumvent the three-wire power cord provided with each unit; plug it into a three-wire outdoor extension cord, and plug that into a grounded outlet. Most units have a safety switch that automatically shuts off electric power when the unit is being disassembled, even if the line cord is left plugged in.

Recommendations

If you think it extravagant to spend $69 to $140 (even if reduced by a discount) for a device that kills common backyard-variety insects, consider the money spent on patios, porches, and landscaping to make backyard areas pleasant. In that context, the purchase of a bug killer makes some economic and recreational sense.

Among the models in Consumers Union's test, however, some make considerably less sense than others. Models with incandescent bulbs seem implausible choices; they don't attract enough insects.

The most sensible choices are the five top-rated bug killers. All of them perform competently.

A bug drowner

The **Pestolite Patio RDIP,** $70, doesn't work the same way as the other outdoor bug killers. It attracts insect victims with an ultraviolet lure light, but instead of electrocuting them it drowns them.

The 4½-pound unit is closed to view on all sides except the front, which has a window of transparent plastic sheet. Inside, an 8-watt ultraviolet fluorescent bulb beckons through the window, while a small fan runs overhead. A slide-out tray full of water, mixed with a bit of detergent to reduce surface tension, lies at the bottom of the trap.

For the **Pestolite**'s method to work, night-flying insects must conform to a predictable pattern of behavior. They fly toward the lure light, bump into the transparent window, are swept off their wings by downdraft from the fan, drop to the water, and drown.

Apparently, bugs can be depended on to behave that way. Practical tests resulted in dense catches of all kinds of night-flyers, including mosquitoes and gnatlike pests so tiny that they can walk with impunity on the electrocution grids of other bug killers.

You will like the silent, unobtrusive method of the **Pestolite**—the absence of zapping noises and the purple dimness of the lure light. The unit is designed to be mounted by bracket to a fairly flat vertical surface, or it can simply be set down on a table.

But you won't like the attention the **Pestolite** requires. Every few days, the tray has to be dumped and a new detergent-water solution prepared.

Ratings of electric insect killers

(As published in a June 1982 report.)

Listed, except as noted, in order of estimated overall quality. All can be hung outdoors or indoors. Unless otherwise indicated, all have a 15-watt "black-light" fluorescent bulb, an open or partly open base to let insect fragments fall through, and a grounded metal external screen. Dimensions are in order of height and width or diameter. Prices are suggested retail. (Prices of replacement bulbs are in parentheses.)

- *The following 2 models were judged approximately equal in overall quality. Listed alphabetically.*

FI-SHOCK FS3000, $92 ($15). 15¹/₂x12¹/₄ in. 8 lb. Very good corrosion resistance. Plastic external screen. Judged difficult to disassemble.

SEARS Cat. No. 1402, $100 ($17) plus shipping. 22¹/₄x12³/₄ in. 11 lb. 20-watt fluorescent bulb. Fluorescent-light output decreased more slowly than with most. Good corrosion resistance. Judged difficult to disassemble.

- *The following 3 models were judged approximately equal in overall quality. Listed alphabetically.*

FLINTROL XL100, $110 ($18). 19¹/₄x10³/₄ in. 9³/₄ lb. 20-watt fluorescent bulb. Fluorescent-light output decreased more slowly than with most. Excellent corrosion resistance. Edge of base became loaded with insect fragments. Grid discharge judged noisier than with most. No mounting instructions.

FLOWTRON BK2000, $100 ($13). According to the company, this model has been discontinued, but may still be available in some stores. 16¹/₂x10³/₄ in. 8³/₄ lb. Excellent corrosion resistance. Bulb removal judged easier than with most. Grid discharge judged noisier than with most. Cannot be disassembled. Optional bait tray, $5.

VANDERMOLEN BUGKIL E85, $100 ($13). 13¹/₄x12 in. 9³/₄ lb. Good corrosion resistance. Grid discharge judged noisier than with most. Judged easier to disassemble than most.

- *The following 2 models were judged lower in overall quality than those preceding.*

WEBER PATIO WEB 23520, $89 ($14). 13³/₄x12¹/₄ in. 8¹/₂ lb. Fair corrosion resistance. Wide openings in external screen admitted more large insects than with other models. Grids sometimes tended to clog. Grid discharge judged noisier than with most. Judged easier to disassemble than most.

UDO EL ZOPPO 1500, $110 ($6). 20¹/₄x9¹/₂ in. 6 lb. Poor corrosion resistance. 20-watt fluorescent bulb. Bulb removal judged easier than with most. Judged difficult to disassemble. Lacks safety switch to keep unit turned off when disassembled.

- *The following 4 models were judged lower in overall quality than those preceding, primarily because their grids tended to clog more frequently.*

EMERSON EBK 15, $80 ($12). 15³/₄x8¹/₄ in. 6¹/₂ lb. Excellent corrosion resistance. Bulb removal judged easier than with most. Cannot be disassembled.

CHARMGLOW RID-O-RAY 6153, $90 ($20). According to the company, this model has been discontinued, but may still be available in some stores. 13¹/₂x6³/₄ in. 7¹/₂ lb. Poor corrosion resistance. Bulb removal judged easier than with most. Judged easier to disassemble than most. Lacks safety switch to keep unit turned off when disassembled.

WARDS Cat. No. 2152, $77 ($15) plus shipping. Essentially similar to **Charmglow Rid-O-Ray 6153**, preceding.

HALL AMERICAN 10206 IDT100, $140 ($24). 19¹/₂x14¹/₄ in. 9 lb. 2 8-watt fluorescent bulbs. Poor corrosion resistance. Closed base became loaded with insect fragments. Judged easier to disassemble than most.

■ *The following 2 incandescent-bulb models were judged lower in overall quality than those preceding because they were judged to attract significantly fewer insects.*

FI-SHOCK FS2500, $69 ($2.75). According to the company, this model has been discontinued, but may still be available in some stores. 15¹/₂x12¹/₄ in. 7¹/₄ lb. 75-watt incandescent bulb; consumed more power than any other. Very good corrosion resistance. Plastic external screen. Bulb removal judged very difficult. Judged difficult to disassemble.

HALL AMERICAN 10200 IDT2, $75 ($2.50). 12¹/₂x9¹/₂ in. 4¹/₄ lb. Poor corrosion resistance, 60-watt incandescent bulb; consumed more power than most. Closed base became loaded with insect fragments. Judged easier to disassemble than most.

The homely art of ironing

At a loss at the ironing board? Here are some tips from CU's home economist:

Iron garments that require the lowest temperature setting first. If you start with high-temperature items, you'll have to wait for the iron to cool down when you go on to items that need lower settings.

With fabrics blended of different fibers, use the setting for the fiber that requires the least heat. If a label has come off and you're not sure of fiber content, test a small area—a seam or an inside hem, say—before ironing. Watch out for synthetic fabrics that look as though they're made of natural fibers.

When ironing any garment, start with small areas such as collars, cuffs, and sleeves; then go on to the larger areas. To avoid distorting fabric, always iron with the fabric's lengthwise grain—up and down a shirt, for instance, not across it.

To avoid putting a shine on silks, acetates, rayons, or dark fabrics, turn the garment inside out; press its inner side.

Press velvets, velours, corduroy, or other nappy fabrics on the wrong side; don't let the iron rest on the fabric. If the nap is flattened, bring it up with a shot of steam from a few inches away—or hang the fabric in the bathroom with the door closed while you shower.

Steam-press woolens on their wrong side, and use a woolen presscloth. Never press wool completely dry.

Don't sprinkle, spray, or steam silks, taffetas, or other fabrics that may water spot. If you're uncertain about the fabric, check an inconspicuous area first.

A steam iron may not produce enough steam to remove all the wrinkles from fabrics such as linen. Or, on big jobs, you may find it a nuisance to keep pumping a steam or spray button. One solution: sprinkle the item, roll it up, put it in a plastic bag, and let it sit a few hours to become evenly damp. If the weather's hot and humid and mold is a possible problem, put the bag in the refrigerator. Alternatively, remove problem items from your dryer before they are completely dry.

Never touch plastic buttons or zippers with a hot iron.

If the bottom of the iron gets scratched, iron a piece of waxed paper. That will coat the soleplate and make the iron easier to push.

Citrus juicers

There's a lot to be said for commercial orange juice. The best frozen concentrates come close to the taste of freshly squeezed orange juice. Frozen juice is far more convenient to prepare than fresh juice. Frozen juice contains all the vitamin C you'd need in a day. And frozen juice can cost only about half as much as juice you squeeze yourself.

For those who are convinced that nothing quite matches the taste of freshly squeezed orange juice, there are a number of squeeze-it-yourself juicers available: electric, lever-type manual, and the old-fashioned manual reamer.

Squeezing

With a hand reamer you push half an orange or lemon or grapefruit down onto a ridged cone, then twist, turn, and squeeze the fruit. The juice then flows into a moat surrounding the cone.

An electric juicer works the same way, but you hold the fruit still. The cone, not the fruit, does the turning. The ridged cone of a typical electric juicer starts to turn when you push the fruit down on it and stops when you lift off the hollowed rind.

After being released by the revolving cone, the juice flows through a strainer, which holds back the seeds and some but not all of the pulp. There may be a plastic scraper affixed to the cone. It's meant to keep the strainer clear of pulp to aid the flow of juice. But a scraper

isn't essential.

The result with most models is a moderately pulpy juice.

The moat around the cone of a manual reamer is usually shallow, so it doesn't hold much juice. Electric juicers often have a built-in bowl or pitcher to collect the juice.

Consumers Union testers prefer the models with their own pitcher to those that channel the juice to a spout, under which you place a glass. With the latter, they find, at least a little juice continues to drip from the spout to the counter after the glass is removed.

Most juicers with a spout can accommodate an eight-ounce glass up to 4¾ inches tall. A two-cup measuring pitcher should fit nicely under the spout.

Other factors

A juicer should be easy to maintain and unobtrusive:

Ease of cleaning. The juicers are simple machines and are generally easy to clean. The best arrangement has the cone, strainer, and juice container as a single unit. Less convenient are models with several pieces that have to be taken apart, washed, dried, and put back together.

Check manufacturers' instructions about washing in a dishwasher.

Storage. The citrus juicer is relatively small and light in weight. None takes up much counter or cabinet space. A few juicers have a cord that might be too short to reach from counter to outlet in some kitchens. Models with a longer cord have a cord-storage feature, so the cord won't snake all over a counter.

Manual juicers

A lever-type citrus juicer and an old-fashioned reamer require only your energy to make citrus juice. To use a lever-type juicer, you put half an orange into the hopper and push down on a lever. You can squeeze fruit just as fast with one as with the electric citrus juicer, albeit with greater effort and perhaps with some bitterness (from squeezing the rind).

A simple reamer requires much more effort than the lever-type juicer. How much juice you extract from fruit squeezed on a manual reamer depends on how hard you work. If your hand doesn't tire out, you can get as much juice from an orange as you would if it were done in an electric juicer. But it takes longer.

Recommendations

If you're not devoted to freshly squeezed juice, you needn't consider an electric juicer. A manual reamer will be adequate for an occasional glass of fresh citrus juice for one or two people. When you start squeezing many oranges regularly, however, you'll need strong hands as well as time. That's when you might think of a juicing machine.

For a few dollars more than a lever-type juicer, you can buy an electric.

Ratings of citrus juicers

(As published in a November 1982 report.)

Listed in order of estimated overall quality. Except where separately rated by bold rules, closely ranked models differed little in overall quality. Except as noted, all made moderately pulpy juice. Dimensions and weight include cover, if any. Prices are suggested retail, rounded to nearest dollar; + indicates shipping is extra. Except for mail-order models, discounts are generally available.

Legend: ● ◑ ○ ● — Better → Worse

Brand and model	Price	Height x diameter (in.)	Weight (lb.)	Capacity of juice container (fl. oz.)	Cord length (in.)	Performance	Convenience	Ease of cleaning	Advantages	Disadvantages	Comments
PANASONIC MJ50P	$30	8x7½	3¼	20	41½	◉	◑	●	B,D,E,G,H	—	F
DAZEY FJ28	40	7¼x8¾	3½	22	74	●	◑	◑	A,D,G	—	F
SANYO SJ60E	25	8x8½	3¼	22	44	●	◑	◑	A,C,D,G	—	F
KRUPS PRESSA C252	45	6¾x8¾	2	16	47½	●	◑	○	A,D,G,H	c	A,F
DOMINION/HAMILTON BEACH 2109	20	7½x6	2¾	9	34	◑	◉	◑	—	d,e	C
WARING JC1108	31	8¼x5½	2¾	—	36	○	◑	◑	D,F,G,H	a,b,g	E
OSTER AUTOMATIC 368	26	8½x7½	3¼	—	27½	○	◑	◑	F	a	F,H
PROCTOR-SILEX J11C	33	9¼x7¼	4¾	—	25	○	◑	◑	—	a,f	B,G,K
SEARS COUNTERCRAFT Cat. No. 83488	20+	9x7¼	4¾	—	25	○	◑	◑	—	a,f	B,G,K
BRAUN CITROMATIC 2 MPZ-4	29	6¼x7	2½	8	41	◑	◑	◑	A,D	d,e	D,J
RIVAL 966	26	7½x7¼	3¾	—	32	◑	○	◑	F,H	a,b,d,e	E,I

SPECIFICATIONS AND FEATURES

Except as noted, all: ● Operate when fruit is pressed down on cone. ● Operated quietly. ● Have plastic motor housing, 1 plastic cone, plastic pulp strainer, opaque plastic bowl or pitcher.

KEY TO ADVANTAGES

A – Has handle or grip on juice container.
B – Has clear-plastic juice container marked in fl. oz. and ml.
C – Has clear-plastic juice container.
D – Has provision for cord storage.
E – Has brush for cleaning.
F – Better than most at juicing limes.
G – Has plastic dust cover.
H – Quieter than most.

KEY TO DISADVANTAGES

a – Lacks juice container. Juice flows into glass; may drip on counter when glass is removed.
b – Requires slightly shorter glass than most.
c – Reversal of cone direction can be disconcerting.

d – Requires somewhat more pressure than most.
e – Requires more effort than most to juice grapefruit.
f – Worse than most at juicing lemons.
g – Spatters somewhat more than most.

KEY TO COMMENTS

A – Juice container locks on to base.
B – Juice container moves slightly during operation.
C – Juice container, strainer, and cone are single unit.
D – Cone and strainer are single unit.
E – Has second, larger cone, which fits over small cone.
F – Has plastic scraper, attached to cone, to clear pulp.
G – Ceramic cone; metal base and strainer.
H – Tall container surrounding cone may interfere with use by someone with large hands.
I – Has on/off switch.
J – Made less-pulpy juice than others.
K – Made pulpier juice than most.

Juice extractors

If your taste tends toward the extraordinary, you probably won't be satisfied with a mere citrus juicer. A juice extractor provides all the mystique of a juice bar in your own kitchen. It can make juice out of a carrot or a beet in seconds. It can turn pears into instant nectar.

Unfortunately, the extractor has one failing. All the citrus fruits—and other thick-skinned fruits—have to be peeled before juicing. That's a major inconvenience. Large pits, as in peaches, also have to be removed.

Juice extractors tested in Consumers Union's laboratories look a bit like food processors. They have a feed tube, into which you put the fruits or vegetables. The produce has to be cut in chunks small enough to fit into the tube. Like food processors, the extractors come with a plastic pusher to force the food through the tube.

At the base of the tube, the food meets a metal grating disk. When the extractor is turned on, the disk spins, shredding the food. Centrifugal force separates juice from shredded food fibers. In a few models the fibers are thrown against the walls of the strainer basket. In others, the fibers are thrown into a separate compartment. However it's separated, the juice passes through a strainer basket surrounding the grating disk and then flows out a spout, either into a container that comes with the machine or into a glass you've placed under the spout.

Juicing

CU testers juiced a variety of fruits and vegetables in the machines and tasted the results. The taste experience for the uninitiated can be strange—the flavor of celery without its crunch takes some getting used to.

Here's a rundown of some of the things extractors can do:

Carrot juice. It takes about a minute to turn a pound of carrots into seven or eight ounces of sweet and carroty juice (and two cups of pulp). Most extractors make a moderately thick juice with some grainy texture.

Celery juice. This will come out pale, green, and bland.

Tomato juice. Tomato juice from a can is generally rather thick, bright red, and spicy. Tomato juice made in a juice extractor is moderately thick and very smooth. But its color is pink and its flavor bland. A dash of lemon or lime juice and some spices can bring it to life.

Orange juice. Once the oranges are peeled, making the juice goes very quickly. But it's not typical orange juice. It's unusually smooth, aerated, creamy, or frothy. Some extractors turn out orange juice with a head, like beer. The juice tends to separate rather quickly.

In an informal taste test involving CU staffers and some children, no one preferred the extractor-made orange juice to conventionally squeezed juice.

An extractor will pulverize small seeds such as those in oranges, as well as bits of the fruit's white membrane that may not be completely removed when the orange is peeled. Processing these can give an extractor's juice a bitter flavor.

Exotic juices. Once you've caught juice fever, you may decide to experiment a bit with off-beat juices and combinations. Extractor-made cantaloupe juice, for example, will be moderately thick and sweet, with a mellow cantaloupe flavor. Broccoli juice comes out thin, dark green, and overpowering.

The Consumers Union adventurers brought together two carrots, an apple, and a stalk of celery. Another combination drink was made from two oranges, half a banana, and a few strawberries. That drink resembles a thick shake.

Separating juice and pulp

With most extractors, you have to provide a glass or serving container to catch the juice as it pours out the spout. Most spouts will accommodate a 4¾-inch-tall, eight-ounce glass.

An extractor can hold a limited amount of pulp. When one reaches capacity, you have to empty it before you can make any more juice.

With most extractors, it is easy to remove the upper housing, which holds the pulp so you can quickly get back to producing juice.

The upper part of most extractors is clamped on with the machine's handle or with a couple of latches.

Other factors

An extractor is more complex than a citrus juicer and has more problems.

Ease of cleaning. A juice extractor is a nuisance to clean. Most have five or six parts to be removed, washed, dried, and replaced. Most manufacturers don't recommend putting the parts in a dishwasher.

A plastic filter strip on the **Krups** and the **Braun** is helpful in removing the bulk of the pulp between batches of juice and when the machines are cleaned. However, the pulp still escapes the filter strip and was difficult to remove from the strainer-basket.

Most come with a toothbrushlike tool for cleaning the basket. The brushes are also handy for cleaning the grating disks, which are a mass of sharp edges. A real toothbrush would work, too.

An extractor will spatter the counter. At times, you may feel like a tomato target.

Noise. A juice extractor is far noisier than a citrus juicer, but it still isn't much of a problem since it's generally used for so short a time.

Safety. There aren't likely to be any significant hazards, but these machines can become unbalanced by what you put into them and start to vibrate. Frequent cleaning helps, and so does slower feeding.

Underwriters Laboratories requires the feed-tube opening of an extractor to be no more than three inches across, and the grating disk to be at least four inches below the top of the feed-tube opening. But even with the four-inch clearance, you can, if you try, reach through the feed tube and touch their grating disk. Don't do it.

Extractors don't have an interlock mechanism to prevent operation with the lid off. Without an interlock, it's possible to remove an extractor's lid while the unit is operating. That's hazardous, so be careful.

Storage requirements. An extractor, particularly a larger rectangular one, takes up a good bit of storage space.

Recommendations

An extractor is noisy, hard to clean, and space-consuming. But it can make beverages that citrus juicers and blenders can't make. Fans of fresh apple juice, carrot juice, or fruit nectars may consider an extractor worth the investment.

Don't consider an expensive extractor if you're primarily interested in citrus juices. Peeling the fruit before processing is time-consuming and inconvenient. And the frothy result may be less appetizing than conventionally squeezed juice.

There are no nutritional advantages to drinking a glass of juice instead of eating the whole fruit or vegetable. Consider the carrot: a carrot provides dietary fiber and exercise for the teeth and gums that a glass of carrot juice doesn't provide. Carrot juice may have more vitamins and minerals than whole carrots—because you get the essence of many carrots. But, in any case, one large carrot provides about twice an adult's RDA for vitamin A.

Consumer Reports
Savings Certificate

Subscribe to Consumer Reports and get the unbiased facts you need on hundreds of products and services. Every month you'll get up-to-date product Ratings and recommendations... expert advice on health and money matters... warnings against unsafe merchandise. And with a subscription, you can save up to $25.40 off the newsstand price.

☐ **YES,** please enter my subscription to Consumer Reports and send me 11 regular monthly issues plus the 1987 Buying Guide Issue when published—all for $16. **I save $9.70 off the newsstand price.**

☐ **I WANT TO SAVE EVEN MORE!** Please enter my TWO-YEAR subscription and send me 22 regular monthly issues plus both the 1987 and 1988 Buying Guide Issues when published—all for $26. **I save $25.40 off the newsstand price.**

Name _____
(please print)

Address _____ State _____ Zip _____

City _____

☐ Payment enclosed.   ☐ Please bill me later.

Please allow 6 to 8 weeks for delivery of first issue.
Rates for delivery in the U.S. only. All other countries add $4.50 per subscription year.

You save up to $25.40
when you order a 2-year subscription

5C6C7

Consumer Reports
Savings Certificate

Subscribe to Consumer Reports and get the unbiased facts you need on hundreds of products and services. Every month you'll get up-to-date product Ratings and recommendations... expert advice on health and money matters... warnings against unsafe merchandise. And with a subscription, you can save up to $25.40 off the newsstand price.

☐ **YES,** please enter my subscription to Consumer Reports and send me 11 regular monthly issues plus the 1987 Buying Guide Issue when published—all for $16. **I save $9.70 off the newsstand price.**

☐ **I WANT TO SAVE EVEN MORE!** Please enter my TWO-YEAR subscription and send me 22 regular monthly issues plus both the 1987 and 1988 Buying Guide Issues when published—all for $26. **I save $25.40 off the newsstand price.**

Name _____
(please print)

Address _____ State _____ Zip _____

City _____

☐ Payment enclosed.   ☐ Please bill me later.

Please allow 6 to 8 weeks for delivery of first issue.
Rates for delivery in the U.S. only. All other countries add $4.50 per subscription year.

You save up to $25.40
when you order a 2-year subscription

5C6C7

BUSINESS REPLY MAIL
FIRST CLASS   PERMIT NO. 1243   BOULDER, CO

POSTAGE WILL BE PAID BY ADDRESSEE

NO POSTAGE
NECESSARY
IF MAILED
IN THE
UNITED STATES

Gift Subscription Department
Box 2475
Boulder, CO 80321

Ratings of juice extractors
(As published in a November 1982 report.)

Listed in order of estimated overall quality. Except where separately ranked models differed little in overall quality. Except as noted, all made moderately thick carrot juice. Prices are suggested retail, rounded to nearest dollar; + indicates shipping is extra. Except for mail-order models, discounts are widely available.

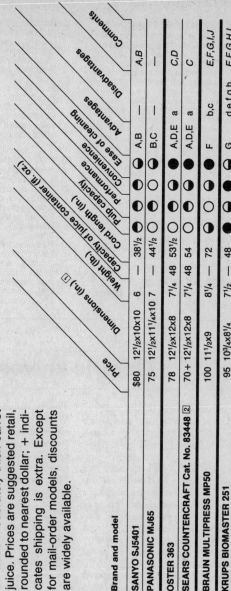

Legend: Better ◉ ◖ ○ ◗ ● Worse

Brand and model	Price	Dimensions (in.) [1]	Weight (lb.)	Cord length (in.)	Capacity of juice container (fl. oz.)	Pulp capacity	Performance	Convenience	Ease of cleaning	Advantages	Disadvantages	Comments
SANYO SJ5401	$80	12½x10x10	6	—	38½	◖	◖	◖	◖	A,B	A,B	—
PANASONIC MJ65	75	12½x11¼x10	7	—	44½	○	◖	◖	○	B,C	—	—
OSTER 363	78	12½x12x8	7¼	48	53½	◖	◖	●	●	A,D,E	C,D	a
SEARS COUNTERCRAFT Cat. No. 83448 [2]	70+	12½x12x8	7¼	48	54	○	●	●	●	A,D,E	C	a
BRAUN MULTIPRESS MP50	100	11½x9	8¼	—	72	○	○	◖	●	F	E,F,G,I,J	b,c
KRUPS BIOMASTER 251	95	10¾x8¼	7½	—	48	●	●	◖	◖	G	E,F,G,H,I	d,e,f,g,h

[1] Height x diameter, if 2 dimensions given; height x width x length, if 3 dimensions given.
[2] 1985 catalog price

SPECIFICATIONS AND FEATURES

All: ● Have plastic motor housing, feed tube with pusher, lid that clamps or latches, on/off switch, and attached cord. ● Lack an interlock system to prevent operation when lid is removed. ● Require peeling of thick-skinned fruits and removal of large pits before juicing.

Except as noted, all have: ● Clear-plastic lid. ● Separate pulp container. ● Metal strainer basket with attached grating disk. ● Single operating speed. ● No juice receptacle. ● Brush for cleaning. ● Fairly noisy operation.

KEY TO ADVANTAGES

A – Has provision for cord storage.
B – Spout cover decreases spraying of juice.
C – Has device for cleaning strainer basket without disassembling unit.
D – Has clear-plastic juice container marked in fl. oz. and ml.
E – Has helpful instruction guide.
F – Has automatic overload feature with reset button.
G – Quieter operation than others.

KEY TO DISADVANTAGES

a – Upper housing and strainer basket sometimes difficult to remove.

b – Spatters somewhat more than most.
c – Strainer basket must be emptied occasionally to prevent vibration when juicing soft foods; has brake that stops unit immediately if vibration occurs.
d – Requires slightly shorter glass than most.
e – Feed tube smaller than others.
f – Pulp collects in strainer basket, which must be emptied frequently to prevent vibration with soft foods.
g – Juice sprays from spout.
h – Metal clamping handle hard to release.

KEY TO COMMENTS

A – Has 3 speeds.
B – Upper housing locks on to base.
C – Has 2 speeds.
D – Carrot juice thicker than others.
E – Opaque plastic lid.
F – Releasing clamp before disk stops can cause disk to shave edge of feed tube.
G – Has filter for use inside strainer basket.
H – Carrot juice thinner than others.
I – Lacks brush for cleaning.
J – Has more removable parts than others, including plastic strainer basket and separate grating disk.

Electric knives

The skilled carver deftly delivers neat, uniform slices of meat to waiting plates before it has cooled. Such a carver probably doesn't need an electric knife.

But if your skill is just average, you may well find that a good electric knife can help. It will also come in handy for a wide variety of kitchen and table chores other than carving—for example, slicing tomatoes and eggplants without mashing them, cutting onions fast enough to stay dry-eyed, slicing bread or pineapple.

An electric knife has two serrated blades, linked near the tip, that fit into a handle containing an electric motor. When the knife is turned on, the blades move to and fro. The basic job is then just to guide them. The carver needs to exert less pressure than with a regular carving knife, and almost no back-and-forth motion; hence, there is less risk of causing a roast to skid and splash juices about. These knives do not operate silently, however, and the noise from some of them is enough to intrude on table conversation.

In general, plug-in models perform better than cordless. Still, many buyers consider freedom from a hobbling power cord well worth a small sacrifice in performance. Properly charged, the batteries on the cordless knives should store enough power to handle the largest roast. You could safely take a cordless knife on a picnic without fear that you might have to serve seconds in chunks.

This plus for the cordless knife is a minus for models with a cord. The power cord may not reach from the table to an outlet in a large dining room. You could, of course, use an extension cord, but then you may increase the risk of tripping one of the diners. And a cord may accidentally trail through the meat juices.

Carving

Most electric knives won't slice straight down all the way through a piece of meat. Either your hand or part of the handle hits the table first, unless you set the carving board right at the edge of the table or tilt the blade to let the tip cut along the bottom of the slice. To eliminate this interference, the blades of some models are inserted at an angle to the handles.

Lightness (not very much more than a pound) and good balance make a knife comfortable.

It would make no sense to cancel out the real advantages an electric knife can offer by choosing one that's so bothersome to take out or use that it will gather dust most of the time. It pays, then, to give some thought to convenience difference.

Take blades. Two of them—not one—will have to be removed after every use for washing. If you can't insert and remove them easily, you may not be very happy with your knife.

If you accidentally press the switch and start a knife up as you're handling the blades, there's a real chance you'll cut yourself. So unplug any cord knife first. It would also be wise to guard against accidental start-ups between uses, even between slicing the first and second helpings. Many cord knives have a locking device for the on/off switch so you needn't pull the plug.

Check the clearance if you plan to keep your knife on a countertop under an overhanging cabinet. With the blades removed, you probably won't have any trouble—the tallest handle, in its stand, is only 14 inches high. If you insist on keeping the blades in the handle, you may need as much as 22 inches of clearance.

Recommendations

Manufacturers are sometimes remiss in quality control; so try to check any knife for defects before you buy. In the store, insert the blades and heft the knife for balance and comfort. Make sure the controls are within reach and easy to use.

Don't forget to insert the blades in the handle—to make sure that there are no gaps between them where food can lodge and that they slide easily. At home, use the knife as soon as possible. If you buy a top-rated model and find that it doesn't slice well, you may have dull blades that should be exchanged. It's important to have sharp blades right from the start and to take good care of their edges—you'd probably have trouble finding someone to resharpen them.

Generally, the less expensive blades are stainless steel and the more costly ones stainless steel with tungsten carbide edges. While the latter will probably stay sharp longer, cutting efficiency also depends on blade design, so replacing all-stainless blades with tungsten-edge blades may not improve performance.

Blades will stay sharper longer if you don't use them to cut frozen foods, if you keep them from striking bones, and if you store them in their sheaths. Carve on a wooden carving board to avoid damage to both blades and platters. Wash the blades separately from other cutlery, and don't dunk the handles in water; they aren't sealed. Wipe the handles with a damp cloth.

Microwave cookware

As you become involved with microwave cooking, you'll find yourself changing a lot of old habits. One of them is what you cook in. You'll make hot chocolate in the same cup you drink it from. You'll heat dinner right on your dinner plate.

Your favorite stainless-steel saucepan will be left hanging on the wall. Metal pots and pans don't work in a microwave oven because metal reflects the microwave energy. If you use a metal pot in a microwave oven, you'll have, at best, a cold meal. At worst, you'll have a damaged oven.

You no doubt already own several items suitable for microwave cooking. Some may surprise you. Paper towels are good for cooking bacon. Foods in some plastic freezer containers can be defrosted right in the container, if you remove the lid and make sure the food doesn't get too hot. You can even heat food in straw serving baskets.

You probably own casseroles and baking dishes of ceramic or glass, which you use in your conventional oven. Most of those utensils can also be used in a microwave oven.

But if you do a lot of microwave cooking, you may need additional cookware. Should you buy cookware marketed specifically for use in microwave ovens?

Dozens of manufacturers make such utensils. You can buy a full line of equipment for microwave cooking: roasting racks, soufflé dishes, baking pans, bacon racks, pie plates, muffin pans, ring molds—just about every form of pan you would find in conventional cookware. But glass and ceramic cookware—plain or fancy—has long been sold. And many of the microwave utensils are designed for cooking things microwave ovens don't cook particularly well.

Versatility

Since microwave ovens are really accessories to conventional ovens and ranges, cookware that is usable in both is a better investment than cookware usable only in microwave ovens.

A further measure of versatility is whether a utensil can be used as a freezer-storage container and whether it is dishwasher safe.

The shape of a pan affects its performance when it is used in a microwave oven. Round pans are superior to rectangular ones for baking and for cooking casseroles and meat loaf. In a rectangular pan, food tends to get overcooked in the corners. Food also cooks more evenly when spread out in a shallow pan than when heaped up in a deeper dish.

The material of the utensil—glass, ceramic, plastic—doesn't seem to have an effect on how well food cooks in a microwave oven. In conventional ovens, the material sometimes does affect the cooking process.

Most glass and ceramic utensils can be used in a conventional oven if the oven temperature doesn't exceed 375° to 400°F.

According to their instructions, some plastic utensils have temperature and use restrictions similar to the glass and ceramic cookware: when the plastic gets too hot it can cause an obnoxious odor, and impart a plastic flavor to the food.

Clever design gives many utensils another kind of versatility. The lids of many casseroles, for example, can be used as cooking vessels.

Some open roasting racks double as bake trays. Racks may be reversible, one side having a smooth surface for baking, the other side ridged for roasting. Or a rack may come with a trivet to use when roasting and remove when baking.

A few racks come with a cover to help keep food from drying out and spattering during cooking.

Or the cover can be used as a cooking vessel. Such a cover also has an opening at the lip so you can insert the oven's temperature probe.

Browning dishes are an attempt to solve the problem microwave ovens have in cooking meat. Unlike other microwave cookware, they have a coating on the bottom that allows them to absorb microwave energy. A browning dish requires preheating in the microwave so it becomes hot enough to brown the food placed in it. Unfortunately, the dishes are more likely to "beige" food rather than brown it.

Cooking something on a browning dish can take twice as long as cooking it in a plain dish. Hamburgers can actually take more time to cook on a browning dish than on a range. Foods cooked on a browning dish need a lot of attention—turning and rotating—for even cooking.

However, the results when you use a browning dish in a microwave are better than not using a browning dish, even though the food is still not as appealing as food cooked in the traditional way. Grilled-cheese sandwiches turn out pale and unevenly cooked. The cheese melts before the bread shows much color.

Other factors

Testers at Consumers Union also checked for heat absorption and breakage of the utensils. They also evaluated ease of cleaning.

Heat absorption. One reason that microwave cooking is so much faster than conventional cooking is that the energy heats just the food. It wastes no time

heating the container. At least, that's how it's supposed to work. But some materials absorb more energy than others.

You can do a test at home to see which of your nonmetal utensils can be used in a microwave. You don't have to make precise temperature measurements—just check to see if the dish is fairly cool. If it is, the dish is probably okay to use in the microwave oven.

It's a good idea to carefully examine a new ceramic utensil for cracks the first few times it's used. If the dish cracks, try to return it.

Remember that even the coolest cookware can become hot to the touch when it's used for cooking something that takes a long time. It picks up heat from the food it holds. Keep potholders handy.

Breakage. Ceramic and glass utensils are likely to be breakable if dropped.

And even plastic utensils may sometimes crack.

Cleanup. Except for the browning dishes and the crevices on some trivets, cleaning the cookware should be easy. Some of the plastic utensils have a nonstick finish. That's not really necessary, since sticking food is seldom a problem in microwave cooking. The nonstick finishes are probably a drawback because they are easily scratched and quickly look worn.

Because browning dishes sear food, they accumulate a fair amount of burnt-on soil and can be difficult to clean.

Recommendations

Before buying a supply of special cookware, take stock. See what **Pyrex, Corning Ware** (except for **Centura** and some **Corelle)** or other nonmetal cookware you have.

Don't consider using your good china. And don't use dishes with a decorative metal trim. Be careful of pottery: it may have metal in the glaze or impurities in the clay.

Do consider what you'll be cooking. The Consumers Union survey of microwave-oven owners indicated that many don't like roasts made in microwave ovens. If you come to the same conclusion, you won't have much use for a roasting rack. And since microwave ovens are usually poor at baking, baking utensils may not be good, although some can be used for heating foods.

For the utensils you decide you need, use ceramic or glass, since those materials can also be used in a conventional oven. But some of the plastic utensils are handy.

If you don't mind microwave meat, you'll probably need a roasting rack (although you can improvise one with a dish and an inverted saucer or casserole lid). Buy the biggest one that you can use in your oven. And, since you're spending the money, try to get a rack that also functions as a baking tray.

Because of the bother involved in cooking with browning dishes and the mediocre results, a browning dish is apt to be disappointing. Your range may still be best for cooking pork chops, hamburgers, fried eggs, and the like.

Hardware and variety stores still sell the same sort of ceramic and glass cookware they've sold for years—for quite a bit less money than many of the dishes specifically designed for microwave use. In one of those stores, you might find an inexpensive "old-fashioned" heat-resistant glass dish that suits your needs as well as an expensive microwave utensil.

Microwave ovens

Microwave ovens are great for fast meals. Microwave ovens can turn dreary leftovers into a meal that tastes freshly cooked. They can bake a potato in five minutes. They don't heat up the kitchen. They're clean and they save energy.

If you've thought about buying a microwave oven, you've heard all this from friends who own one and from salespeople. If you own one yourself, chances are you'll agree. But microwave ovens cannot be all things to all foods.

Microwave ovens don't really do justice to meat. They won't win a pastry chef's accolades for baked goods. Making full-course meals from scratch requires donning running shoes. And simply learning how to use the features on some microwave ovens is no small accomplishment.

Microwave cooking

In conventional cookery, you might begin cooking a roast by preheating the oven. When the oven is hot enough, you put in the roast, in its pan. The heat in the oven warms the pan and the surface of the meat. During cooking, the heat penetrates the meat until it's cooked through. It takes time.

Microwave cooking is faster. Microwave ovens don't heat up. The energy passes through glass, ceramic, plastic, and paper, the materials of utensils commonly used in microwave cooking. Microwaves work on the object of concern—the food.

Microwave energy penetrates into the food. The energy causes moisture molecules in the food to vibrate. That results in a kind of internal friction— and the friction produces heat.

The energy is piped into the oven cavity from the magnetron (the source of microwave energy), usually entering the cavity through the top. Most models have a rotating reflector for distributing the microwaves around the cavity so food will cook evenly. Despite that, the pattern of energy distribution within an oven is not easily predictable. As a result, some areas can be overexposed to microwaves while other areas are underexposed.

But uneven cooking is not necessarily the oven's fault. How food turns out also depends on the type of food being cooked, the quantity of food, the amount of turning and stirring you do, the cooking time, the power level, and where the food is placed within the oven. Successful microwaving, therefore, also requires practice and patience—and your willingness to put up with the occasional disaster.

Some models move the food around, within the energy field, on a rotating turntable. A turntable presumably exposes food evenly to the microwaves as it rotates and provides more predictable results. With a turntable model you will probably get best results if you line potatoes for baking around the outer rim of the turntable. Asparagus will be best cooked with tender tips toward the center, tougher stems toward the rim.

Dry foods absorb less energy than liquids. If you're warming a casserole, you can end up with steaming hot liquid and tepid noodles. To prevent that, microwave cookbooks recommend rotating the dish and stirring the food periodically during cooking. With turntable-equipped models, dish rotation isn't necessary, but stirring might be.

Obviously, you can't stir a veal chop or lasagna. You can promote more even cooking of chops by turning them periodically. And lasagna heats more evenly if, instead of being placed whole in the microwave oven, it's cut into smallish serving pieces that are well spaced on the cooking surface.

Microwaves work well with liquids. If you're making a pudding or warming some soup, the results should be fine—though you must remember to stir such a dish a couple of times during cooking so it heats uniformly. Actually, heating liquids in a microwave oven doesn't offer much advantage over using a pot on a conventional surface burner.

In dense foods, such as roasts, microwaves are absorbed primarily by the outer section. The inner section picks up heat from the outer (in the same manner that meat cooks in a conventional oven). But if the outer portion cooks too quickly, it may end up overcooked, with the inside undercooked. That's where variable power settings come in.

Large or small

Smaller sizes have solved one of the biggest problems consumers have had with microwave ovens—too much oven in too little kitchen. The new, smaller models take up a lot less counter space than full-sized ovens. Some, designed to be mounted on the wall or under cupboards, take up no counter space at all.

The smaller ovens save money as well as space. While an elaborately equipped, full-sized oven sells for $600 or more, some of the smallest, most basic models sell for as little as $150.

When appliance manufacturers talk about "compact" or "subcompact" ovens, they're referring to the size of the cooking compartment. Depending on how much kitchen space you have, the exterior dimensions of the oven may be far more important.

To a certain extent, an oven that's small on the inside is small on the outside. But some ovens are designed to use space more efficiently than others.

Aside from their size, the most important difference between a large microwave oven and a small one is the amount of microwave energy that they produce. All microwave ovens cook fast, but a big one cooks faster than a small one, from 25 to 33 percent faster.

Controls

Differences in controls often result in differences in price. Deluxe models come with relatively fancy electronic touch-pad controls. You can save $100 or more in some brand lines by buying a simpler version—one with mechanical, not electronic, controls. (Buying a model with mechanical controls might save money down the road, too, since mechanical controls are apt to be cheaper to repair.)

Electronic touch pads do offer some advantages over knobs and dials. Because they're flat, they're easier to clean. They can be set more precisely. And many models with touch-pad controls let you program a sequence of cooking steps, so you can cook dishes that require defrosting first, or a final "simmer" stage.

Touch pads can be single function—one pad for time, another for power level, and so on. Or they can be multifunction—providing different information depending upon the mode (time, power, or temperature) that you're setting. A digital display shows the numbers you've entered and whether they've been entered as time, power, or temperature.

Most ovens have separate pads numbered from 0 to 9. To enter the cooking time, you enter the digits from left to right: 4 followed by 5 for 45 seconds, for example. A few have count-up systems that require tapping a "tens" pad four times and a "ones" pad five times to arrive at a cooking time of 45 seconds.

Programming a microwave oven is not laborious, but it can be confusing without good instructions. Some models make it even easier by building instructions into the controls. The display on some models provides prompts, in words, that tell you what information has been entered and what information to enter next.

Most ovens further aid the entering of information by beeping when each touch on the pad has been registered. Without that feature a hesitant touch could result in a timing or power error.

Every microwave oven, no matter what kind of controls it has, needs to know two things: the power level to use and how long to operate. To operate a typical oven at full power, for instance, you simply program the time and push the "start" pad. Some ovens make you do more, such as first turning it on or selecting the power level.

Power levels. Different cooking chores require different levels of microwave energy. Four or five power levels are sufficient for handling any task—from melting chocolate and warming bread, to defrosting stews and roasting meat, to cooking bacon and vegetables.

The number of power levels often tracks with the price of the model. Fancy models have at least 5 power levels; some models have as many as 10 or even 100 levels, which is 5 or 95 more than you need. Less-expensive models may have only 2 or 3 power levels—one of the minor sacrifices you make in return for paying less.

So far, power levels haven't been standardized, which has led to a lot of confusion. A conventional oven temperature of 350°F is 350°F no matter how large or small the oven. But power level 5 on one microwave oven might not be the same as power level 5 on another. That makes it risky to use a cookbook other than the one that comes with the oven you buy.

Defrosting foods

Consumers Union's reader survey indicated that more than half of the microwave-oven owners responding use their ovens for quick-defrosting much of their food. But that function proved to be wanting in CU's laboratory tests. Instructions say to break up moist frozen foods such as casseroles and stews with a spoon a number of times

during the defrost cycle. If you do that, the exterior portion of a deep bowl of stew, for example, may begin to cook while the interior remains frozen solid. (Freezing stews and casseroles in shallow containers should make the defrosting chore easier and faster.)

There are some foods that you can't break up to hasten their defrosting—

pork chops, for example. You have to turn such foods over now and then.

Dry foods—breads and cakes—defrost in just a few minutes. And they can taste bakery fresh.

Temperature probes

One of the problems with cooking from scratch the microwave way is that you can't rely on the signs of doneness familiar to you from conventional cooking. Meats don't crisp, or brown much. Baked foods don't brown, either. Casseroles don't have a golden, crisp crust.

Use of a temperature probe can help in determining when food is cooked. A temperature probe is like a meat thermometer, but more versatile. A probe can work with liquids and casseroles as well as with meats. And, because the probe plugs into the oven, it can stop the cooking process when the food's desired internal temperature is reached.

A cord connects the probe to a receptacle in a wall or in the ceiling of the oven's cavity. The probe itself is inserted into the thickest part of the roast, the thigh of the fowl or the center of the casserole, much like a meat thermometer. Then the oven is set to the appropriate power level and to the desired internal temperature of the food, typically 100° to 200°F.

As the food cooks, the oven's display shows the current internal temperature, usually changing the display with every five-degree change in temperature.

Most probes and displays are accurate within five degrees in measuring and reporting the internal temperature of the food.

All ovens sound off to let you know when the food has reached the desired temperature. Many shut off at that point. Models that have a "hold" or "keep warm" feature continue cycling on and off to keep the food hot. During that time, some models simply state "hold" on their display. The more helpful ovens tell how long the food has been on hold.

Delayed start. Adding a delayed-start feature to an appliance that has a timer is fairly easy. Such a feature has been found on conventional ranges for years. Adding it to a microwave oven, however, doesn't make a lot of sense, considering the speed of microwave cooking. Having foods standing at room temperature for hours isn't a good idea anyway.

Nevertheless, the delayed-start capability is a common feature.

Programmed cooking. Many microwave ovens that have electronic controls also allow programmed cooking. One kind of programmed cooking is to cook in more than one stage—defrosting a casserole at a low power level, for instance, then cooking it at a high level. Many models let you cook in two stages; some allow three or even four. Some models have a few programs built in for added convenience.

Living with a microwave oven

In day-to-day use, little things can mean a lot. Here are several:

Door. Most microwave ovens have a door that is hinged at the left and opens with just a tug on the handle. Some open instead at the touch of a push button (large enough to be operated with your elbow). The free-swinging door can be an advantage when your hands are full; it can be a disadvantage if you're standing in the way as it swings.

Window. The window found in just about every microwave-oven door is backed with a screen that prevents microwaves from going through the glass. On most models the screen is coarse and the glass dark, so it's hard to see what's happening to food inside the oven. Models that have fine screening and a brightly lit interior provide a good view.

Light. When you operate a microwave oven or open its door, an interior light goes on. Some lights are brighter than others. A more serious difference is how easy it is to replace the light bulb. With a number of models you have to disassemble the cabinet to get at the bulb. That, in our judgment, is a job for a professional. Such an oven either has to be taken to the shop or you'll need a service call.

Shelf. A few ovens come with a removable wire shelf. While such a shelf can increase the oven's capacity for cooking small items, it can complicate cooking if one dish shields another from the microwaves or if the foods cook at different rates. All in all, a shelf

is not a feature on which to base a buying decision.

Clock. On an oven with a digital timer, you can generally use the timer as a clock when the oven isn't in use. It can usually be used as an all-purpose kitchen timer as well (basically by just programming the oven to "cook" without power).

Noise. The assorted beeps you hear as you program a microwave oven and as it signals the end of cooking are generally loud enough to be heard but not loud enough to be annoying.

Cleaning. Since microwave ovens stay relatively cool when cooking, food doesn't have the opportunity to bake onto oven surfaces as it does in a conventional oven. Spills and spatters are easy to wipe up.

Portability. Some manufacturers advertise their smaller models as "portable." But even those weigh anywhere from 40 to nearly 80 pounds.

Installation. The new, smaller microwave ovens are countertop models in that they have a fully finished, enclosed case and feet. But most can be built into a wall or cabinet if you purchase the manufacturer's trim kit. The kits are designed to provide an attractive installation and the ventilation the ovens require.

Speed and energy efficiency

For some jobs, microwave ovens are fast and relatively energy-efficient. A five-pound roast beef cooked in an energy-efficient electric range might take 1½ hours and use one kilowatt-hour (kwh) of energy. In a microwave oven, a roast of the same size might take about half the time and use half the energy.

There would also be measurable differences in cooking foods smaller than roasts—hamburgers, for instance. But

the differences would not always be large enough to be meaningful.

Fixing a cup of tea in a microwave oven is the ultimate convenience. You heat, brew, and drink with the same cup. But boiling water on your range would be about as fast. And if you're making tea for two, the range is faster.

Cooking four quarter-pound hamburgers is faster in a microwave oven than in a frying pan on the burner of an electric range. In the oven, it took 6 minutes in a Consumers Union test, including standing time, and used about 95 kwh of electricity. In the frying pan, it took 8 minutes and 180 kwh. But a dozen hamburgers would be another matter: all could be cooked at once in the broiler of a conventional range. But it is an unusual microwave oven that can easily cook twelve hamburgers. In fact, most manufacturers recommend that no more than six hamburgers (or a total of one pound of hamburger meat) be cooked at a time.

While a roast is cooking in a regular oven, you can add a few potatoes for baking without affecting the roast or using a lot of energy. Doing that in a microwave complicates things. The potatoes rob energy from the roast and change the cooking time.

It's easy to wait until the roast comes out of the microwave oven before putting in the potatoes. A solitary potato cooks in about five minutes. Four potatoes take nearly four times as long. Still, that's only a third of the time the same task takes in a regular oven. And cook-

ing roast and potatoes separately in a microwave still uses less energy than cooking them together in an electric oven.

The cookbooks that come with a few of the ovens give directions for cooking entire meals, all at once. That doesn't work too well.

Most of the owners who responded to a Consumers Union survey expected to save energy when they bought their microwave ovens. But how much energy can you really save with a microwave oven? That depends on how often you use the oven and on *how* you use it. Since a microwave oven cooks foods faster and more directly than your conventional oven, it saves energy. But unfortunately, defrosting in the microwave oven uses energy that isn't used in ordinary defrosting in the refrigerator.

The cost of the energy you use for cooking is only a small part of your total energy bill. It's unlikely that you could save enough energy with a microwave oven to justify its cost.

Food aesthetics

No matter how fast or how convenient microwave cooking is, speed and ease are meaningless unless the food looks appetizing and tastes good. Consumers Union's home economist cooked some typical foods in a microwave; here are the results:

Roast beef. Most are pink and rare on the inside. Some are well done toward each end. They are juicy and tender, but the lack of a crisp, dark-brown outside is a disappointment. And it is not possible to make a rich brown gravy out of the drippings. Microwave cooking is really no way to treat a fifteen-dollar roast—cook it in a conventional oven instead.

Chicken. Unstuffed roasters taste very good. The meat is tender and juicy. But the yellowy, cooked fat looks unappetizing. The appearance may not bother you, but if you like crisp brown skin, roast your bird in a conventional oven.

Meat loaf. Without a dark-brown outer crust, meat loaf isn't a pretty sight, and microwave-cooked meat loaf tends to come out gray-brown. No matter what shape pan is used, the meat loaves have several hard, overcooked spots (although a **Pyrex** ring mold gives better results than the others). But the meat loaves are juicy and flavorful.

Stew. The microwave-oven recipe for

stew doesn't require braising the meat—just putting meat, tomatoes, and seasoning into a casserole dish and cooking it for 45 minutes. Then the vegetables go in. Naturally, the meat isn't a braised brown, but in a stew that's acceptable. The tomatoes give some color. All the ingredients cook to an appropriate degree of doneness, and the flavors blend nicely. Survey respondents were generally happier about their microwave ovens' performances with stews and casseroles than with other types of main dishes.

Hamburgers. Microwaved burgers did not work in the Consumers Union tests. They emerged from the oven gray-brown and oozing, without a crisp brown outside, and with a steam-table taste.

Quiche. By comparison with quiche prepared in a conventional oven, microwaved quiche looks pale and wan. But its flavor and texture are good.

TV dinners. Some microwave cookbooks say you can use the aluminum tray that a TV dinner comes in to heat the dinner in the microwave (most say it's okay as long as the tray isn't more than ¾ inch deep). Though you might think the microwave oven was born to do justice to a TV dinner, test results are disappointing. Some of the foods in the dinners (the vegetables) will cook before the others (the meat). You can get much better results in a regular oven.

Baked potatoes. A microwave oven cooks a single baked potato in record time. And many of the owners who responded to the CU survey praised their oven for its prowess with baked potatoes. But microwave-baked potatoes may not be the kind you're accustomed to. The texture of the skin is softer than that of a regular baked potato—more like the skin on a potato that has been baked in foil.

Vegetables. Microwaved vegetables are a pleasure, with natural color and good flavor.

Cakes. Since baked goods won't brown, a white cake comes out very white. Unless you plan to disguise cakes with topping or icing, stick with chocolate or spice cakes, which have some color to begin with. We find that pineapple upside-down cake looks as good as it tastes; it is light and moist. A rum cake is not as successful: spots are overcooked, so hard that the rum poured over the cake cannot penetrate them.

Puddings. A batch of vanilla pudding, stirred twice during its six-minute cooking time, comes out smooth, creamy, and tasty. What's more, it is cooked in a glass bowl that goes right into the refrigerator—no gopey pan to clean. On the other hand, some moist foods require constant stirring when cooked in a microwave and are best done on the rangetop.

Leftovers. The CU survey respondents reheat most of their leftovers in the microwave oven and are satisfied with the results.

Overall, the biggest drawback to microwave cooking is the lack of browning. Browning trays don't seem to work very well. Commercial browning agents that you paint or sprinkle on meats tend to be unsatisfactory, too. They give an artificially browned look to food and, in many cases, an artificial, salty flavor.

Installation

A microwave oven needs extra room around its vents (at the side, the rear, or the top). If the vents are blocked, moisture may condense inside the oven. But more importantly, the oven's working parts can't be kept appropriately cool, which may affect durability.

Some manufacturers offer kits for building-in the microwave. A logical place to do that might be above the conventional range. But it may not be the best place. The heat and humidity that are generated by the range may cause premature breakdown. Only install microwave ovens designed for that location. Some of those models also incorporate a venting fan.

You may want to avoid locating a microwave oven near a radio or TV, particularly if you already have poor reception. The closer the oven is to a receiver, the greater the interference is likely to be. Interference will probably increase, too, if the radio or TV and the microwave are hooked up to the same circuit. Some manufacturers make "filter" kits that could help to reduce interference.

Since ovens draw up to 1,600 watts, they're best operated on a separate 15-amp circuit. If you can't give an oven a circuit of its own, it may be wise to choose a lower-wattage oven that draws about 1,300 watts or less. Avoid putting it on a circuit with other appliances that use a lot of energy, or you may find yourself constantly replacing fuses or resetting circuit breakers. A few lights on the same circuit will probably do no harm.

Recommendations

A microwave oven is a cooking accessory, like a blender or an electric mixer. But its price puts it into the major-appliance category. Before you buy a microwave oven, then, you should carefully examine your reasons for wanting one.

Microwave ovens aren't very convenient for cooking complete meals from scratch. If you plan to use the oven that way, plan to spend a lot of time at it. And that largely offsets the oven's biggest advantage—speed.

If you plan to use the oven for convenience foods or for quick workday meals, you'll probably be happy with it.

If you cook in quantity and freeze the leftovers, you'll have a great new toy.

If kitchen cleanup is the bane of your existence, a microwave oven will suit

Pork in microwave ovens

The U.S. Department of Agriculture issued a warning about cooking pork in microwave ovens. Uneven cooking (a typical characteristic of microwave ovens) might leave parts of a pork chop or roast insufficiently cooked. The danger is that the larvae of the trichinosis-causing nematode might not be killed. Although trichinosis is rarely encountered in the U.S. today, the USDA recommends the following precautions: Rotate the dish for more even cooking; allow the pork to stand in aluminum foil several minutes after cooking to ensure thorough heating; and use a meat thermometer after cooking is complete to make sure the pork has reached 170° throughout.

A separate turntable

Microwave ovens don't take all the work out of cooking. In fact, to a certain extent, they *make* work. A rotating turntable inside a microwave oven can save some of that work—namely, manually turning the food while it's cooking to make sure it cooks evenly.

Only a few microwave-oven brands come with a turntable. And most of those turntables must be used all the time, which can lessen the effective capacity of the oven.

For ovens without a turntable, there's the $40 **Micro-Go-Round** by Nordic Ware. It can be used when needed, removed when not. After you wind it up (like an inexpensive kitchen timer), the plastic turntable makes a complete turn every two minutes or so, for nearly an hour. However, the **Micro-Go-Round** has a significant amount of metal in it and shouldn't be used in ovens where manufacturers warn against use of any metal whatsoever.

you. If you want a microwave oven to keep your kitchen cool in warm weather, you're likely to be pleased, too. But if you purchase one to save energy, you could be disappointed.

Above all, don't expect a microwave oven to revolutionize your cooking. If you do, you may be disappointed.

In particular, since a microwave oven doesn't present meats at their best advantage, you may not want to cook large cuts of meat in it.

A relatively small-sized model comes pretty close to the ideal compromise between exterior size and interior capacity. It's small enough to avoid overwhelming the average kitchen and large enough to handle most common microwave-cooking chores.

While a built-in turntable can contribute to cooking uniformity, it can also encroach on the overall capacity of an oven. You may like the idea of a removable turntable such as the **Micro-Go-Round**.

Whichever microwave oven you choose, don't expect to fall in love with it at first use. Microwave cooking is quite different from conventional cooking. If you find yourself thinking of the new arrival as an unmastered monster lurking in the kitchen, you aren't alone. Many respondents to the Consumers Union questionnaire spoke strongly in favor of taking cooking lessons. Adult-education programs and county extension services usually offer them for a modest fee. Manufacturers offer courses, too. Their lessons are generally free.

Microwave safety

Like radio waves or visible light, microwaves are a form of electro-magnetic energy. Each form is identified by its frequency—the num-ber of times the wave oscillates per second. Microwaves oscillate faster than radio or television signals, but slower than infrared or visible light.

Microwave radiation isn't the same as nuclear radiation, which is a far more intense type of electromagnetic energy. X rays and gamma rays oscillate millions of times faster than microwaves do and can create electrically charged, or "ionized," molecules in substances they penetrate. Ionizing radiation can harm living cells even at rela-tively low levels.

Although microwaves don't carry enough energy to be ionizing, they can still be dangerous. The well-known heating effect of intense microwave radiation can cook food or damage human tissue, but research has turned up no convincing evidence of hazard from microwave ovens.

Extensive evidence from human and animal studies indicates that even extended microwave exposure in a range from 0.1 to 1 milliwatt per square centimeter (0.1 to 1 mW/cm^2) has no apparent harmful effects on humans. Occupational exposure within that range is con-sidered safe even if extended over the whole working day.

Safety standards for occupational exposure are commonly less stringent than those for the general public, which includes children, pregnant women, sick people, and others who may be more vulner-able than a healthy adult worker. Still, that level of exposure is con-siderably greater than any you're likely to experience near a micro-wave oven.

How much leakage?

Radiation leakage from today's microwave ovens is low. A federal safety standard allows leakage of 1 mW/cm^2 at two inches from a newly purchased oven and 5 mW/cm^2 in older ovens (to allow for some deterioration of the door seals over the oven's lifetime). In Consumers Union's tests, which paid particular attention to areas around the door, vent, and controls, leakage didn't exceed 0.5 mW/cm^2 in any model.

It's important to note that the amount of leakage an instrument measures a few inches from the oven is not the same as the exposure a person using the oven actually receives. Radiation intensity declines sharply as distance from the oven increases. Moreover, few people are likely to stand in one fixed spot while the oven is on. They move around the kitchen, approach the oven window to check the cooking, leave the room, and the like. Thus, a person's exposure depends on how much the oven leaks.

Under ordinary conditions, such exposure will normally be far below the 0.1-to-1 mW/cm^2 level judged safe for occupational exposure. And most exposures last but a minute or two, versus the eight-hour workday in the occupational limit.

In tests done at various distances and angles from ovens in a typical kitchen environment, regardless of the angle, exposure declines substantially with distance from the oven. At four feet, a reasonable distance for someone working in a kitchen, radiation levels averaged roughly 0.01 mW/cm^2. That's about one hundred times less than the maximum leakage level permitted by the federal standard for new ovens at a two-inch distance. Accordingly, anyone who wishes to reduce any possible risk from leakage can do so by keeping a reasonable distance from the oven when it's on.

For an added margin of safety, don't make a habit of peering through the oven's window for long periods. If you're pregnant, don't linger near the oven while it's on. To keep the risk of excessive leakage low, pay particular attention to the door seal. Make sure that food residue doesn't build up around it, and don't let anything get caught between the sealing surfaces. Also, don't use a swing-down door as a shelf. If the door, its seal, or its hinges become damaged, don't use the oven until it's repaired.

Measuring leakage

It isn't clear whether leakage meters advertised for home use are sufficiently reliable. In their principal microwave-leakage tests, Consumers Union engineers used an expensive, laboratory-grade instrument. It seems highly unlikely that any of the simple devices using a fluorescent bulb would have the necessary precision.

Ratings of microwave ovens

(As published in a November 1985 report.)

Listed in groups by size, based on exterior dimensions. Within groups, listed in order of estimated overall quality based on performance and convenience. See also features table opposite. Prices are mfr. suggested retail; * indicates price is approx.

Better ←——→ Worse

Small ovens

	Price	Dimensions (HxWxD), in.	Weight, lb.	Usable capacity, cu. ft.	Heating speed—large load	Heating speed—small load	Power levels	Cooking stages	Advantages	Disadvantages	Comments
GE JEM31E	$295*	11³/₄x23³/₄x13¹/₄	44	0.7	○	○	5	2	A,B,C,D	d	F,I,L
LITTON 1455	259	11¹/₂x21³/₄x14¹/₄	40	0.7	○	○	10	2	—	d,g,i	D,M
WHIRLPOOL MW3500X	269	12³/₄x20¹/₂x16	42	0.7	◐	●	10	2	F	d,f,i	B,D,P

Medium ovens

■ *The following ovens were judged about equal in quality. Listed alphabetically.*

FRIGIDAIRE MC400A	189	14x22x17	57	0.8	○	9	2	B,G,H	j	G,T,P
HOTPOINT RE86	255*	13x20½x18½	57	0.6	○	5	3	D,G,H	k	I,K
J.C. PENNEY 5645	200	13¾x21x17¼	43	1.0	○	10	3	B	e	D,G,J,P
MAGIC CHEF M41A6P	269	15¼x22x16	62	1.0	○	10	1	B,C,D,H	h	D
MONTGOMERY WARD 8135	228	14x21¾x15½	46	0.7	○	5	3	D	j	I,J,K,P
PANASONIC NE6960	360	13¾x21x17¼	43	0.6	○	6	3	E	a,g	J,O
QUASAR MQ6674XW	330	13½x21x17¾	43	1.0	○	6	3	B	—	D,G,J,P
SAMSUNG RE620TC	250	13x20½x18½	56	0.6	◐	10	3	—	k	G,I,J,K,N
SANYO EM2520S	300	13¾x20½x18½	48	0.6	○	100	1	—	j,l	E,K
SEARS 87750	[1]	13¾x20½x18	48	0.6	◐	100	2	D	j	E,K,P,Y
SHARP R4850	319	14¾x22x16	47	0.6	◑	5	3	E	a,j	J,O,P
TAPPAN 56247310	249	15x21½x15½	51	0.7	◑	10	2	—	c,g,j	C,D,H,P
TOSHIBA ER175BT	299	14x22x17¾	57	0.8	○	9	2	B,G,H	j	G

Large ovens

AMANA RR920	365*	15¼x22¾x18¼	68	1.0	◉	10	2	B	g	A,C
MAYTAG CME500	300*	15x25x18¾	78	1.0	◑	10	1	B	b,d	C,G

[1] *Price varies from store to store; CU paid $270.*

SPECIFICATIONS AND FEATURES

All: ● Have touch-panel controls.
Except as noted, all: ● Come with a removable glass tray. ● Beep when a touch pad is activated. ● Operate on full power unless otherwise programmed. ● Cannot be set to start cooking automatically at some later time. ● Will not easily hold a medium turkey. ● Have a probe that allows temperature-controlled cooking with settings in 1-degree increments ranging from a low of 90° to 110°F to a high of 190° to 200° (displayed in 5-degree increments). ● Shut off when set temperature is reached. ● Provide a dim view of food cooking. ● Can be built into a wall or cabinet using optional hardware.

KEY TO ADVANTAGES

A – Design uses space most efficiently.
B – Turkey easily fits inside oven.
C – Has removable metal shelf that increases capacity for small items.
D – Exceptionally clear control display.
E – Comes with turntable; provides predictable microwave distribution during cooking.
F – Comes with particularly clear operating instructions.
G – Window offers relatively clear view.
H – Fan is fairly quiet.

KEY TO DISADVANTAGES

a – Holds turkey, but turntable can't rotate.
b – Does not beep when touch pads are activated; judged more prone to programming errors than models that beep.
c – Separate on/off switch must be activated to operate oven.
d – End-of-cycle signal is barely audible.
e – Needs extra step to run on full power.
f – LED display is relatively dim.
g – Fan is fairly noisy.
h – Inconvenient, recessed door handle.
i – Door didn't always latch when closed.

j – Cabinet must be disassembled to change light bulb, requiring service call.
k – Sharp edges on sheet metal surround light bulb.
l – Glass tray slides forward too easily.

KEY TO COMMENTS

A – Has highest heating rate, over 600 watts.
B – Has lowest heating rate, under 450 watts.
C – Draws close to 1500 watts; more likely than most to overload a circuit.
D – Does not come with removable glass tray.
E – Temperature-probe range relatively narrow, 115° to 185°.
F – Temperature probe can be set down to 80°.
G – Temperature can be set only in 5- or 10-degree increments.
H – Temperature-probe setting less accurate than others, off by more than 10° at low- and middle-range settings.
I – Temperature is displayed in 1-degree increments.
J – Opening door requires pushing button or level.
K – According to mfr. literature, installation hardware not available.
L – Comes with hardware for under-cabinet installation and very good installation instructions. Hardware for installing in cabinets or on wall not available.
M – Hardware for installing under cabinets or on wall available as option.
N – Has alarm that can be set to repeat at the same time each day.
O – Instead of temperature probe, has moisture-sensor system, which worked well.
P – Model discontinued at the time the article was originally published in *Consumer Reports*. The test has been retained here, however, as a guide to buying.

Microwave-oven features

Key features on tested
and similar models.
Prices are suggested
retail; * = approx.
● = model tested.

Brand and model	Price	Controls [1]	Maximum timer setting, min.	Power levels [2]	Cooking stages	Other [3]	Temperature probe	Clock	Keep warm
AMANA RR720	$285*	M	99	10	1	—	✔	✔	✔
RR820	325*	T	99	10	1	—	✔	✔	✔
●RR920	365*	T	99	10	2	—	✔	✔	✔
FRIGIDAIRE ● MC400A	189	T	99	9	2	—	✔	✔	—
GE JEM10E	235*	M	35	Inf.	1	—	—	—	—
JEM20E	265*	T	99	5	1	—	—	✔	—
●JEM31E	295*	T	99	5	2	—	✔	✔	✔
HOTPOINT RE83	215*	M	35	2	1	—	—	—	—
●RE86	255*	T	99	5	3	—	✔	✔	✔
J.C. PENNEY ● 5645	200	T	99	10	3	B,D	✔	✔	✔
LITTON 1435	229	M	35	10	1	—	—	—	—
●1455	259	T	99	10	2	—	✔	✔	✔
1465	269	T	99	10	2	B,D,P	✔	✔	—
MAGIC CHEF M41A3	210	M	35	Inf.	1	—	—	—	—
M41A6S1	250	T	99	10	1	—	—	✔	—
●M41A6P	269	T	99	10	1	—	✔	✔	✔
MAYTAG CME300	225*	M	30	Inf.	1	—	✔	—	—
●CME500	300*	T	99	10	1	—	✔	—	—
CME700	345*	T	99	10	4	D	✔	✔	✔
MONTGOMERY WARD ● 8135	228	T	99	5	3	D	✔	✔	—
PANASONIC NE6660	260	M	30	Inf.	1	—	—	—	—
NE6765	290	T	99	4	1	—	—	✔	—
●NE6960	360	T	99	6	3	D,M	—	✔	—
QUASAR ● MQ6674XW	330	T	99	6	3	D	✔	✔	✔
SAMSUNG RE610D	200	M	35	2	1	—	—	✔	—
RE610TC	230	T	99	10	4	—	—	✔	—
●RE620TC	250	T	99	10	4	B,D,P	✔	✔	✔

Brand and model	Price	Controls [1]	Maximum timer setting, min.	Power levels [2]	Cooking stages	Other [3]	Temperature probe	Clock	Keep warm
SANYO EM2500	270	T	99	100	1	—	—	—	—
● EM2520S	300	T	99	100	1	—	✔	✔	✔
SEARS ● 87750	[4]	T	99	100	2	D	✔	✔	✔
SHARP ● R4850	319	T	99	5	3	D	✔	✔	✔
TAPPAN ● 56247310	249	T	99	10	2	—	✔	—	✔
TOSHIBA ER145BT	249	M	45	Inf.	1	—	—	—	—
ER155BT	279	T	99	9	1	—	—	—	—
● ER175BT	299	T	99	9	2	—	✔	✔	—
WHIRLPOOL ● MW3500XM	269	T	99	10	2	D	✔	✔	✔

Programmable settings

[1] *M = mechanical; T = touch pad.*
[2] *Inf. = infinite settings.*
[3] *D = Delayed start; B = Built-in recipes;*
P = Programmable recipe memory;
M = Moisture sensor.
[4] *CU paid $270.*

Microwave/convection ovens

An appliance that has evolved from the microwave oven is the microwave/convection oven. It's sometimes claimed to be "the best of both worlds." The combination unit is said to offer the speed of microwave cooking and the browning and crisping of convection cooking. A combination oven does let you cook with microwaves, by convection, or by combining the two methods.

Microwave cooking can be fast. Substituting convection heat for part of the microwave cooking slows the process down. But the circulating hot air of convection cooking is what gives meats and baked items their traditional color.

Combination ovens look like ordinary microwave ovens, but larger. They have to accommodate the magnetron, the key part of a microwave oven, and other microwave hardware—plus a heating element and circulating fan, the basics of a convection oven. Combination ovens require substantial counter space and shouldn't be mistaken for portable appliances; they weigh in at upward of 75 pounds.

Microwave ovens have power-level settings rather than temperature settings. Convection units combine microwave power with conventional temperature settings. Depending on the model, they may alternate rapidly between microwave and convection cooking during the entire cooking process. The two modes cycle on and off so rapidly that it seems as though they operate simultaneously.

Or combination cooking may be achieved by your setting the temperature and time (convection), then power and time (microwave).

Cooking

Consumers Union's testers cooked meat, vegetables, and poultry, and baked bread and cake in the microwave/convection ovens and in a conventional oven. Here's what they found:

Roast beef. The objective was roasts rare inside, crisp and brown outside. The roasts were almost as good as the conventionally roasted beef, although not quite as evenly cooked inside, in general a big improvement over pale, microwave-only roasts.

The combination ovens can produce rich, brown juices for gravy—something you could never achieve with an ordinary microwave oven.

Meat loaf. Microwave ovens make mediocre, grayish-brown meat loaf. Combination ovens make very good, dark-brown meat loaf.

Chicken. Chickens cooked in regular microwave ovens aren't much to look at, though they may taste just fine. Combination cooking in a microwave/convection oven produces nicely browned skin, as long as you turn the chickens while cooking.

Turkey. All the ovens' cookbooks give instructions for roasting turkey, but the size of turkey you can cook depends on the oven's volume. You may not be able to roast a bird that weighs more than twelve pounds or so, if that much.

The birds may taste very good, but probably will not look as handsome as turkey cooked in a conventional oven.

Hamburgers. These ovens turn out very good hamburgers, with browned exteriors and rare to medium-rare interiors. The outsides, however, won't be really crisp, as they would be if cooked conventionally.

Steak. T-bone steaks come out moist, tender, and juicy inside when broiled conventionally and also when cooked in a combination oven. But the outsides look different. Those cooked in a combination oven aren't as brown or crisp as a conventionally broiled steak.

Vegetables. Ordinary microwave ovens do a very good job with fresh and frozen vegetables. Combination ovens do, too, cooking the microwave way.

Bread and cake. The combination ovens can do the job, but probably not really well enough.

Leftovers. A microwave oven is convenient for reheating leftovers. But different foods have different needs. When meat, potatoes, and vegetables are heated on the same plate, they're not likely to be evenly heated. A combination oven performs the same way when used in the recommended microwave-only mode.

Timing

Unless you're cooking a lot of things at once, microwave ovens can be fast, no doubt about it. When used on microwave only, a combination oven is just as fast. It can bake a potato in about five minutes.

When these ovens are used in the combination mode, cooking is slower than straight microwave cooking. Still, they are faster than a conventional oven for many tasks. For example, baking a meat loaf takes one-half to two-thirds the time of a conventional electric range oven.

Other factors

Cooking in an ordinary microwave oven requires an assortment of non-metal cookware. You'll need a fairly good assortment for a combination oven too.

For most cooking chores, ceramic or glass cookware is the safest choice for combination cooking. Metal could interfere with the microwaves and damage the oven's magnetron. Plastic might soften under the heat of the convection heating element.

Microwave ovens are very easy to clean because spattered grease doesn't bake on the surfaces of the oven. But convection heat does bake on grease.

Energy use. A microwave oven isn't a big energy saver. It's unlikely that you could save enough energy with a microwave oven to justify its cost. The same is true of combination ovens. In fact, they may use more electricity than a conventional oven for some cooking tasks.

Space. Combination ovens take up lots of room, not only because they are large, but because they need room at the top and back so their vents can work properly (otherwise, some parts of the ovens could overheat).

Some models, according to their installation instructions, shouldn't be in hot or damp places. That could rule out a location near the stove, where heat or steam could affect the oven.

Heat. Microwave ovens don't get hot, but convection ovens do. Some exterior surfaces of all the ovens are hot to the touch when operating in the convection mode. Dishes in microwave ovens aren't supposed to get hot, but they can when they pick up heat from the food that's in them. Dishes in convection ovens, of course, get very hot.

Microwave/convection ovens are covered by the same federal standards for microwave emissions as plain microwave ovens (see pages 200–201).

Noise. During all cooking modes, there's some noise from the blowers that cool the oven's working parts. It's not likely to be very loud or annoying.

Recommendations

If you own a microwave oven now, don't feel compelled to upgrade it. For the things a microwave oven does well—defrosting, warming, cooking vegetables—an ordinary microwave oven with a few power levels is all you need. It makes a nice adjunct to the conventional range oven.

Even a combination microwave/convection oven will not replace the range oven. It is still a better all-around baker, a more even roaster, and a crisper broiler. But if you're starting from scratch, a microwave/convection model can go a long way toward correcting the major shortcoming of a microwave—the lack of browning and crisping—and perform many other cooking chores very well.

What microwave-oven owners think

A few years ago, 23,000 readers of *Consumer Reports*, owners of microwave ovens, filled out a questionnaire about how they used their oven and how well they liked it.

The questionnaire respondents characterized the microwave ovens as everything from a necessity of life to an "expensive bottle warmer." Those who said they love the appliance, however, were in the majority. Some said, "Take away my dishwasher, but not my microwave."

The microwave oven seems to be a standby for people who have little leisure time. More than half of the respondents were from a two-income household.

Many owners had great expectations when they bought their microwave oven. Nearly all of them expected that the oven would save time. Many also hoped for a future of less oven cleaning and fewer pots and pans to scrub.

The microwave ovens lived up to those expectations: 72 percent of the respondents said they were very satisfied, overall, with their oven. Virtually no respondent checked "very dissatisfied" on the questionnaire.

Most of the respondents owned ovens with more features than basic models—larger capacity, programmable cooking, more power settings, longer cooking time, a temperature probe to sense doneness. The respondents were especially pleased with two of the features on their ovens: the variable power levels and the programmable cooking.

Of those who reported having an oven with a temperature probe, very few said they use it often. A small percentage of the respondents had a browning feature in their oven. Many more had a browning tray. The level of satisfaction with both features was fairly low.

Only 13 percent of the respondents had an oven with a turntable. But most of them were very satisfied with this feature, particularly with its effect on evenness of cooking.

Though most of the respondents had big expectations when they went out to purchase a microwave oven, 17 percent said they didn't expect food cooked in it to taste as good as food cooked by traditional methods. But that expectation, apparently, didn't scotch the sale.

As it turned out, the owners felt that some foods were actually better when cooked the microwave way. They especially liked what the ovens can do with vegetables and with leftovers. Fifty-six percent said vegetables cooked in the microwave were better than vegetables cooked any other way. They praised the vegetables' color, texture, and flavor. More than 80 percent said leftovers reheated in the microwave oven were better than leftovers reheated any other way.

Fifty percent of the respondents said frozen foods come out better with the microwave oven.

The classic microwave-oven miracle is the fast baked potato. A microwaved potato has a less crisp skin than a regular baked potato. That didn't bother the respondents. About half of them said that baked potatoes were better when cooked in the microwave.

The questionnaire didn't ask about baking cakes, breads, and cookies, but many people volunteered the information. Most said baked goods were unsatisfactory. One reader said brownies came out like "hockey pucks." Another claimed reasonable success by adding extra eggs and milk to favorite recipes.

More than half the respondents said they use the microwave oven for at least 50 percent of their cooking—mostly for snacks, quick meals, convenience foods, and leftovers.

Cooking a meal from scratch requires skill and endurance. Yet 21 percent of the respondents said they do most of their cooking from scratch with their microwave ovens. Generally, those hardy cooks own an expensive model with many features. Those people may do a lot of cooking from scratch because the many features on their oven allow them to do so. Or they may do it to justify the cost of the fancy appliance. Or they may do it simply because they like to cook and are willing to take the time to learn the microwave way. Whatever the reason, that group of owners expressed more satisfaction with their oven than those who use it mainly as a reheating device.

Ranges (electric and gas)

There are a great many choices to make when you buy a range—more substantive choices than you'd have with most major household appliances.

Some of the basic decisions—whether the range should be electric or gas, for instance—will probably be made by your kitchen's layout. Either a gas line or a 240-volt circuit is already in place, and switching, if it's possible, would require extra work and expense. The cabinetry and floor plan of the kitchen will probably decide the range's width (21, 24, 30, 36, and 40 inches are the usual sizes) and whether the range is one that slips in between counters, is built into the countertop and cabinets, or stands by itself.

Even with those decisions taken care of, there are more to make. Do you want one oven or two? Should the oven be a self-cleaning model, continuous-cleaning, or do-it-yourself cleaning? Have you considered a microwave oven as part of the range, either above the cooktop or as part of the regular oven? Or do you want a range with a convection oven? How about a model with a modular cooktop that lets you substitute a rotisserie, grill, griddle, or deep-fat fryer for two of the cooktop elements?

Some of those decisions may be made by your kitchen, too, or by how much money you've budgeted. A range that has a microwave oven built into its regular oven, for instance, can cost something like $1,000, while a range that has a microwave oven above and a self-cleaning oven below will cost about $1,600.

Many people decide on a 30-inch-wide, freestanding range with a self-cleaning oven. Such a range is still not cheap: The suggested retail prices run anywhere from $400 to $1,100. There are still lots of choices to make, although they are of a less significant nature. Do you want an oven window? Lots of indicator lights or few? A regular clock, a digital clock, or no clock at all? Electronic push-pad controls or knobs? And the list goes on.

Such niceties add more to the appliance's price than they are perhaps really worth. The most important considerations should be a range's versatility at cooking and how easy it is to clean.

The big feature in today's electric ranges is a self-cleaning oven—that is, a special high-heat cycle that incinerates any accumulated food. Not all major range manufacturers, however, market gas ranges with a self-cleaning oven. Many gas ranges come with a so-called continuous-cleaning oven, in which a porous interior coating is supposed to slowly dissipate soil as you cook.

Many of the new gas ranges do not use a pilot light. Electric igniters, used instead of pilot lights, can greatly decrease a range's gas consumption. In some states, ranges with pilot lights are no longer sold.

Energy use is not a reason to choose one brand of range over another. Ranges don't consume a lot of energy, certainly not as much as a water heater. At average national utility rates (7.75 cents per kilowatt-hour for electricity and 61.7 cents per therm for gas), an "average" amount of cooking would cost about $60 a year with an electric range and about $35 with a gas range.

Cooktops

The usual cooktop layout for a 30-inch electric range comprises two large elements (about 8 inches across) and two small elements (about 6 inches across), each with a removable bowl underneath to catch spills and drips. Gas ranges usually have four equal-sized burners. Sometimes their drip bowls are removable, and sometimes those bowls are just a depression in the porcelain-enamel cooktop.

Any range should be able to hold four pots of various sizes at one time, though the pots may not fit comfortably. For example, a 12-quart stockpot may not fit on the center of the rear elements because the backguard protrudes too much. On a gas range, big pots on the back burners may crowd pots on the front burners.

Even though gas ranges provide instant heat under a pot and electric elements have to warm up, an electric range generally manages to boil water faster than a gas model.

With an electric range, the shininess of the drip bowls makes no difference in the speed. What is important to cooking fast and efficiently on an electric range is your choice of pots. Their bottoms should be flat, and their diameters should match those of the elements.

On a gas range, the shininess of the drip bowl does make a difference. That's because gas burners rely partly on radiant heat, which can be enhanced by reflections, whereas electric elements cook mainly by contact with the pan.

Overall, the difference in energy costs between the fastest and the slowest ranges will probably amount to no more than a few dollars a year.

Many people prefer cooking with gas because it's easy to control the heat of a gas burner. That's especially important in, say, Chinese cooking, where you want short, concentrated bursts of heat, or when you want to cook at very low temperatures, as you would when making a delicate sauce.

Electric elements achieve a low temperature by cycling on and off at full current, not by using a steady, small amount of electricity. It may take some practice, therefore, to discover which setting on a range will gently simmer a white sauce and which setting will burn it.

Gas ranges are pretty straightforward—the bigger the flame, the more the heat. The control knobs on some models (generally the more deluxe models in a line) make it very easy to adjust the flame because they turn nearly a half-circle. On many gas ranges, however, the knobs turn only a quarter-circle. With those, you may need a delicate touch to set a very low flame.

Oven cooking

An oven window seems like a useful feature, but it can be difficult to see through, even with the interior light on.

Most ovens have two wire racks and four sets of supports. In a continuous-clean gas oven, be aware that the special coating can make loaded racks hard to slide. And, if you leave the racks of a self-cleaning oven inside the oven during the cleaning cycle, the racks may no longer slide easily.

In addition, check to be sure that wire oven racks aren't so springy that they won't support a heavy roast or casserole.

You'll also discover that the apparatus and insulation required to make a self-cleaning oven reduces the space inside the oven, especially in a gas range.

The useful space in an oven is the room left with one of the racks at the lowest position. The average self-cleaning gas-range oven is only about $1\frac{1}{2}$ cubic feet. That's big enough to hold a large turkey, but there's not a lot of room left over.

To test oven performance, we baked four layers of white cake per oven, two per rack. The layers provided a map of each oven's heating uniformity. An oven that is very uniform turns out four nearly identical layers, each with an evenly browned top and a fine, bubble-free texture.

The layers done in other ovens do not cook very evenly. As far as cake baking is concerned, just about any range oven should bake to most people's satisfaction.

The accuracy of the oven's thermostat can vary from sample to sample, and it can change over time. An oven thermometer is still the best way to be sure of oven temperature. (There's an article about oven thermometers on page 266, and another one about meat thermometers on page 263.)

Many gas ranges use electric ignition to light their ovens. Some use spark igniters similar to those that light the burners. Others use a "glow-type" igniter—a piece of ceramic-based material that heats up enough to ignite the gas. In the past, the design of many glow-type igniters was such that they could break very easily. For the most part, present designs have been substantially improved.

Broiling

A self-cleaning range has the broiler conveniently located at the top of the oven. Continuous-cleaning gas ranges, however, may have a burner at the bottom, so the broiler is in a drawer below the oven, where it's not convenient to get at.

Ranges come with their own broiler tray. The sizes of broiler trays vary: they can hold as few as 12 or as many as 20 or more hamburgers.

In Consumers Union's tests, one electric range in particular, the **White-Westinghouse,** had a broiling feature that sounded like a good idea at first but turned out to be a lot of trouble. The feature, called "Speed Broil," cooks on both sides at once. To use it, you have to plug in an extra heating element, let the elements heat up, then slip a rack holding the food between two red-hot elements. To catch the drippings, you fill the broiler pan with water and put it on a lower rack. The feature cuts actual cooking time by about half. But the time and care required to prepare the oven—positioning the food and the broiler pan correctly, pulling out the food rack and the drip pan simultaneously so the oven door stayed clean—turn broiling into a chore.

Controls

Electric ranges generally have their controls on the backguard, where they are easy to see and out of the way of spilled food. On some models, however, the control for a cooktop element can be blocked when a tall pot is on the element.

Controls of gas ranges on the backguard means trouble, since you might have to reach across an open flame to get to them. So most gas ranges have the controls at the front of the cooktop. To help keep you (or a curious child) from turning on the range accidentally, the cooktop knobs usually have to be pushed in before they'll turn.

The control markings on the electric ranges are reasonably easy to read. Less attention has been paid to clearly marking the controls on the gas ranges, perhaps because people tend to look at the size of the flame rather than the position of the dial.

A light that indicates when the oven or an element is on is necessary with an electric range, since a hot element can often look much the same as a cold one. Any electric range should have such a light and it should be easy to see. Gas ranges don't have a burner-on light—it's not really necessary. And only a few have an oven-on light.

A clock can be a useful feature, and a digital clock is generally easier to see than a dial clock.

Two types of timer are often associated with the clock. One is a regular kitchen timer, usually with an hour maximum. The other is a timed-bake feature, which typically lets you turn the oven on and off at some time in the future. On some models, timed-bake will start only at the present, not in the future.

Cleaning the cooktop

The simpler the design of the cooktop and its surroundings, the easier the range is to clean. Ideally, there shouldn't be any decorative trim to catch dirt, and the meeting of the cooktop and the control panel on the backguard should be a seamless, easy-to-wipe curve of porcelain enamel. Try to avoid models with dirt-catching gaps between the top and the side panels.

Range knobs pull off so you can clean the panel underneath. The panels most likely to need cleaning are on gas ranges whose knobs are mounted horizontally.

On many ranges, you have to be careful about scrubbing around the control panel, for the markings can be rubbed off with steel wool or scouring powder. The markings some ranges have on a glass panel stand up best to scouring.

Cleaning under gas burners is easier than cleaning under electric elements. You simply remove the burner grate and wipe. To get to the drip bowls of the electric ranges, you have to first unplug the element (on some inexpensive ranges, they are merely hinged).

With most drip bowls, excess drippings will overflow through the hole in the middle and dirty the "floor" under the cooktop. Most ranges let you raise or remove the cooktop to clean. But some electric ranges have a fixed cooktop; to clean below on those models, you have to poke your hand through the burner holes.

Ranges with a fixed cooktop may partially solve this problem by making three of their four drip bowls solid, without a hole in the middle (the fourth drip bowl has a hole to allow venting of the oven). The hole-less drip bowls do contain large spills. But if the large spill is grease and you don't turn the element off, you could start a grease fire.

Electric elements are all "self-cleaning" since spills burn off quickly. In fact, if you soak an electric element in water, you can damage it.

Cleaning the oven

Self-cleaning ovens are easy to clean. They work by heating up to more than 800°F, hot enough to turn the most stubborn spills into a powdery gray ash residue. After the cycle has run, you merely wipe off the residue. A wiping solution of vinegar and water works well.

Because an 800° oven is nothing to fool with, self-cleaning ranges have a light to warn that the cycle is in progress, and the oven door locks when the oven goes higher than about 550°. On some ranges, the door locks automatically. Many ranges, however, have a door latch that you have to lock.

The self-cleaning cycle produces smoke and fumes, which exit through a vent that's on the backguard of gas models or under a rear element on electrics. If there's a loose duct from the oven to the rear element that's not in place during the cycle, a lot of hard-to-clean gunk can be deposited under the cooktop.

The smoke and fumes of self-cleaning are not only unhealthy to breathe, they can redeposit oven dirt on the walls and ceiling of your kitchen. So it's important to ventilate the kitchen during the self-cleaning cycle. You should also be careful about touching the stove during the cycle, for parts of the exterior are apt to be hot enough to cause a burn.

The doors and frames of self-cleaning ovens usually still need some scrubbing outside the door seal, where some vaporized soil leaks through. (Avoid scrubbing the gasket itself, for it can be damaged.)

Since self-cleaning heats up the oven to such high temperatures, you might think it would use a lot of electricity or gas. It actually uses relatively little. We figure that cleaning the oven once a month would cost perhaps $4 a year for electric ranges and $2.50 a year for gas ranges, at average utility rates. That's less than what you'd pay for a year's supply of oven cleaner.

Continuous-cleaning ovens are no match for self-cleaning ovens in cleanability. The porous finish of the continuous-cleaning oven is supposed to dissipate light dirt gradually at normal cooking temperatures. But major spills won't go away—they have to be wiped up right after they happen. Minor spills are slowly eliminated, partly because they spread out on the finish, which is mottled and thus helps disguise the patches of dirt. And you can't scrub the oven or use conventional oven-cleaning chemicals—either would wreck the special finish.

A sensible compromise is to have an oven floor and the inside of the door—the areas most likely to be dripped on—made of scrubbable porcelain.

Recommendations

If you can choose between an electric and a gas range, choose the electric. Electric ranges are cleaner in operation, cheaper to buy, and tend to bake more evenly than comparable gas ranges. Electric ranges' controls are on the backguard, where they are easy to see and out of the way of spilled food.

Self-cleaning electric ovens are usually roomier than gas ones. In addition, electric ranges usually have a storage drawer below the oven; and it's generally easier to clean beneath them than beneath gas ranges.

Gas ranges do have their advantages, though. They are cheaper to operate in much of the country—enough to offset their greater initial cost, over time—and their burners give you greater control. But if you want a self-cleaning oven in a gas range, you have to pay a premium. Self-cleaning gas ranges can be priced as much as $200 more than comparable electrics. And you'll probably have to sacrifice some oven capacity.

Although continuous-cleaning ranges will never have a sparkling-clean oven, they are worth considering. They usually have a larger oven than a self-cleaning range, and they cost a lot less. (Of course, a range with a clean-it-yourself oven costs even less, and the oven can be scrubbed until it's shiny.)

As for how many and what kind of features to choose, added features mean added expense, and a valuable feature is often available only on models with a lot of other not-so-valuable features. You'll generally get the best value for your money by avoiding the frills.

Ratings of ranges (electric and gas)

(As published in a January 1984 report.)

Listed by types. Within types, listed in order of estimated overall quality based mainly on oven capacity, performance, and ease of cleaning. Models in brackets were judged equal; listed alphabetically. All models are 30-in. wide, free-standing, and with a single oven. All similarly sized electric elements performed similarly; all electric ovens used about the same amount of energy. Performance of gas burners and energy use for gas ranges varied slightly. Prices are approximate retail; + indicates plus shipping.

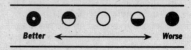

Brand and model	Price	Depth with door open (in.)	Baking performance	Broiler performance	Oven capacity	Advantages	Disadvantages	Comments
Electric ranges with self-cleaning oven								
WHIRLPOOL RJE3750	$650	46¾	○	◐	○	C,E,Q	j,p	—
AMANA ARR402	735	43½	◐	○	○	S	j,o,p,t,u	C,E,P,S
CALORIC ESR348 [1]	730	47	○	◐	○	F,Y	e,g,u	B,P
FRIGIDAIRE REG36H	529	46	○	◐	○	I,P,S	g,j	C,G,I
GENERAL ELECTRIC JBP26D	590	44¾	◐	○	○	E,F,Q	d,f,o,p,r,s,u	H,Q
HARDWICK EPC979KA650	752	44½	○	○	○	D,F	k,o,p,s,t	A,B,F,N,P
HOTPOINT RB737D	575	44½	○	○	○	F,Q	f,o,p,r,u	H,Q
KELVINATOR REP306A	629	46	○	◐	○	I	g,j	C,G,I
MAGIC CHEF 38CCW [1]	649	43¼	○	◐	○	D	f,r,s	C,D,E,G
MAYTAG CRE600	590	44½	◐	○	○	D	k,o,p	A,F,S
WHITE-WESTINGHOUSE KF535D	650	46	◐	◐	○	J,O,S	c,f,j,o	C,K,N
ROPER 2382	659	45¼	◐	○	◐	F	e,g,s	B,I,J,N,P
SEARS Cat. No. 93731	580+	46	◐	○	◐	Q	e,g,s,u	B,J
O'KEEFE & MERRITT 317442	710	42¼	◐	◐	◐	—	s	A,B,G,I,J,P
TAPPAN 312432	750	42¼	◐	◐	◐	Q	f	A,B,G,I,J,P
Gas ranges with self-cleaning oven								
SEARS Cat. No. 73931	$780+	44	○	◐	○	B,K,M,R	k,l,m,n	A,B,N,P
MAGIC CHEF 342KCW [1]	699	43¼	○	◐	◐	B	a,m	C,D,E,G
CALORIC RSR364	930	44	○	○	◐	M,N	l,m,p,q	B,L,N,P
ROPER 1393	839	44¼	○	◐	◐	M,R,U	e,i,n	B,N,P
O'KEEFE & MERRITT 308482	930	44¾	○	○	●	A,B,L,M,U	f	B,G,J,N
TAPPAN 302602 [2]	840	44¾	○	○	●	B,V	f	A,G,I,J

Brand and model	Price	Depth with door open (in.)	Baking performance	Broiler performance	Oven capacity	Advantages	Disadvantages	Comments
Gas ranges with continuous clean oven								
HARDWICK CKC9641KW650A	745	41	○	◗	◗	M,T,U	b,d,h	*B,M,N,O,P*
MAYTAG CRG400	520	41	○	◗	◗	S,U	b,d,h	*S*
WHIRLPOOL SF3300EK	475	41	○	◗	◗	S,U	b,d,h	—
FRIGIDAIRE GCG38KB	569	44	◗	○	○	H,K,R	d,n,u	*B,N,O,P*
WHITE-WESTINGHOUSE GF732C	580	44	○	○	○	G,K,T	d,n,u	*B,N*

1 *Model discontinued or about to be discontinued at the time the article was originally published in* Consumer Reports. *The test information has been retained for its use as a guide to buying.*

2 *According to manufacturer, this model was replaced by* **302603.**

SPECIFICATIONS AND FEATURES
All electric ranges take 120/240 volt, 3-wire service; all gas ranges are for use with natural gas but can be converted to LP gas.

KEY TO ADVANTAGES
A – Burner speed and energy efficiency best of all gas models—if drip bowls are kept clean.
B – Oven used less energy than most other gas models.
C – After self-cleaning cycle, oven judged cleaner than any other.
D – Areas outside door gasket stayed cleaner than most during cooking and self-cleaning.
E – Design of cooktop and surrounding made range easier to clean than most.
F – Control markings are on glass and were most resistant to scrubbing off.
G – Burners are porcelain-coated instead of aluminum, so were somewhat easier to clean than the others.
H – Has porcelain finish on oven bottom and inside door instead of continuous-clean finish so oven is easier to clean thoroughly than other continuous-clean ovens.
I – Drip bowls are porcelain enameled, so can be cleaned in oven during self-cleaning cycle.
J – Side panels are porcelain enameled.

K – Lower front panel is scratch-resistant.
L – Indicators for cooktop controls are behind windows, so are easier to see and clean than on other gas models.
M – Cooktop controls allow finer tuning of flame than the others.
N – Allows access to floor under range without moving range (but access panel is flimsy).
O – Control markings very clear.
P – Controls have additional indicator (bright orange marking) that element is on.
Q – Manufacturer makes it easy to reset oven temperature control to match actual temperature.
R – Unlike most gas ranges, has "oven on" indicator light.
S – Oven window easy to see through.
T – Timer has 4-hr. maximum.
U – Cooktop less crowded than on other gas ranges.

KEY TO DISADVANTAGES
a – Cooktop speed and energy efficiency worse than most other gas models.
b – Oven used more energy than most other gas models.
c – With heavy load, oven rack bent enough when pulled out to dump food.
d – Oven racks don't slide easily when loaded.
e – Door-locking lever for self-cleaning cy-

cle can pinch hand if door is closed carelessly.

f – Broiler tray may have very sharp edges.

g – Storage drawer may have sharp edges.

h – Oven light is unguarded and protrudes, easy to replace but susceptible to breaking.

i – Broiler igniter is fragile and easy to bump.

j – Part of door exterior is easily scratched paint.

k – Oven light apt to be left on accidentally because there's no window or automatic turn-off device.

l – Self-cleaning cycle is fixed at about 3 hr., less flexible than most others.

m – Oven took longer to self-clean than most.

n – Cooktop controls are on horizontal shelf in front, slightly safer but harder to keep clean than those mounted on a slant or vertically.

o – Cooktop is fixed, somewhat difficult to clean underneath.

p – Oven door is not readily removable.

q – Cooktop is recessed, with dirt-catching gaps at sides.

r – "Element-on" indicator dimmer than most.

s – Controls easy to block with large pots.

t – 12-qt. stock pot couldn't be centered on either rear element.

u – Broiler tray would only hold or cook 12 burgers.

KEY TO COMMENTS

A – Has no oven-door window.

B – Oven door covered with black glass panel.

C – Door locks automatically for self-cleaning.

D – Relation between door lock, timer, and indicator light for self-cleaning cycle poorly explained in owner's manual.

E – Shield must be raised over door window before self-cleaning.

F – 3 of 4 drip bowls have no hole in middle, so cleaning is easier but chance of grease fires is increased.

G – Some care required when cleaning under cooktop to keep liquid from running into oven insulation.

H – Duct that vents self-cleaning gases from oven is loose and easily misaligned.

I – Has fast-preheat setting for oven.

J – Oven-rack supports are removable.

K – Extra broiling element allows broiling on both sides at once, which saves time but was inconvenient; element must be removed before self-cleaning oven to prevent damage.

L – Broiler tray largest of any model, but hamburgers hardly cooked at edges.

M – Has "cook and hold" feature to keep food warm (not tested).

N – Has backguard lighting.

O – Timed-bake feature that starts from present time only.

P – Clock is digital.

Q – Frequency-of-Repair data indicates brand less likely to require repair than most others.

A look at the range of ranges

(As published in a January 1984 report.)

The following list of ranges should cook about the same as the ones tested. Although dozens of variants are listed, not all are included. Among gas ranges, there are more real choices than among electric ranges. For instance, only gas ranges with electric ignition were tested, although every manufacturer also makes a version with pilot lights. A few gas models are listed with an oven of plain porcelain; apart from that, those ranges appear to be the same as the continuous-cleaning gas ranges tested, and they're quite a bit cheaper. Of course, you have to clean the oven yourself.

Within their types, the ranges are listed by brand in alphabetical order. Within brands, they're listed in order of approximate retail price—a price that is often discounted. (In the price column, + means that shipping is extra.) Features are noted only when they differ from those of the model tested.

Manufacturer and model	Price	Comparison to model tested
Electric ranges with self-cleaning oven		
AMANA ARR401	$ 640	A
ARR405	800	D,G,I
CALORIC ESR343	600	A,C,L
ESR347	660	A
ESR375	800	G
FRIGIDAIRE RE36H	499	A,R
REG36HB	559	D,R
REG38H	699	G,R,M
GE JBP23	495	A,T,U
JBP24G	529	A,D,T
JB500D	685	G,M,V
JB500GD	715	D,G,M,V
JB600GD	889	D,G,M,W
HARDWICK EP976K430A	671	C,H
EP979K430A	705	H
HOTPOINT RB734	495	A,U,X
RB735G	525	A,D
RB747D	649	G,M,V,X
RB747GD	679	D,G,M,V,X
KELVINATOR REP303A	529	A,Z
REP309A	729	D,G,M
MAGIC CHEF 383CXWON	559	D
383CW	575	AA
384CXW	729	D,G,M
MAYTAG CRE700	640	D,G,M
ROPER 2352	439	A,C,H,L,R,Z,DD
2372	599	A,H,R
SEARS 93631	600+	AA
93831	680+	G,M,Y
TAPPAN 312542	890	B,G,J,M
WHIRLPOOL RJE3600	575	A,U
RJE385P	675	D,M
RJE395P	725	D,G,M
WHITE-WESTINGHOUSE KF335D	550	A,F,H,U
KF435D	560	A,D,H
KF500D	590	A
KF735D	800	D,M
KF835D	940	D,M,W

Gas ranges with self-cleaning oven

CALORIC RSR350	780	A,C,H,K,N,P,BB
RSR359	820	A,L,P,BB
RSR369	1090	EE,FF
MAGIC CHEF 343CKW	739	G
343CKWON	749	D,G
344CKXW	789	D,G,M
ROPER 1353	729	A,H,L,R
SEARS 73331	680+	H,L,P,BB
TAPPAN 302632	970	D,G,GG
302662	1140	B,D,G,J,BB,GG

Gas ranges without self-cleaning oven

WARDS 2460	590+	A,C,K,P,R,S,HH
FRIGIDAIRE G32K	399	A,C,H,K,N,P,Q,S
G36K	449	C,L,P,Q
HARDWICK C9642M53OR	569	C,H,L,P
CK9642M64OR	651	C,P
CK9641K630A	718	L
MAYTAG CRG300	480	Q
CRG500	580	D,G,Q,BB,CC
CRG600	630	D,G,M,BB,CC
WHIRLPOOL SF33100EK	370	Q
SF333PEKT	530	D
SF350PEK	610	D,G,M,O,CC
WHITE-WESTINGHOUSE GF630C	520	C,Q
GF832C	650	M,BB,DD

KEY TO COMPARISON

A – Oven door has no window.
B – Oven door has window.
C – Door has plain finish.
D – Door has black or brown glass panel.
E – Has no storage drawer.
F – Has no 2-sided broiling.
G – Backguard has light.
H – Backguard doesn't have light.
I – Has more indicator lights.
J – Color selection is wider.
K – Has no clock.
L – Has dial clock.
M – Has digital clock.
N – Has no timer.
O – Has timed-bake feature.
P – Doesn't have timed-bake feature.
Q – Oven interior is plain porcelain.
R – Color selection is limited.
S – Has no oven light.
T – Control panel isn't glass.
U – Has 1 large and 3 small cooktop elements.
V – 1 cooktop element heats to several diameters.
W – Some controls are electronic touch pads.
X – All or some cooktop elements don't unplug.
Y – Control panel is glass.
Z – Cooktop elements have lower wattages.
AA – Trim is different.
BB – Burner controls or indicators are different.
CC – Has cook-and-hold feature.
DD – Timed-bake feature starts at present only.
EE – Has rotisserie for oven.
FF – Has meat probe for oven.
GG – Has shiny burner bowls.
HH – Has plain burner bowls.

Gas-range emissions

Gas ranges are inherently more dangerous than electric ranges. They cook with open flames, and their fuel can be explosive. A less well-known danger is their contribution to indoor air pollution. While natural gas is a relatively clean fuel, its combustion does release gaseous by-products, some of which can be hazardous to health. Of particular concern are carbon monoxide and nitrogen oxides.

Gas ranges have been standard equipment for decades. But there are two reasons for the recent concern. In the past 20 years, a great deal has been learned about the health effects of pollutants in indoor air. Scientists now suspect that the same comparable levels of pollutants in indoor air can be just as hazardous as those outdoors. In addition, indoor pollutant levels are apt to be higher now than in the past. Many houses used to be drafty, but as people have tightened up their homes to conserve energy, they have also helped to seal in any contaminants that are emitted indoors.

Some studies have suggested an increased incidence of respiratory diseases such as colds and flu in children who live in homes with gas ranges, while other studies have failed to find such a correlation. The elderly and those who suffer from chronic lung disease are also likely to be sensitive to the effects of gas indoor pollutants.

In our judgment, the evidence so far suggests that emissions from a gas range do pose a risk—though probably not a major one—of impairing the health of some people in some homes. If you are buying a new range and can choose between electric and gas, that fact, added to other advantages of electric ranges, may make you choose an electric one. If you now own a gas range, should you consider junking it? Probably not, unless someone in your family has chronic respiratory problems. But make sure your kitchen is well ventilated for cooking. The best method is a range hood that vents to the outside. An exhaust fan about ten feet from the range would ventilate the kitchen adequately, but even an open window will help.

Refrigerators

Most top-freezer, "no frost" refrigerators have a fairly similar array of features and conveniences. In fact, without their nameplates, it's hard to tell one refrigerator from another.

Since refrigerators are major energy users, one important difference between them is energy use. Another is how well and how uniformly the refrigerators maintain "ideal" temperatures in both the freezer and the main compartment (see page 269 for thermometers that can help to check and adjust temperature even in your old refrigerator).

How much electricity?

Today's refrigerators typically use 15 percent less electricity than comparable models did five years ago. For the group of refrigerators tested most recently by Consumers Union, yearly electricity costs would run from about $82 to $105.

CU's estimates of energy consumption are higher than the estimates headlined on the "Energy Guide" labels manufacturers put on refrigerators. The discrepancy is due in part to differences in test methods, in part to the fact that some labels use out-of-date electricity rates.

The cost of operating a refrigerator is lower if you use the "energy-saver" switch controlling electric heaters that keep water from condensing around the door in humid weather. In dry weather you can save even more if you turn the heaters off.

Some models come with a "hot-tube" system instead. The system fights condensation by channeling waste heat from the condenser to areas around the doors. But the drawback is that the hot-tube models tend to cost more to operate than units with electric heaters.

How they perform

A good refrigerator should be able to keep food at 37°F in the main compartment and 0° in the freezer. And it should maintain those temperatures uniformly, through all kinds of room-temperature conditions. Consumers Union tested for:

Temperature balance. At a temperature of 70°, most models are able to hold, or stay within a degree or two of, the ideal 37°/0° temperature balance. Most will do the same with an "outside" temperature of 90°.

Temperature uniformity. Refrigerators are good at keeping temperatures uniform in the main compartment and in every part of the freezer but the door shelves; those are usually substantially warmer. It's not advisable to store frozen foods on door shelves for long periods.

Compensation for temperature change. When your kitchen's temperature goes up during a heat spell in summer, or when it goes down in the winter because you've lowered the thermostat, the refrigerator's temperature balance should remain stable without manually having to change control settings.

Most models won't vary their temperatures by more than a few degrees over the course of that temperature rise. But the testers noted an average variation of 5° in the freezer of some models; to compensate, you need to adjust the controls yourself during a heat spell.

Reserve capacity. Without this, a refrigerator that works well at normal temperatures may balk if the temperature soars. Reserve capacity is also an index of how a refrigerator will work after compressor, door seals, and other components have aged.

To measure reserve capacity, CU's testers raised a test room's temperature to a torrid 110°. Most models adjusted nicely, holding the ideal balance.

Controlling condensation. Door heaters or hot tubes help to resist condensation around the door when the temperature and humidity are high.

Most refrigerators will have some outside moisture despite the heaters' assistance.

Meat-keeping. It's good to have a control that lets you feed cold air into the meat-keeper through a duct; that feature will let you keep the meat at from 30° to 35°, a temperature range considered ideal for the purpose.

Vegetable-keeping. Crisper drawers keep fruits and vegetables from drying out.

Most models have at least one crisper with rubber flanges or gaskets; those seals generally help keep in moisture better than crispers without seals.

Crispers that are full retain moisture better than crispers that are only partially filled.

Noise. The new, no-frost refrigerators are much noisier than the old manual-defrost models. In addition, a no-frost model's more energy-efficient compressor runs for longer periods.

Convenience

There are a number of differences in features and design that affect a refrigerator's convenience:

Main-compartment shelves. It's hard to find a refrigerator without adjustable shelves in the main compartment—some shelves as wide as the compartment, some half its width. Half shelves allow more options for stowing things—and more chances of accidentally knocking things off. Some models have heat-resistant, tempered-glass shelves; others have wire shelves. A glass shelf is easy to clean but breakable, heavier than wire, and somewhat of a chore to remove or adjust.

Some shelves are so narrow that getting even a two-liter bottle in and out can be awkward. With a few, the door-

shelf retainers aren't high enough to keep a two-liter bottle from falling out if you open the door abruptly.

Freezer shelves. Most models have a shelf in the freezer; the shelf is often adjustable.

Usually, there are also shelves in the freezer door; some are extra deep, a natural repository for packages of frozen food (but for short-term storage only, given the warmer door temperatures we measured). The retainers on some door shelves let packages fall through.

Egg storage. To preserve freshness, it's best to store eggs where they aren't jostled. That means in a container in the main compartment. Egg shelves in the door are really second best, even though they may be convenient.

Dairy-product storage. A covered compartment in the door is generally at a temperature slightly higher than in the main compartment. The butter stays spreadable—and shielded from odors. Many models also have a covered utility compartment. It's best to have butter and utility compartments whose door stays up when raised, since that permits you to get something out with only one hand.

Fruit and vegetable storage. One big drawer is handy for large items, but in most refrigerators, each crisper is about half as wide as the compartment. Make sure the main compartment door opens wide enough to permit removing the full-width crisper or the crisper next to the hinge.

Controls. Temperature adjusters and other controls should be easy to reach and easy to operate, and their functions should be clear.

The best location for controls is in the front top section of the main compartment. A control at top center of the compartment tends to obstruct tall items on the top shelf. And the inconvenience of a freezer control hidden at the back of the freezer speaks for itself.

Lighting. Any refrigerator should have a light bulb that provides adequate light. The bulb should be easy to change and it should be protected by a shield.

Door. You should order your new refrigerator with the door mounted to open in the appropriate direction for your particular kitchen arrangement.

Cleaning, maintenance

A textured steel door effectively hides fingerprints; but some people complain that it's more rust-prone, especially in humid oceanside areas.

The condenser coil is easy to overlook when you clean. The coil, which helps remove heat, is outside the cabinet, where it tends to collect dust. Dust brings the refrigerator's efficiency down and its operating cost up. Some manufacturers suggest cleaning the coil annually or semiannually.

Cleaning the back-mounted coil on a few models is a simple matter, once you pull out the refrigerator. With most others, the coil is mounted in a compartment underneath the cabinet, where it

easily picks up dust. To clean most bottom-mounted coils, use a condenser-coil cleaning brush—available at many hardware and appliance stores for about $6—and a vacuum cleaner's crevice tool. Most manufacturers tell you to clean the coil from the front. That's difficult if the coil is under a shield and toward the refrigerator's back. Cleaning the coil from the back after you remove the cardboard "service-access" cover is a bit easier.

Most models have a seamless, easy-to-clean plastic liner in the main compartment. Others are easy-to-clean porcelain-on-steel, with plastic molding at the front.

Ratings of refrigerators

(As published in a July 1985 report.)

Listed in order of estimated over-all quality. Prices are average (and range) for white models, as quoted to CU shoppers in an 11-city survey.

Legend: Better ◉ ○ ◐ ● Worse

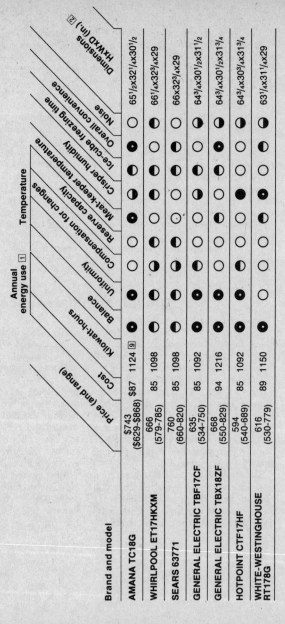

Brand and model	Price (and range)	Annual energy use [1]		Temperature					Crisper humidity	Ice-cube freezing time	Overall convenience	Noise	Dimensions HxWxD (in.) [2]
		Cost	Kilowatt-hours	Balance	Uniformity	Compensation for changes	Reserve capacity	Meat-keeper temperature					
AMANA TC18G	$743 ($629-$868)	$87	1124 [9]	●	○	○	○	◐	◐	●	●	○	65 1/2 x 32 1/4 x 30 1/2
WHIRLPOOL ET17HKXM	666 (579-785)	85	1098	◐	◐	◐	○	○	○	○	○	◐	66 1/4 x 32 3/4 x 29
SEARS 63771	760 (660-820)	85	1098	◐	◐	◐	○	○	◐	○	○	○	66 x 32 3/4 x 29
GENERAL ELECTRIC TBF17CF	635 (534-750)	85	1092	◉	◉	◐	◐	○	◐	●	●	◐	64 3/4 x 30 1/2 x 31 1/2
GENERAL ELECTRIC TBX18ZF	668 (550-829)	94	1216	●	●	◐	◉	○	○	◉	●	◐	64 3/4 x 30 1/2 x 31 3/4
HOTPOINT CTF17HF	594 (540-689)	85	1092	◉	◉	○	◐	●	◐	○	○	○	64 3/4 x 30 3/4 x 31 3/4
WHITE-WESTINGHOUSE RT178G	616 (530-779)	89	1150	●	○	◐	◐	●	◐	◉	●	◐	63 1/4 x 31 1/4 x 29

FRIGIDAIRE FPE18TM	676 (589-799)	82	1052	◓	○	◒	○	◓	○	○	◓	◒	65x31 1/4x30 3/4
KELVINATOR TMK180AN	657 (549-895)	82	1052	⦿	◒	○	○	◒	○	○	○	◒	64 3/4x31 1/4x31 1/2
HOTPOINT CTX18GF	634 (550-849)	94	1216	⦿	⦿	○	○	◓	●	○	○	◒	64 3/4x30 1/2x31 3/4
SEARS 64861	684 (600-760)	96	1244	⦿	◓	◒	◒	○	○	○	○	○	66x29 3/4x29 3/4
MAGIC CHEF RB17C3A	574 (530-675)	103	1327 [10]	◓ [11]	⦿	◓	○	◓	⦿	◒	○	◒	65 1/4x29x29 1/2
MONTGOMERY WARD 1772	566 (500-740)	103	1327 [10]	◓ [11]	⦿	○	○	○	⦿	◓	○	◒	64 3/4x29 1/4x29 1/4
ADMIRAL NT17B8	631 (529-760)	103	1327 [10]	◓ [11]	◓	◓	○	◓	⦿	◒	○	◒	65x29 1/4x29 1/4
WHIRLPOOL ET18MK1L	695 (595-819)	105	1360	⦿	◓	◒	◒	○	◓	○	◓	○	66 1/4x29 3/4x29 1/2
ADMIRAL À LA MODE ICNT18E9	915 (699-1050)	100	1284 [12]	●	◓	◓	○	◓	⦿	⦿	◓	◒	66x31 3/4x29 1/4
GIBSON RT17F9WM	672 (579-799)	89	1150 [12]	●	◓	○	○	⦿	⦿	○	○	◒	63 1/4x31 1/2x29 1/4
TAPPAN 95-1773	600 (530-680)	103	1327 [10]	◓ [11]	⦿	○	○	—	⦿	[14]	●	◒	65x29 1/4x28 3/4

[1] *Costs are to nearest dollar, based on rate of 7.75¢/kwh. Estimates based on switchable door heaters, if any, turned on for 4 mo., off for 8 mo.*

[2] *Height may vary slightly depending on setting of leveling legs; depth with door closed includes door handles.*

[3] *Side clearance required when door is opened 90 degrees or more to allow removal of crispers.*

[4] *Usable capacity, discounting space around crispers, etc.*

[5] *All models have plastic, removable crisper tops that serve as shelves. Number of other shelves is listed; unless noted, all are 1/2 or partial width, adjustable, and made of tempered glass.*

[6] *Number of shelves is listed. On all models: at least 1 shelf holds items as large as 2-liter soda bottles; door has butter compartment. Unless noted, all shelves are nonadjustable, full width, made of plastic, with fixed retainers.*

[7] *Number of shelves is listed; count doesn't include floor of freezer. Unless noted, all shelves are removable, nonadjustable, full width, made of wire.*

[8] *Number of shelves is listed. Unless noted, all are full width, made of plastic, with fixed retainers.*

[9] *Measured using higher of 2 heater settings.*

Ratings of top-freezer refrigerators. CU-measured capacity footnote [4]; measured shelf/clearance footnotes [3], [5]; shelf description footnotes [6], [7], [8].

Brand and model	Depth, door open 90 degrees (in.)	Clearance, opened door (in.) [3]	CU-measured cap. — Refrigerator (cu. ft.)	CU-measured cap. — Freezer (cu. ft.)	CU-measured cap. — Total (cu. ft.)	Mfr. claimed total capacity (cu. ft.)	Mfr. claimed shelf area (sq. ft.)	Refrigerator shelf area (sq. ft.) [5]	Shelves — Refrigerator	Shelves — Refrigerator door [6]	Shelves — Freezer [7]	Shelves — Freezer door [8]	Advantages	Disadvantages	Comments
AMANA TC18G	59¾	0	10.0	3.9	13.9	17.8	23.0	25.0	3,1s	2ad,2af,1ap,1g	1c	1,1q	D,F,H,I,M	j,n	E,K
WHIRLPOOL ET17HKXM	58¾	0	9.7	4.2	13.9	17.2	21.3	22.7	3,1s	1k,1o	1	2	—	s	A,F,H,N
SEARS 63771	58¾	12	10.2	4.2	14.4	17.1	22.3	23.3	3t,1s	1,1z,1o	1	2	—	s	A,F,H,N
GENERAL ELECTRIC TBF17CF	58¾	15	9.8	4.1	13.9	17.2	22.2	24.5	2,1b	1,1z,1o	1c	2	D,K	c,f,j,o	H,I,J,K,L,O
GENERAL ELECTRIC TBX18ZF	59	15	9.5	4.3	13.8	17.7	22.5	25.1	2,1b	1,1gnz,1gq	1c	2w	D,E,F,G,H,I,J,K	c,f,o,q	G,H,J,K,O
HOTPOINT CTF17HF	59	17	9.9	4.2	14.1	17.2	21.7	23.5	3t,1s	1,1z,1o	2a	2	—	c,f,j,o	I,L,O
WHITE-WESTINGHOUSE RT178G	57¾	0	10.1	3.8	13.9	17.0	22.5	24.1	4	1e,1z,1aep	1	2ae,1q	B,F,I	h,k,l,m,n,s	F,H,J
FRIGIDAIRE FPE18TM	60	0	12.1	3.8	15.9	18.0	25.0	25.9	3,1b	2a,1am,2ho	1	4a	B,C,F,I,J,K	l,m,n,s	E,H,J,L,O
KELVINATOR TMK180AN	60¼	0	11.7	3.6	15.3	18.0	21.8	22.9	4	2ek	1c	1e,1q	B,D,I	h,l,m,n,s	F
HOTPOINT CTX18GF	58¾	15	9.8	4.4	14.2	17.7	21.9	24.3	4	1,1gnz,1go	2a	2w	E,I	c,f,o	L,O
SEARS 64861	56½	16	9.6	4.4	14.0	17.7	21.6	22.1	3t,1s	1,1m,1p	1	2	A,I	a,n,s	A,F,H,N

MAGIC CHEF RB17C3A	55½	9	8.9	4.2	13.1	16.6	21.1	23.0	3,1s	1n,2o,1ao	1acsv, 1ac	D,F,I	b,c,i,j,n,p	C,E,J,O
MONTGOMERY WARD 1772	55¾	9	8.9	4.2	13.1	16.6	23.7	25.7	2bt,1r	1n,2o,1ao	1c	2w	C,D,I	b,i,j,n,p,t	C,E,H,L
ADMIRAL NT17B8	55½	9	8.9	4.2	13.1	16.5	21.1	23.0	3,1s	1n,2o,1ao	1acsv, 1ac	D,F,I	b,c,i,j,n,p	E
WHIRLPOOL ET18MK1L	56¼	16	9.4	4.4	13.8	17.6	21.1	22.4	3,1s	1,1fz,1o	1	2	H	a,j,n,p,s	A,H,M, N
ADMIRAL À LA MODE ICNT 18E9	58	9	9.4	4.1 / 3.3[13]	13.5 / 12.7[13]	18.1	20.5	23.2	3,1s	1ak,1hmn, 2hnz,1ap	1w, 1ac,1u 1aw, 1a	D,F	b,c,h,j,n,o	B,C,D, E,H,L,O
GIBSON RT17F9WM	57¾	0	9.5	4.1	13.6	17.0	24.0	25.6	3b	1e,1ez,1x	1,1e	B,C,F,I	h,l,m,n,s	E,H,O
TAPPAN 95-1773	55¾	9	9.0	4.3	13.3	16.6	20.8	23.0	3bt	1n,2o,1ao	1y	2w	C	b,c,d,e,g, i,j,n,p,r	C,E,F

[10] Average of 4 samples; 1 was significantly higher; another, whose freezer couldn't be depressed to 0°F, had low energy consumption.

[11] Temperature balance varied from ○ to ● depending on sample.

[12] Freezer temperature couldn't be made cold enough. Unit would probably use more energy if freezer temperature could be made lower.

[13] Measured with ice-maker in place.

[14] Longer than with similar tested Wards because freezer couldn't be depressed to 0°F.

KEY TO DISADVANTAGES

a – Door heaters less effective than most in reducing condensation.
b – Condenser very hard to clean.
c – Cabinet hard to level.
d – Wheels not included (CU paid $37 for optional wheels).
e – Doors open more than 180 degrees; may hit adjacent cabinets.
f – Doors do not open wide enough for easy removal of crisper next to hinge.
g – Small, flimsy freezer rack instead of built-in shelf.
h – Smaller door-shelf area than most.
i – Refrigerator door-shelf retainers too low to keep 2-liter bottles from toppling.
j – Freezer-door retainers couldn't keep small packages in.
k – Retainers on juice-can dispenser inadequate to store cans of different sizes.
l – Refrigerator control obstructively located at top of main compartment.
m – Freezer control inconveniently located at rear of freezer.
n – Light bulb unshielded, a potential safety hazard.
o – Crisper drawers difficult to slide.
p – Crispers smaller than most.
q – Crisper/meat-keeper controls may shift if drawer is slammed.
r – No meat-keeper.
s – Butter-compartment door won't stay open by itself.
t – Flimsy egg tray.

KEY TO COMMENTS

A – Brand Frequency-of-Repair record better than most.
B – Through-the-door water and ice-cube dispenser, automatic ice-maker, automatic ice-cream maker, fold-out wine rack.
C – "Hot-tube" model; has no electric door heaters.
D – Door is not reversible, but models can be ordered with either right- or left-opening door.
E – Butter compartment but no utility compartment.
F – Egg or utility bins.
G – Bin with molded bottom for eggs.
H – At least 1 crisper has seal to retain moisture.
I – Has smooth steel doors.
J – Ice-cube-tray rack or compartment.
K – Meat-keeper has seals.
L – Egg tray.
M – Has porcelain-on-steel crispers and meat-keeper.
N – Has porcelain-on-steel liner.

O – Model discontinued at the time the article was originally published in *Consumer Reports*. The test information has been retained here, however, for its use as a guide to buying.

SPECIFICATIONS AND FEATURES

Except as noted, all have: ● Controls that can be set to maintain 37°F in refrigerator, 0° in freezer. ● Plastic, seamless liner. ● Bottom-mounted condenser. ● 2 plastic crispers and 1 uncooled plastic meat-keeper in main compartment. ● 2 ice-cube trays in freezer. ● Light in main compartment (not in freezer). ● Provision for optional ice-maker. ● Reversible, textured-steel doors. ● 4 rollers.

KEY TO SHELF ARRANGEMENTS

a – 1/2 or partial width (more or less than 1/2).
b – Full width.
c – Adjustable to 2 positions.
d – Adjustable shelf retainers.
e – Removable retainers.
f – Adjustable door shelf.
g – Adjustable door-shelf separators.
h – Adjustable and removable partial-width bin/shelf.
k – 2-liter soda bottles fit very tightly.
m – Holds 1/2-gallon milk container and 1-lb. coffee can.
n – Holds 6-pack of 12-oz. beverage cans.
o – Holds 12-oz. beverage cans.
p – Small items only.
q – Cans only.
r – Full width; half-wire, half-plastic.
s – Plastic.
t – Wire.
u – Ice-cube-maker, removable, takes up 1/3 of freezer.
v – Part of floor is partitioned into ice compartment whose top can serve as shelf.
w – Extra-deep freezer-door shelf.
x – For eggs only.
y – 1 small rack only.
z – Holds 1-lb. coffee can.

KEY TO ADVANTAGES

A – Reached lower freezer temperature than others.
B – Rear condenser coil; requires less cleaning than bottom coil.
C – Main compartment has more shelves than most.
D – Adjustable freezer shelf.
E – Adjustable "snuggers" in refrigerator-

door shelves keep small items from
falling horizontally.
F – Meat-keeper has adjustable cold-air
duct to control temperature.
G – Convertible crisper/meat-keeper.
H – At least 1 crisper has adjustable humidity control.

I – Has ice-cube bin.
J – 1 crisper is larger than other.
K – Ice-cube-tray rack conveniently lo-
cated at top of freezer.
L – Light in freezer.

Quality, repairs

Most refrigerator parts and labor are covered under a one-year warranty. The sealed refrigerator system generally has a five-year coverage.

Consumers Union's Frequency-of-Repair data disclose that the repair record for **Whirlpool** and **Sears** refrigerators has been better than for most brands.

Repairs to an appliance require a house call and are a costly nuisance. To help avoid unnecessary repairs, Whirlpool has a toll-free number you can call for advice on the proper operation of its appliances. Besides providing an easy-to-follow service manual, GE and Hotpoint offer a "Quick Fix System" for do-it-yourselfers bold enough to tackle a major appliance, and a 24-hour, toll-free number.

Recommendations

Shop carefully for the best price. There tend to be enormous price spreads from one store to another.

Take your time and check out as many features as possible. After all, a new refrigerator is an appliance you and your family may use dozens of times a day.

If you live in an area with high electricity costs and you expect to own a refrigerator for its entire lifetime (perhaps fifteen years), it would certainly pay to buy a model that doesn't use much energy.

But the odds are that you won't own your refrigerator nearly that long. Statistically, the average family moves every six or seven years, and the refrigerator often stays behind. In that case, electricity consumption would be less significant, and shelf arrangements or other features would become more important. If your family is at least as mobile as the average, you'll be better off buying a less expensive model that's fairly energy-efficient.

Bathroom scales

You may be interested in knowing your weight precisely, but a bathroom scale is not a laboratory instrument— nor need it be. Small errors are tolerable, provided the scale is consistent. You use the scale principally to monitor **changes** in weight. Therefore, consistency is more important than accuracy.

In addition to being consistent a scale should be easy to use and read. The conventional squarish platform with the dial is still around, as is the upright, clinical-looking "doctor's" scale. But now there are digital readout models competing with the dials and balance-beams, and there are high-tech-looking instruments. There are even scales that talk to you, with a built-in memory to remind you what you weighed the last time.

Weighing in

Besides being reasonably accurate and consistent, a scale's dial should be easy to read, the zero easy to reset, and the scale's bottom shouldn't mar the floor.

Accuracy and consistency. Most scales are accurate to within a pound or two; the best are accurate to within three-quarters of a pound, up to their maximum capacity. There is a tendency for some units to lose their accuracy at the high end of the dial, with even a good scale being off by about 1½ pounds.

Carpeting. Unfortunately, the accuracy of even the best scales is likely to be affected by using them on a soft surface. So if your bathroom is carpeted, you'll have to arrange for a hard,

smooth place to put the scale.

Readability. Your scale should have large, bright numerals that are easy to read in any light. In general, most digital dials and "speedometer"-type dials are easy to read, as are the workings on beam-balance "doctor's" scales.

Setting zero. Every once in a while, a scale's zero may drift from its setting and have to be readjusted. On scales with conventional dials and speedometer dials, it is just a matter of locating and turning a thumbwheel or knob. Any adjuster should be conveniently located and easy to move.

A few of the digital scales have a finger-operated device for resetting their zeros. The rest will zero in automatically.

Features

Platform digital scales have a switch, button, or lever that you must push with your toe before you can get a reading. If you've always had a dial-type scale, you'll probably find that a bit inconvenient at first. A few upright digital scales are automatically activated when

you step on the platform; the others have a push button on their display panel.

Once activated, most digital displays will stay on for as long as you stand on the scale, or for an extra second— which is more than sufficient time for reading the bad (or good) news.

Unlike dial scales or scales that use the beam-balance, a digital scale takes batteries. Most last at least several months, but even that may be an inconvenience. The majority are powered by 9-volt batteries, but some take three, four, or six AA cells or four C cells.

Dial scales have conveniences and inconveniences. To activate a dial scale, all you do is step on. And, of course, a dial scale doesn't need batteries. But a dial scale is more apt to need adjustment for zero.

A conventional beam-balanced upright is fairly easy to use once you get the knack. But the weighing operation takes a bit longer than with a digital or dial scale. The method requires you to balance a long beam by sliding two weights (called "poises") along its length. The traditional way to use a beam-balance is to stand on the platform and move the large poise to the nearest 50-pound-multiple setting below your weight, and then slide the small poise till you find the exact quarter-pound mark that will cause the right end of the beam to settle in the center of the slot, which indicates the beam is balanced, and then you must add up your weight.

Some digital scales have interesting features. There are display panels that can be mounted on the wall at eye level.

There's a scale that literally talks back. It displays your weight and simultaneously announces it—in a fairly intelligible male voice, loud enough to be heard in an adjacent room.

Worse for some people than a talking scale might be one that remembers and can store a number of past weights that can be individually recalled at the touch of a memory button on the display panel.

If you've ever wanted to weigh something while you're holding it—say a baby or a pet—a few digital scales make it especially easy: you step on the platform and, by pushing a button, zero out your own weight. You then step off, pick up what you want to weigh, step on the platform again, and press the start button. The weight displayed on the readout will be that of baby or pet.

Recommendations

A high-rated scale will give consistent measurements. You can rely on any of them to reveal quite precisely how much weight you've gained or lost. It will tell you what you weigh at the moment as truthfully as you need to know.

There are big price differences, but the extra money goes largely for such niceties as digital readout, or a beam balance—and for a more stylish object in the bathroom.

Ratings of bathroom scales

(As published in an August 1983 report.)

Listed in order of overall quality based primarily on consistency and accuracy. Models judged approximately equal in overall quality are bracketed and listed alphabetically. Prices are suggested retail; + indicates shipping is extra.

Better ● ◐ ○ ◐ ● Worse

Brand and model	Price	Type	Claimed capacity (lb.)	Readout interval (lb.)	Consistency	Accuracy	Readout clarity	Ease of setting zero	Style notes	Comments
NOVUS NSC22	$70	Platform, digital	300	1/2	●	●	●	●	"Modern"; square; vinyl surface; hang-up display.	A,B,C,D,E
KRUPS FIT CONTROL MEMORY 801	150	Platform, digital	286	1/10	◐	●	●	●	"Modern"; rectangle; vinyl surface.	B,E,F
SEARS Cat. No. 6442	70+	Platform, digital	300	1/2	●	●	○	●	"Modern"; square; vinyl surface.	A,C,D,E,U
HEALTH-O-METER 230	190	Upright, beam-balance	350	1/4	●	●	◐	◐	"Clinical"; rectangle; vinyl surface; metal pedestal and display.	A
SEARS Cat. No. 6450	120+	Upright, beam-balance	350	1/4	●	●	◐	◐	Essentially similar to preceding model.	A
KRUPS 821 WEIGHT MINDER	50	Platform, speedometer	260	1	◐	◐	○	○	"Modern"; oval; vinyl surface.	A,G,V
DETECTO CELEBRATION K200, A BEST BUY	15	Platform, dial	280	1	◐	◐	○	○	"Traditional"; square; vinyl surface.	E,H
HEALTH-O-METER 720	130	Upright, digital	300	1	◐	●	●	◐	"Clinical"; rectangle; vinyl surface; metal pedestal and display.	I,J,K
HEALTH-O-METER 130	180	Upright, speedometer	300	1/2	◐	●	◐	●	"Clinical"; rectangle; vinyl surface; metal pedestal and display.	A

Product		Type								Surface description	Notes
HEALTH-O-METER 134	140	Platform, speedometer	300	1/2	◑	●	◑	◑		"Clinical"; rectangle; vinyl surface.	A,L
GENERAL ELECTRIC COMPUTERSCALE EDS3	44	Platform, digital	300	1	●	●	○	◑		"Modern"; square; raised dot, plastic surface.	J,K
COUNSELOR PIN STRIPES 111	25	Platform, dial	300	1	◑	●	◑	●		"Traditional"; square; padded, vinyl surface.	I
COUNSELOR FIBER OPTIC 800	30	Platform, dial	260	1	●	○	○	●		"Traditional"; square; wicker surface.	M
DETECTO K620	59	Platform, digital	280	1/2	◑	●	◑	●		"Clinical"; rectangle; padded, vinyl surface	I,N
HANSON 9950	150+	Upright, digital	300	1/2	◑	●	◑	●		"Clinical"; rectangle; vinyl surface; plastic pedestal and display.	I
HEALTH-O-METER 250	120	Upright, beam balance	300	1/4	○	◐	◑	●		"Clinical"; rectangle; vinyl surface; metal pedestal and plastic display.	O
COUNSELOR VICTORY 2000	60	Platform, digital	300	1	◑	●	◑	◐		"Modern"; rectangle; vinyl surface; raised and angled display.	N
HANSON W64	13	Platform, dial	280	1	◑	◑	◑	◑		"Traditional"; square; vinyl surface.	E
TERRAILLON 1020	26	Platform, dial	260	1	○	○	◑	○		"Modern"; square; vinyl surface.	E
SEARS COLORCARE Cat. No. 6428	17+	Platform, dial.	280	1	◑	◐	●	○		"Traditional"; square; vinyl surface.	E
SOEHNLE 6613	40	Platform, dial	285	1	○	◑	◐	○		"Modern"; square; grooved, plastic surface.	E
COUNSELOR 7170	45	Platform, digital	300	1	◑	◑	◐	●		"Traditional"; oval; wicker surface.	P,Q
DETECTO K600	89	Upright, digital	300	1/2	●	●	◑	●		"Clinical"; square; padded, vinyl surface; metal pedestal and plastic display.	N,W
HEALTH-O-METER 700	50	Platform, digital	300	1	◑	◑	○	–		"Modern"; rectangle; grooved plastic platform.	J,K,R
HEALTH-O-METER 1022	14	Platform, dial	300	1	◑	◑	●	○		"Traditional"; square; vinyl surface.	
TERRAILLON 146	40	Platform, dial	260	1	◑	◑	◑	○		"Modern"; rectangle; wood surface and plastic base.	E

Brand and model	Price	Type	Claimed capacity (lb.)	Readout interval (lb.)	Consistency	Accuracy	Readout clarity	Ease of setting zero	Style notes	Comments
BORG 7210PM	15	Platform, dial	270	1	○	○	●		"Modern"; square; vinyl surface.	Q
COUNSELOR PEDESTAL II 8100	90	Upright, digital	300	1	●	●	◉		"Clinical"; rectangle; padded, vinyl surface; plastic pedestal and display.	N
SEARS Cat. No. 6435	30+	Platform, digital	300	1	●	●	◉		"Traditional"; square; vinyl surface.	B,S
SOEHNLE 6960	22	Platform, dial	285	1	●	◑	○		"Modern"; square; vinyl surface.	E
TALK WEIGHT TE1000	100	Platform, digital	[1]	1	●	●	○		"Modern"; rectangle; padded, vinyl surface.	J,K,T
BORG 9100PM	35	Platform, digital	300	1	◑	○	○		"Modern"; square; grooved vinyl surface.	K,U
COUNSELOR MONTEREY R931	20	Platform, dial	300	1	◑	◑	◑		"Traditional"; oval; wicker surface.	P,Q
J.C. PENNEY Cat. No. 2403	20+	Platform, dial	300	1	◑	◑	◑		Essentially similar to preceding model.	P,Q
BORG 2533PM	20	Platform, dial	270	1	◑	●	●		"Modern"; oval; vinyl surface with pattern.	A,P,U
COUNSELOR 7140	23	Platform, dial	300	1	◑	●	◑		"Traditional"; oval; vinyl surface.	P,Q

[1] Claimed maximum capacity, 270 lb.; measured to about 264 lb.

SPECIFICATIONS AND FEATURES
Except as noted: ● All have steel base and platform. ● None worked well on carpet. ● All digital models have an activating switch and measure minimum weight of from 1 to 6 lb.

KEY TO COMMENTS
A – Unaffected by carpet.
B – Display mounts on wall.
C – Unlike most other digital scales, can be used to weigh objects.
D – Also works on AC with optional adapter.
E – Also has kilogram scale.
F – Has memories for 7 weights.
G – Pointer did not consistently return to zero.
H – Bare metal feet may scratch wood floors or leave rust marks.
I – Only slightly affected by carpet.
J – Minimum weight measured was from 20 to 28 lb.
K – Unlike most digital scales, works without activating switch.
L – Dial housing above platform may interfere with feet.
M – Required 40-lb. weight minimum to light up.
N – Display stays on for extended time, slowing repeat weighings.
O – Unusual weighing technique can easily produce error.
P – Tended to tip unless weight was well forward on platform.
Q – Has undersized platform.
R – Zero is not set but readout is raised or lowered to reflect known accurate weight.
S – Raised readout window frame on platform can stub toes.
T – Also announces weight.
U – Reflection of overhead lights made display hard to read.
V – According to the company, a later designation is **Family Data 881.**
W – According to the company, a later designation is **K750.**

A closer look at two diet scales

A desire to lose weight is enough motivation for some people to buy a kitchen scale. And the manufacturers know it: most of the small-capacity scales tested by CU have names that suggest weight loss, with words such as "diet," "dietetic," and "skinny." Some come with calorie charts and other nutritional information on a variety of foods.

The **Compucal Personal Diet/Kitchen Computer PDC1R,** a medium-capacity scale, goes much further (it should, for a price of $135). It has an electronic memory that holds nutritional data on six components of several hundred foods: protein, fats, and carbohydrates (in grams); sodium and cholesterol (in milligrams); and calories. Each food has a three-digit key number. You look up the key number in a booklet that comes with the scale, then you punch in the number on the **Compucal**'s membrane keyboard, put the food on the platform, and press a button. The scale gives you a digital readout of any of the six components. The scale automatically calculates nutrient content of foods on the basis of their weight in grams or ounces.

The **Compucal** also lets as many as nine people keep track of their individual food intake. You simply punch in your personal code number—1 through 9—before keying in the food-code number. The scale keeps a cumulative record of your food intake; to cancel the record you enter your personal code and follow it with "000."

With food weighing two pounds or less, the scale is accurate to within about three-fourths of an ounce—an adequate performance. But with loads ranging between two and five pounds, the maximum average error is a bit more than two ounces.

You can operate the scale with six C-size batteries, or, as with the other digital scales, you can use a special adapter. But even with the adapter, the scale needs three C-size batteries to operate its memory, according to the manufacturer. The batteries fit in their recess almost too snugly; batteries tend to pop out before the cover can be put in place.

The expensive **Compucal** delivers a lot of information about a lot of foods. It can be useful for a single dieter or a family of dieters, or for people following a special regimen, such as a sodium-restricted or low-cholesterol diet.

A much cheaper approach for calorie counters comes with the $30 **Terraillon D200.** It has a set of ten cards that give nutritional information—mainly calorie counts. You insert the proper card in a recess behind a window on the scale's housing, read the weight that shows on a dial next to the window, and consult the card for the value of that weight of food. The cards give information on about one hundred foods, including some unfamiliar ones such as hake, celeriac, and salsify.

Kitchen scales

Dieters are instructed to use a scale so that calorie intake can be monitored closely, but a kitchen scale can be handy for ordinary cooking, too. In Europe, where cookbooks traditionally specify ingredients by weight instead of volume, a scale is a routine kitchen utensil.

The kitchen scales in this chapter cover the entire size range that's useful in the kitchen, from dieter's scales small enough to tuck in a drawer to scales big enough for canning or mass feeding. For convenience sake, the scales are divided into three groups: small (diet) models, with a capacity of 2 pounds or less; medium-capacity models, whose range runs up to 11 pounds, handy for most kitchen tasks; larger-capacity models, which can handle loads up to 26 pounds—the capacity you'd need for a Thanksgiving turkey.

The scales vary widely in design, from the old-fashioned dial or beam-balance types to models that are sleekly modern with digital readout.

Most of the tested scales are table models. The **Eva, Terraillon 3000,** and **Soehnle 1140,** three medium-capacity models, are designed to be wall-mounted and thus save counter and storage space.

Accuracy

You shouldn't expect laboratory precision in a kitchen scale, but you do have a right to expect reasonable accuracy.

Nearly two-thirds of the models are very accurate, off by less than a half-ounce over their entire weight range.

Maximum errors often come at the high end of a scale's range. For example, with weights between 4 and 12 pounds, the average error for the large-capacity **American Family 3000** is 5 ounces; with weights of more than 20 pounds, the scale would be off by almost 3 pounds. The **Soehnle 1140** errs by at least 3 ounces, on average, at the top of its weighing range.

You can make sure your scale is accurate by testing it in your kitchen with a range of loads of a known weight (be careful not to overload). Commercially packaged foods such as sugar, butter, and sliced processed cheese are handy for the test; their net weight, marked on the package, is labeled.

All the scales should respond to small changes in the load when under light or heavy loads. But readings will tend to be a touch high or low unless you are careful to center the weight on the scale's platform or in the weighing container that comes with some models. Inaccurate readings are also likely unless you use a table scale on a smooth, flat surface. Wall-mounted scales are very sensitive to verticality; they require perfectly upright, tilt-free installation.

Ease of reading

Accuracy aside, reading a scale should be quick and easy.

Digital scales are very quick and easy to read. The liquid-crystal display of the check-rated **Hanson 7200** is not quite as bright nor as easy to read as the light-emitting diode of the other digitals. Unlike the other digital models, that **Hanson** won't shut off automatically after use. You must switch it off to avoid battery drain.

With some nondigital models, clearly identifiable weight graduations—neither too few nor too many, and properly spaced—make it easy to get precise readings.

Even with small weight graduations, an accurate scale won't be very useful if the graduations are narrowly spaced or hard to identify. Some scales have a pointer that's too wide or too far from the numbers for accurate reading. Or the dial position may be poor—hard to read without stooping when the scale is on a countertop.

Among the nondigital scales, the easiest to read are three small-capacity models—the **Soehnle 1301,** the **Soehnle 8600,** and the **Terraillon D200.** All three can easily be read without stooping.

If wall-mounted at eye level, the medium-capacity **Eva, Terraillon 3000,** and **Soehnle 1140** models are fairly easy to read, with no bending—but their weighing pans all sit in front of the dial. A pan piled high with food will block your view of the pointer.

Most of the scales with a circular dial or vertical index have a plastic window over the pointer to protect the pointer and keep the dial clean.

Other convenience features

Platforms and stability. Some scales' weighing platforms lift off for cleaning. Some platforms are circular, but most are rectangular—and big enough, in some cases, to hold a dinner plate. A number of small- and medium-capacity scales have platforms that are really too small for practical kitchen use; for most foods, you have to use the bowl that comes with the scale.

Container. Many models come with at least one lightweight container or bowl, useful for weighing liquids and loose food. The medium-capacity **Krups, Soehnle 1230,** and **Terraillon BA5000** come with a container you can fit over the scale like a cover, for more compact storage.

Zero setting and holding. By adjusting a dial or sliding a weight, you can get each model to register zero, as it should when there's nothing on the platform. The zero-adjustment control should be easy to reach, and it should respond to your touch. Zero adjustment is easiest—essentially automatic—with the digital model.

Tare compensation. You'll want to know a given portion's weight minus tare (the weight of the container) without having to do arithmetic. Some models have a mechanical adjustment for tare: You put the container on the platform, set a dial at zero, and add food to the container. Tare-free readings are very easy with the digital scales—you simply push a button or move a switch. Surprisingly, most of the tested scales lack tare compensation, though you can use the zero adjustment as a clumsy substitute.

Inconvenience. The medium-capacity **Terraillon BA5000** has an open area around the weighing platform; a spill could find its way through the opening into the scale's innards. The medium-capacity **Soehnle Digita S 8003** has a stiff on/off switch; so using the scale on a smooth countertop is a two-handed job, requiring one hand for the switch and the other to steady the scale. During tests of the large-capacity **Hoan Balance,** the beam behaved erratically when being balanced.

Recommendations

If you plan to buy a kitchen scale, start with a clear idea of what you'll use it for. That will help you decide what the scale's capacity should be.

Choose a model that has both customary and metric readings, since metric weights are used in some cookbooks and diet plans.

Among the small-capacity scales, the **American Family Tanita Electronic** and the **Hoan Electronic Diet/Kitchen** are high in quality. Essentially similar models, they are accurate and easy to use. But be aware that their prices match their modish style—$80 and $90, respectively. You could spend a good deal less, with no sacrifice in accuracy, on the small-capacity **Soehnle Dietetic 1301** ($22) or **Terraillon D200** ($30).

If you want a portable scale, the compact **Soehnle Dietetic 8600** may be just the ticket at $24; folded, it will fit in your pocket. It doubles as a postal scale.

Among the medium-capacity scales, the $90 **Hanson 7200** and the $85 **Soehnle Digita S 8003** are standouts.

Wall-mounted models should be welcome in a kitchen where counter space is at a premium—and the medium-capacity **Eva 28510,** at $38, is the best of the wall scales.

If you're unsure of the capacity you need, consider the large-capacity (22 pounds) **Terraillon T565,** $48, for its versatility. Though it can handle a turkey, it can also handle the very small weights that would normally be entrusted to a diet scale. That **Terraillon** is a beam-balance model, so it does require some patience in fiddling with its calibrated weights (and some mental arithmetic). Note that its weighing platform has very sharp edges. You should file those sharp edges smooth.

Ratings of kitchen scales

(As published in an August 1984 report.)

Listed by groups according to capacity; within groups, listed in order of estimated overall quality. Models in brackets were judged approximately equal in overall quality and are listed alphabetically. Prices are suggested retail; + indicates shipping is extra.

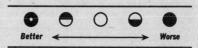

Better ← → Worse

Small-capacity (up to 2 lb.)

Brand and model	Price	Readout [1]	Claimed capacity (lb.)	Readout interval (oz.)	Accuracy	Readability	Ease of setting zero	Comments
✔ AMERICAN FAMILY TANITA ELECTRONIC COOKING/DIET 3500	$80	DD	2	$1/10$	◉	◉	◉	A,B,C
✔ HOAN ELECTRONIC DIET/ KITCHEN 21200	90	DD	2	$1/10$	◉	◉	◉	A,B,C,GG
SOEHNLE DIETETIC 1301	22	D	$1\,1/8$	$1/4$	◉	◒	○	E
SOEHNLE DIETETIC AND LETTER 8600	24	D	$1/2$	$1/8$	◉	◒	○	F,G
TERRAILLON DIET D200	30	VI	1	$1/4$	◉	◒	○	C,G,H,I
AMERICAN FAMILY 3120	11	VI	1	1	◉	◔	○	C,EE
HANSON DIET CALCULATOR 185	11	VI	1	$1/2$	◒	◔	○	C
HANSON FOOD PORTION 1523	40	D	2	$1/8$	◉	◔	○	E,G,O
AMERICAN FAMILY DIET 3100	7	VI	1	1	◉	◔	○	C,J,K,L,M
COOKS TOOLS DIET 21679	7	VI	1	1	◉	◔	○	C,J,K,L,M
IDL SKINNY 33300	7	VI	1	1	◉	◔	○	C,J,K,L,M
HOAN DIET WATCHER'S 410	8	VI	1	$1/2$	◉	◔	○	J,N,HH

Medium-capacity (up to 11 lb.)

Brand and model	Price	Readout [1]	Claimed capacity (lb.)	Readout interval (oz.)	Accuracy	Readability	Ease of setting zero	Comments
✔ HANSON PRECISION ELECTRONIC DIGITAL 7200	90	DD	$4\,3/8$	$1/10$	◉	◉	◉	A,P,Q,R,DD
✔ SOEHNLE DIGITA S 8003	85	DD	5	$1/5$	◉	◉	◉	A,C,S,T
COMPUCAL PERSONAL DIET/ KITCHEN COMPUTER PDC1R	135	DD	5	$1/10$	○	◉	◉	A,U,V,W
KRUPS ADDIGRAMM M844	30	D	$4\,3/4$	1	◉	○	○	C,E,X
POLAROID DIGITAL ELECTRONIC KITCHEN	40	DD	10	1	○	◉	◉	A,Y,DD
SOEHNLE 1230	25	D	$4\,1/2$	1	◉	◔	○	C,E,X
EVA 28510 [2]	38	D	11	$1/2$	◉	○	○	—
TERRAILLON BA5000	32	D	10	1	◉	●	○	D,Z,AA,BB
TERRAILLON MURALE 3000 [2]	38	D	$6\,1/2$	1	○	○	○	—
SOEHNLE 1140 [2]	35	D	7	$1/2$	◒	○	○	—
COOKS TOOLS WAYMASTER 21681	20	D	10	1	◒	●	○	C,K,L
HANSON 4400	17	D	10	1	○	●	○	C,K
HOAN 411	20	D	10	1	○	●	○	C,K,II

Brand and model	Price	Readout [1]	Claimed capacity (lb.)	Readout interval (oz.)	Accuracy	Readability	Ease of setting zero	Comments
Large-capacity (up to 26 lb.)								
TERRAILLON T565	48	BB	22	1/16	●	○	○	G,O
HOAN BALANCE 427	35	BB	26	1/4	●	○	○	G,CC,FF
AMERICAN FAMILY KITCHEN/ CANNING 3000	16	D	25	1	●	◖	○	—
SEARS Cat. No. 8886	13+	D	25	1	●	◖	◖	J

[1] *BB = beam balance; D = dial; DD = digital display; VI = vertical index.*
[2] *Wall-hung model; tray folds up when not in use.*

SPECIFICATIONS AND FEATURES
Except as noted, all have: ● Customary and metric calibration readings. ● Dials with transparent protective cover. ● Zero adjustment. ● No tare compensation.

KEY TO COMMENTS
A – Has switch for automatic tare compensation.
B – Uses 4 AA-size batteries; can be used with AC adapter.
C – Comes with 1 container.
D – Comes with 2 containers.
E – Provides for tare compensation.
F – Can be carried folded in pocket or purse; has provision for wall-mounting.
G – Has no metric calibration markings.
H – Has dish that can substitute for platform.
I – Has 10 slip-in cards that align with scale index to give calorie equivalents for a number of foods according to weight.
J – Dial lacks protective cover.
K – Has small platform.
L – Tended to tip over with load on scale.
M – Includes "Food and Drink Counter" calorie guide.
N – Platform judged too small for general use.
O – Has very sharp platform edges; edges should be filed smooth.
P – Manufacturer's instructions warn that exceeding 4³/₈ lb. (or 1,999 g) load might damage scale.
Q – Uses 9-volt battery. Can be used with AC adapter (not tested).
R – Lacks automatic switch-off to conserve batteries. LCD readout relatively dim.
S – Uses 6 AA-size batteries. Can be used with AC adapter (not tested).
T – Push-switch for turning scale on worked rather stiffly, requiring one hand to restrain unit on a smooth surface, a minor inconvenience.
U – Uses 6 C-size batteries.
V – Batteries fit very tightly; some brands may be hard to insert. 3 batteries required for memory function even if AC adapter is used.
W – Has calculator and readout functions.
X – Container inverts over scale for storage.
Y – Uses special **Polaroid** battery. Cannot be used with AC adapter.
Z – Adjuster for tare compensation judged somewhat hard to turn.
AA – Opening around platform edge creates potential for spills inside scale housing.
BB – Comes with 1¼-qt. container that substitutes for tray and inverts over scale for storage.
CC – Erratic balance-beam behavior made weighing difficult.
DD – Model discontinued at the time the article was published in *Consumer Reports*. The test information has been retained here, however, for its use as a guide to buying.
EE – Available as Sears Cat. No. 8889, $9 + .
FF – A later designation of this model was 21061.
GG – A later designation of this model was 21320.
HH – A later designation of this model was 21011.
II – A later designation of this model was 21022.

Electric shavers

In other major industrialized nations, 50 to 75 percent of all men who shave use electric shavers. In the United States, only 25 percent of men who shave do it electrically.

Furthermore, American men who now shave with lather and blade are probably not even potential converts. In a Consumers Union survey of shaving habits a good number of men said that they owned an electric shaver but had returned to lather-and-blade shaving. More often than not, the electric shaver had been received as a present.

That survey dates from years before many of the shavers currently on the market, particularly the Japanese brands, were introduced. And, as they have done with many other products, the Japanese have redesigned the shaver. Today's typical Japanese model is much lighter and more compact than past shavers.

More than one-third of the models tested for the latest Consumers Union report are Japanese-made: the **Hitachis, Panasonics,** and **Sanyos,** the **Mitsubishi,** even the **Sears** models. As it turns out, most of the other shavers tested are foreign-made, too: some **Norelcos** are Dutch; **Eltrons** and **Krups** are German; **Schicks** are Austrian. The American predilection for scraping whiskers off instead of buzzing them off may help explain why **Remingtons** are the only shavers still made in America.

The new competitors and refinements offer a market of wide variety.

Cutting

The head of most shavers comes in one of two designs. Rotary heads, most familiar on **Norelcos,** have two or three spring-mounted guards that shield spinning cutters. Foil heads, now the most common design, have a thin, flexible screen over a cutter that shuttles underneath.

Shavers also vary by how they are powered. Top-of-the-line models, ranging in suggested price from $50 to $100, are often cordless, with built-in rechargeable batteries. Plug-in shavers can range from basic to deluxe, and are priced at $24 to $80.

To evaluate the razors, CU's testers lined up two groups of 25 panelists each. One group regularly used an electric razor; the other customarily shaved with lather and a blade. Every week for ten weeks, each panelist compared a different pair of shavers, using one on the left side of his face and another on the right. During the week, the panelists were asked to make performance and preference judgments about the two shavers. A statistical analysis of those judgments gave the Ratings order.

The **Remington XLR3000,** with a broad, double-curved foil head, turned out the overall favorite. But that's not to say that the **XLR3000** is the inevitable choice for Everyman. Among panelists who usually shave with a foil-head model, the **Eltron 900,** with its slender foil head, is about as well liked as the **Remington** model.

Someone who now shaves with a blade should also consider the **Norelco**

HP1606, the **Panasonic ES827,** and the **Schick F45.**

Men who are dissatisfied with the rotary shaver they're now using might try the **Hitachi RM2530** as a replacement. Rotary users also liked the triple-head rotary **Norelco** models and the foil-head **Remington XLR3000.**

Features

When buying a shaver, there are factors beyond performance to consider:

Trimmer. A cutter that's distinct from the shaving head is handy for tidying the boundaries of beards, mustaches, and sideburns. So most models offer an auxiliary cutting edge at the side or top of the shaver. Some trimmers are fixed in position. Others retract and are moved up or flipped open for use.

Ideally, a trimmer should have its own on/off switch and protrude enough so that you can see what you're trimming. A trimmer mounted on the side of the shaver is likely to be better suited to trimming sideburns than to trimming mustaches or beards—the body of the shaver blocks your view in the mirror when you use one of those trimmers on the front of the face. And a trimmer that's set close to the shaving head tends to trim at inopportune times and in inopportune places. That's especially true if it runs whenever the main cutter does, as it does on the **Remington** models. Still, better to have one than not—the **Norelco HP1620** and the **Mitsubishi** lack a trimmer entirely.

Comfort adjustments. The Norelco HP1606 and **HP1328** have a "closeness/comfort" control that lets you raise or lower the shaving head for a closer or less irritating shave. Our panelists saw no advantage—or disadvantage—to the presence or absence of such features.

Power. A number of models run or recharge only from regular 120-volt AC outlets. Most others offer an extra for world travelers; they can also be powered from the 240-volt circuits commonly found abroad. Some of the models will automatically switch between the domestic and foreign voltages; with others, you must switch by hand.

There's more to shaving abroad, however, than having a razor that handles the local voltage. Plugs differ from country to country as well. The **Sanyo** models come with an adapter for the round-prong receptacles of some foreign countries. You can buy adapters for the others at an electrical-supply or appliance shop. As an alternative, you might want to consider a model powered by throwaway penlight batteries (see page 249).

Power cords on the plug-in models are at least five feet long. On a number of models, the cords are coiled. That's a help for storage, but the coil can exert a mild pull on the razor as you shave.

Recharging. Cordless rechargeable shavers are typically powered by nickel-cadmium ("nicad") batteries. In some models, a small cube right at the plug end of the cord contains the circuitry that converts household current to the low voltage that recharges the batteries. The **Panasonic ES827,** the **Eltron International,** and the **Sanyo SVM530** have their charging circuitry in the shaver.

You charge three models—the **Remington XLR3000,** the **Panasonic ES869,** and the **Mitsubishi**—by placing them in a charging stand that has a cord that plugs into the wall. That arrangement is less handy than it sounds because you can't power the shaver directly from

Ratings of electric shavers

(As published in a November 1984 report.)

Listed in order of overall preference of CU's test panelists. Except where separated by heavy rules, closely ranked models differed little in preference. Specific performance and convenience judgments are noted only where different from average preferences. Prices are suggested retail, rounded to nearest dollar; + indicates shipping is extra.

Brand and model	Price	Head type	Power	Advantages	Disadvantages	Comments
REMINGTON MICRO SCREEN XLR3000	$ 50	Foil	Rechargeable	E,F,H,I	b,o	B,C,J
HITACHI RM2530	50	Foil	Rechargeable	C,J,L	c	—
NORELCO ROTATRACT HP1606	80	Rotary	Plug-in	D,E,K	q	B,E,G
PANASONIC MR. WHISK ES827	43	Foil	Rechargeable	A,C,G,I	—	B
ELTRON 900	80	Foil	Plug-in	K,L	—	—
PANASONIC MR. WHISK WET/DRY ES869	60	Foil	Rechargeable	A,D,G,I	o	B,C
PANASONIC MR. WHISK ES889	24	Foil	Plug-in	A	p	A
SANYO SV3210	40	Foil	Plug-in	H,J	—	—

Model						
REMINGTON MICRO SCREEN XLR1000	40	Foil	Plug-in	F,H,I	b,d,f	A,J
NORELCO ROTATRACT HP1605	64	Rotary	Plug-in	D,K	q	B,G
ELTRON INTERNATIONAL	100	Foil	Rechargeable	B,C,H,K,L	—	B,G,H
SCHICK FLEXAMATIC F45	75	Foil	Rechargeable	C,K	q	I,J
KRUPS FLEXONIC 488	60	Foil	Plug-in	I,J	d,f	G
SCHICK FLEXAMATIC F20	35	Foil	Plug-in	—	p	A,G,I
SCHICK FLEXAMATIC F33	50	Foil	Plug-in	—	p	G,I
HITACHI AV4500	35	Foil	Plug-in	H,J	—	G
NORELCO ROTATRACT HP1328	100	Rotary	Rechargeable	B,C,D,K	q	E,H
ELTRON 660	65	Foil	Plug-in	H,K,L	—	—
REMINGTON MICRO SCREEN XLR800	32	Foil	Plug-in	H,I	b,e,f	A,J
SANYO SVM530	70	Foil	Rechargeable	K,L	o	F
NORELCO SPEEDRAZOR HP1620	26	Rotary	Plug-in	A,K	a,l,q	A,G,J
SEARS ROTOMATIC II Cat. No. 6835	50+	Rotary	Rechargeable	B,C,I	d,j,m,q	A
MITSUBISHI MICRO SHAVER SM600ST	90	Foil	Rechargeable	J,L	a,c,h,i,o	A,D
SEARS ROTOMATIC II Cat. No. 6834	40+	Rotary	Plug-in	I	d,f,g,h,i,k,n,q	G

SPECIFICATIONS AND FEATURES
Except as noted, all: ● Have cutter that will operate independently of trimmer. ● Have retractable trimmer. ● Operate on either 120 or 240 volts AC. ● Have straight power cord. ● Rechargeables give about 1 week's worth of shaves. ● Foil-head models have single head and rotaries have 3 heads.

KEY TO ADVANTAGES
A – Somewhat easier to manipulate than most.
B – Gave more shaves per charge than most.
C – Can operate directly from power cord when battery is discharged.
D – Relatively free of noise and vibration.
E – Somewhat less likely than most to pull and nick skin.
F – Shaved somewhat closer than most.
G – Somewhat easier than most to use around mouth.
H – Has fixed trimmer (but trimmer is less convenient than a retractable trimmer).
I – Has soft case.
J – Has semi-rigid case.
K – Has hard case.
L – Has mirror.

KEY TO DISADVANTAGES
a – Lacks trimmer.
b – Trimmer runs whenever shaver is on.
c – Gave fewer shaves per charge than most.
d – Judged somewhat unbalanced.
e – Judged unbalanced.
f – Somewhat harder to manipulate than most.
g – Somewhat more likely than most to pull and nick skin.
h – Somewhat slower than most.
i – Shaved somewhat less closely than most.
j – Somewhat harder than most to use around mouth.
k – Harder than most to use around mouth.
l – Somewhat less effective than most in trimming sideburns and mustaches.
m – Somewhat more irritating than most.
n – More irritating than most.
o – Will not operate directly from power cord; a discharged battery must first be recharged.
p – Lacks case.
q – Relatively hard to clean completely.

KEY TO COMMENTS
A – Runs only on 120 volts AC.
B – Automatic converter adjusts itself to 120 or 240 volts AC.
C – Has charger stand.
D – Has charger built into case.
E – Has "closeness/comfort" setting.
F – Has cutter-speed control.
G – Has coiled power cord.
H – Has light or lights to indicate when recharge is needed.
I – Model discontinued at the time the article was originally published in *Consumer Reports*. The test information has been retained here, however, for its use as a guide to buying.
J – Has 2 heads.

household current; when the battery runs down, you have to wait for it to recharge before you can shave.

All the other rechargeable models but the **Sanyo SVM530** can work directly off household current. With those, you can use the shaver until it weakens, plug it in to finish your shave, and then let it charge.

Battery capacity. Your shaving habits and the toughness of your beard largely determine the number of shaves you'll get on a single battery charge. On average, you can expect about a week's worth of shaves per charge. But some will do a bit better and others a bit

worse. The **Eltron International** and the **Norelco HP1328** signal their need for recharging by flashing one or two small red lights. Unfortunately, those shavers may slow down unacceptably before the lights ever come on.

Electrical safety. You'd be wise to avoid using a plug-in model over a basin of water. And a shaver that happens to fall into water must, of course, be unplugged before you try to fish it out.

Maintenance. Daily cleaning is no particular problem—you just unclip the blade cover and shake or brush clippings from the cutters and the underside of the head. Once in a while, how-

Electric shaving with soap

In 1931, Schick introduced the electric shaver with the promise that it made shaving "painless" by freeing the user from the need for soap, water, and razor blades. Now Panasonic has a shaver usable with soap and water.

The **Mr. Whisk Wet/Dry ES869** is worth considering purely on its merits as a dry shaver. And its waterproof design lets it be conveniently cleaned under a stream of running water.

As an automated wet shaver, the Mr. Whisk Wet/Dry can shave as nicely wet as it does dry. Using lather as a lubricant can give a pleasantly gentle shave. The only disadvantage is that the shaving cream may make the shaver's handle slippery, and make the shaver seem to skate over the face rather than cut.

Compact shavers for travelers

Going abroad? You can beat power problems with a compact shaver that runs on penlight batteries. Four such models are the foil-head **Eltron Battery 200i** ($50), **Panasonic Mr. Whisk ES811** ($19), and **Sanyo SVM730** ($30), and the rotary-head **Norelco Cordless Speedrazor HP1208** ($27).

The **Panasonic** and the **Sanyo** are particularly compact—about as big as a box of toothpicks. Both are powered by two AA-size batteries. The other two shavers are a bit longer, but still more svelte than most full-sized models. They take three AA batteries. The batteries give about one to two weeks' worth of shaves. In shaving quality, none of the traveling models equals a top-rated standard-size shaver. But they aren't bottom rated, either.

The **Eltron** is a full-featured shaver. The other three lack a trimmer—an important drawback.

All the travel shavers are handy for touch-ups. Kept in a glove compartment, any of them would help a salesman tidy himself up for a meeting, or a playgoer get rid of his five-o'clock shadow en route from work to the theater.

ever, you'll have to face up to a more extensive cleaning job that involves removing, disassembling, brushing, and refitting the cutters and the head. That chore is a bit tougher with the rotary shavers, which have several pairs of cutters and combs, each of which should be kept together and put back as a set. Some manufacturers recommend that you put a touch of light oil on the cutters after a complete cleaning.

So-called preshave preparations are sold to help a shaver glide over the face when the skin is sweaty. They do seem to help, but they make the shaver harder to clean (they'll also void the warran-ty if used with the **Eltron 900** or **Eltron International**). If you use a preshave, you may want to use a shaver-cleaning solution or rubbing alcohol afterward to remove preshave residue from the cutters.

Case. A number of the shavers come without a case. Others sport coverings ranging from a simple plastic shell that fits over their cutters to soft pouches and semirigid "shaving kit" cases and hard cases that open on hinges like a clam shell. A number of cases also provide a metal or glass mirror—usually rather small for convenient shaving but adequate in a pinch.

Recommendations

Not only are more and more foil-head razors entering the market, they are apparently getting better and better. The top-rated **Remington XLR3000,** cited for the closeness of its shave, is a foil-head razor.

The **Eltron 900** ($80) offers sleek European styling and a more compact design than the **Remington** model. The **Panasonic ES827** ($43) and the **Schick F45** ($75) are foil-head shavers particularly liked by panelists who usually shave with a blade.

The **Norelco HP1606** ($80) is the highest-rated rotary-head shaver, a favorite both of panelists used to rotary shavers and of panelists who don't shave electrically.

Given the highly subjective nature of shaving and the uncertainties of panel testing, people may disagree about what constitutes the "best" shaver. That's why it's important to buy from a store with return privileges.

Whichever model you choose, use it with a light hand—especially if it's a foil model. Many of the foils are thin enough to bend or break if pressed too hard. You shouldn't have to jam a shaver into the face to get a good shave; a light-to-moderate touch will produce fewer burns and less irritation. Stretch your cheek with your tongue or your free hand to give the cutters the best advantage with any wayward hairs.

Electric shock

Leakage of electric current from live interior parts to exposed metal parts such as a frame or housing is a common source of shock hazard in consumer products. The leakage may result from inadequate or worn insulation, assembly-line defects, marginal circuit design, or from just plain aging of critical components.

A shock hazard is simply that—a *possible* cause of shock. Consider, for example, the hazard presented by an appliance that has somehow come off the assembly line with a short circuit, so that its outer metal shell becomes as electrically live as the power line to which it is connected. You can touch this shell and feel no shock whatever— so long as no other portion of your body simultaneously touches a ground. A ground is an electricity-conducting material that, at some point, enters the earth, or touches a conductor that enters the earth. Typical grounds are a metal cold-water pipe, gas pipe, or storm drain, or any metallic device connected to any of these, such as a faucet, a metal sink, or a radiator. A bathtub or a sink full of water are both good grounds because water, a conductor of electricity, is in contact with the drainpipe, a potentially excellent ground. A damp concrete basement floor and an outside patio are good grounds, since there are moisture paths through them to the earth beneath. And, of course, if you are standing on damp, bare earth, you are grounded. Only when you are touching a ground and the shock hazard at the same time do you complete the circuit essential for the flow of electric current. Your home may contain numerous shock hazards—lamps and lighting fixtures often develop current leakages—but you haven't gotten a shock simply because you haven't touched a ground and a shock hazard simultaneously.

How severe a shock?

The severity of an electric shock is determined by a number of factors, including the path of the current through the body, the duration of the shock, and the age, sex, and physical condition of the victim. But the most important factor involved is the rate of flow of current through the body, usually measured in milliamps (a milliamp is a thousandth of an amp). The human body is extremely sensitive to electric current. Some persons, although they constitute a minority, can feel a flow of current through them that measures only .2 milliamp. Before leakage current reaches one milliamp, a substantial portion of the population can perceive it as an unpleasant sensation of heat, tingling, or throbbing at the point of contact.

As leakage current increases, human reaction to it becomes more pronounced. A shock only slightly in excess of one's perception level is not likely to be dangerous *per se*. But if it comes unexpectedly, the startled reaction it can produce—jumping against a hot kitchen range or dropping a dangerous power tool—can lead to a serious injury to the person shocked or to people nearby.

When an adult grasps an object and is

exposed to leakage current, the muscular contractions induced by the electricity may be so severe if the current is 5 milliamps or higher (often called "let go" current), that the object can't be released (the metal handle of a power tool, for example). Such continuous shock can quickly lead to fatigue, collapse, and ultimately, due to stopped breathing, death. Children are more sensitive to leakage current than grown-ups are because their muscular strength is not fully developed and they can't "let go" as easily as an adult can. Even a shock of short duration can be fatal; if it is severe enough, it will affect the heart.

On occasion, tests in Consumers Union's laboratories have revealed a shock hazard of 50 to 100 milliamps or more—patent evidence not simply of less-than-adequate insulation in the item being tested, but usually of direct contact between live interior wiring and an exposed metal part. Such high leakage currents constitute a *lethal* hazard.

There is a fair chance that you now have various electrical devices in your home whose electrical safety is questionable.

What you can do

In many circumstances you can eliminate the possibility of electric shock by grounding the equipment. To do this, connect a length of wire between a screw on an exposed metal part of the device and a suitable ground. Always disconnect the appliance while attaching the grounding wire.

You should not attempt to ground every device in your home. In no circumstances should you ground an appliance that has an accessible open-coil heating element. Most toasters and some older broilers fall into this category. In such appliances, the open coil is a live wire (in *some* cases even when the appliance is switched off). Grounding the appliance may actually increase the hazard; if you touched both the heating element with a fork held in one hand and the outer metal shell with your other hand, your body would then serve to complete the electric circuit.

You should ground all electrical devices whose heating elements are *not* exposed, if these devices are within reach of any ground. The rule holds for all electrical appliances used in the garage or basement, on the patio, or any place where you stand on a concrete surface that is in contact with the earth or where you stand on the earth itself. Portable power tools, such as electric drills, will be held firmly in use, perhaps with a sweaty hand, and should be grounded wherever they are used (but see "All-insulated construction," on page 254). Other electrical devices do not need to be grounded if they are used in a room where there are no grounds or if they are kept well out of reach of any grounds that might be accessible.

Finally, there is the problem of safeguarding yourself when using electrical equipment to which it is inconvenient to attach a grounding wire—a portable food mixer, for example, or a vacuum cleaner. The best protection is your own awareness of the hazard. Try to use ungrounded electrical devices only where no grounds or grounded appliances are within reach. (Touching the bare metal of a grounded device is precisely the same as touching any other ground.) It might be a good idea to unplug small appliances when they are not in use. This practice should be followed especially in kitchens or bathrooms, where good grounds are abundant. The bathroom, in particular, is a dangerous place to use electrical devices, unless the appliance is specifically intended for such use (a whirlpool bath, for example); and unless you observe stringent safety rules:

- Do not reach for an appliance that has fallen into water. Unplug it immediately.
- Do not use an appliance while bathing or in a shower.
- Do not place or store any appliance where it can fall or be pulled into a tub or sink.
- Do not place an appliance in or drop it into water or other liquid.

- Always unplug an appliance from the electrical outlet immediately after using.

When you do use an ungrounded electrical device, such as a vacuum cleaner, in the vicinity of a ground, make a conscious effort not to touch the ground while you are holding the ungrounded device. If there's a detachable cord, always plug the cord into the appliance first, then plug the other end into the wall outlet.

Recently built homes are likely to have, as part of the wiring for outdoor receptacles, ground fault circuit interrupters (GFCIs). These devices will open the circuit should there be current leakage that can cause a severe shock (see pages 254–55).

Polarized plugs

If the two prongs on the plug of an electrical appliance are different in shape— one wider than the other—the plug is "polarized," a feature that reduces the chance of a shock.

A polarized plug is designed to fit the slots of a polarized wall outlet only one way. Assuming the outlet is currectly wired, the polarized connection means that the chassis of some TV sets, for example, which might be touched accidentally, are always connected to the neutral wire in an outlet, rather than by chance to the electrically "hot" wire there.

In a lamp with a nonpolarized plug, there is a 50-50 chance that the threaded portion of the bulb socket is live. If it is, you could get a shock when you screw in a bulb: One of your hands might touch the base of the bulb or the edge of the socket while your other hand is grounded—resting on a radiator, for example. With a correctly wired polarized outlet, the hot line of the circuit will always connect to the lamp socket's center contact, which you're not likely to touch accidentally.

But there are problems. Although two-slot polarized outlets have been available since the 1920s, outlets in old houses may not be polarized. (The three-hole grounded outlets in new homes can accept a polarized plug properly.) Also, outlets that appear to be polarized may be wired incorrectly. And even if the outlet is polarized properly, an intermediate plug-in device such as a multiple (or cube) tap, a timer, or an extension cord may not be polarized (polarized two-prong extension cords are rare).

If a polarized plug doesn't fit your outlet, don't defeat the safety benefits of the plug by filing down the wider prong or replacing the plug. And don't try to force the plug to fit; some nonpolarized receptacles (especially those on extension cords or cube taps) will accept a polarized plug with a bit of pushing, and you might be able to force a polarized plug into a polarized socket the wrong way.

Have nonpolarized outlets replaced. If you're not sure a polarized outlet is wired correctly, ask an electrician to check. And when you need an extension cord for an appliance with a polarized plug, use a polarized cord.

Grounding to the outlet box

The best way of meeting the shock-hazard problem with appliances that should be grounded is by the use of a three-wire power cord with a grounding plug; this is the usual two-prong plug with a third, grounding prong. A great many appliances are equipped now with just such a cord and plug. To take best advantage of a three-wire power cord, you should have an electrician convert your outlet boxes (making certain that the boxes are properly grounded), installing in them three-prong receptacles (which, incidentally, will also accept two-prong plugs) if you do not already have such receptacles in your home. The cost of such a conversion will depend on the existing circuitry in your house.

A very much less desirable method is to use an adapter that permits plugging the three-prong plug into a two-prong receptacle. Such an adapter has a pigtail designed to be secured to the screw holding the cover plate in place over the wall outlet in order to provide a connection ground. However, there is just too much of a temptation to use the adapter without securing the pigtail. Furthermore, even with the pigtail secured, the appliance will be grounded only if the outlet itself is properly grounded. And that's a big if.

The act of grounding an appliance can itself serve to uncover dangerous shock hazards. If a grounded appliance blows a fuse or trips a circuit breaker when you plug it in, you have uncovered a potentially lethal shock hazard. Have the appliance repaired before trying to use it again.

All-insulated construction

An important step forward in electrical safety has been the development of "all-insulated" or "double-insulated" tools and appliances: drills, saber saws, sanders, vacuum cleaners, electric toothbrushes. These are so designed that, if insulation fails, no exposed metal part will be electrically live. Double-insulated tools are an exception to the general rule that all portable tools should be grounded—an important consideration since most users will omit this precaution, at least on occasion. But this does not mean that you need exercise no care at all in the use of such a tool. For example, the metal chuck of a drill will become live if you hit house wiring when drilling into a wall. And you cannot safely immerse double-insulated tools in water.

Ground fault circuit interrupters

A ground fault circuit interrupter (GFCI) offers dependable protection against serious shocks caused by electrical leakage. Like a fuse or circuit breaker, a GFCI stops the flow of electricity when abnormal situations develop. Although circuit breakers and fuses prevent high levels of current from overloading a circuit, a GFCI trips at very low levels of current straying from the circuit. More than ten years ago, the National Electrical Code began requiring that GFCIs be used in outdoor receptacles in new buildings. The Code has since been expanded to require GFCIs in bathrooms, garages, and other high-risk areas in new residences. GFCIs would be useful additions to the electrical system in existing dwellings as well.

A GFCI carrying a Class A, Group 1 designation from Underwriters Laboratories should be sensitive enough to respond to leakage as low as 6 milliamps. Speed of response is also important. GFCIs tested by Consumers Union engineers usually worked far more rapidly than the UL standard calls for—response time to 100 milliamps of leakage ranged from 3 to 30 milliseconds. That's enough to eliminate the greatest danger but still does not eliminate the hazard associated with leakage less than the 6-milliamp threshold. Nor do GFCIs eliminate the need for properly grounded appliances and circuits. There should be a test button to let you check to make sure the device is working properly, and some provision for resetting the unit to restore power to a circuit once the GFCI has been tripped. When a GFCI is activated, it generally means there's a faulty circuit or an appliance leaking electricity. The problem must be corrected, because it's impossible to override a GFCI.

Portable models are especially useful when you're working around the house with power tools. You plug the GFCI into any existing outlet (either one end of a three-wire extension cord or a conventional, grounded, wall outlet). You then plug the device you want to use into an outlet on the GFCI. If you take a

GFCI outdoors for short periods, be careful to keep water out of its housing. Some GFCIs can also be permanently attached to an indoor outlet. Portables work only at the outlet they're plugged into, leaving other outlets along a circuit vulnerable. For better protection, you need a built-in model.

Receptacle models. Indoor-type models are similar to conventional wall outlets and are meant to be installed in existing wall boxes. Besides providing protection for their own outlets, they can be wired to supply "feed-through" protection to outlets farther along on the same circuit. Some indoor receptacle models have only one outlet, while the rest have two. Aside from that, it's often hard to tell one brand from another. Outdoor-type models have special housings, gaskets, and other features to make them waterproof, but lack a feed-through provision. Indoor models can also be used outdoors, provided they're installed with a UL-approved weatherproof cover plate.

Combination models combine a GFCI with a circuit breaker in one unit. If you install one in the central electrical service box, it will protect an entire circuit. The model you choose should have mounting hardware that's compatible with the circuit breaker box you now have. In some older homes, deteriorating electrical insulation or dampness within the wiring can bring about fairly high levels of background leakage in the electrical circuits themselves. A combination model might then trip spontaneously, responding to the leakage in the circuit and not to a hazard in an appliance. Such "nuisance tripping" is also possible but somewhat less likely if you've installed a receptacle-type GFCI to provide feed-through protection. Using portable or receptacle GFCIs at key outlets can solve this problem.

Recommendations. Give serious consideration to protecting your garage, bathroom, kitchen, and workshop. You should also protect the wiring in any part of the house that's built on a concrete slab at or below grade, since the slab itself can be a very effective ground. Either a receptacle or a combination model will provide the necessary protection. If you have an electrician install receptacle or combination models, that reduces the chance of making a mistake that would nullify the GFCI's protection or cause other hazards. Local codes may, in fact, require you to hire a professional. Use a portable GFCI when you have power tools or appliances plugged into otherwise unprotected outlets.

Shower massagers

Basically, a shower massager is simply a water-powered gadget that gives you a shower or a "massage," whichever you're in the mood for. The shower may be anything from a sharp, fine spray to a coarse downpour. The massage is created by a pulsating effect in the stream of water.

There are two types of shower massagers—shower heads and shower hoses. Both attach permanently to the shower arm in your bathroom, replacing the ordinary shower head. The hose type is held in the hand or attached to a shower-arm clip or wall mount to deliver a stationary shower or massage.

Wetting down

CU equipped a shower stall in the laboratory so that four massagers could be hooked up simultaneously. That way, the testers were able to switch from one showering unit to another and make comparisons.

Here's what the testers found out:

Shower. The massagers generally produced a regular shower spray, usually rainlike but, from a few, rather forceful and sharp. A number of them could be adjusted to produce a variety of sprays, ranging from a coarse, gentle shower on up to a forceful, sharp one. Since people sharing the same bathroom don't always share the same shower preferences, the versatility can be an advantage.

Massage. You are not apt to become a more sensuous person than before using a shower massager, but the massage can be a pleasant sensation.

All units provide at least one pulsating setting; some provide a selection of such settings, ranging from a slowly pulsating, mild-mannered spray to a rapidly pulsating spray that can really give you a workout. Most of the sprays are forceful and fairly sharp, with well-defined pulsations.

Combination setting. Most models have a combined shower-massage setting. The combination is not always marked on the faceplate, but adjusting the unit between "shower" and "massage" gives you the setting. In most cases, the only purpose it seems to serve is to soften the sharpness of a needlelike shower spray or to reduce the pummeling effect of a forceful massage spray.

Convenience. Most units can be pivoted enough to aim the spray at your calves or point it away so you can put shampoo on your hair.

Shower-head models should swivel easily in the horizontal and the vertical directions, but, set in their shower-arm clip or post, or in their wall mount, some of the hose types may not swivel very far.

Though some of the hoses have drawbacks for stationary showering, all allow more precise spraying than do the shower-head models. Of course, if you happen to let go of the hose while the water is running, you may be thrashed, not massaged. But the handles are easy to hold, which should solve the problem.

Other considerations

Installation. A shower massager can't turn you on if you can't install the thing and turn *it* on. For installation in most bathrooms, you unscrew the old shower head from its pipe-threaded arm and screw on the massager. The post or clip that is the stationary mount for most of the hose models gets screwed onto the shower arm when the hose is installed. Some hose models have a wall mount or mounts; those must be attached with screws, anchors, foam pads, or epoxy glue.

All that sounds easy enough, but installation can get complicated if your shower arm ends in a ball joint rather than in a length of threaded pipe. If you can unscrew the ball-joint arm, you can replace it with a threaded-pipe arm and install any massager we tested. Or you can buy a massager that has a removable ball joint, first making sure that it's the same size as the shower-arm ball joint.

But to install one of those models, you'll still have to unscrew the ball-joint shower arm. If your ball-joint shower arm isn't removable, you have problems: You should be able to get an adapter that changes the ball joint to a pipe thread.

Water consumption. Some manufacturers boast that their shower massager uses less water than those made by the competition. On average, most of them use about the same amount of water as a conventional shower head. In actual use, the rate of water consumption depends upon how you shower. If you turn the faucets on full, you'll use more water in the same period than someone who turns them on only halfway.

Maintenance. Very little is necessary. At most, you might have to remove a filter screen every now and then in some models to prevent clogging. But not all models have filters.

Recommendations

There's really no way to tell in advance whether you'll enjoy the "massage" experience. Try to find a store with a working display. Stick your hand under the spray and decide for yourself.

If you like a sharp, invigorating shower, you may like a unit that delivers a massage spray that's sharper than most.

If you like a gentle shower, investigate one that gives a massage spray that is less sharp. If a number of people will be using the unit, examine one of the models that offers a range of settings.

Whether you choose a shower head or a hose model depends on personal preference.

Smoke detectors

Your best protection against fire is early warning. Smoke is a natural warning, but you may not see or smell it early enough. And, depending on the type of fire, you may have only minutes or even seconds to reach safety.

Most people aren't willing to take that risk. A majority of homes now have at least one smoke detector.

Do smoke detectors really make a difference? Are they worth the frustration of nuisance alarms? Absolutely. In homes with detectors, deaths and injuries are substantially lower and property loss is substantially less than in unprotected homes.

According to a study by the National Fire Protection Association, more than 90 percent of the fire fatalities and more than 80 percent of the fire injuries in a recent year occurred in homes that apparently had no detectors. Residential property loss in unprotected homes accounted for 87 percent of that total.

Homes with detectors are obviously safer. Yet they still account for nearly 10 percent of the fatalities, 20 percent of the injuries, and 13 percent of the property damage. Some of those detectors, it is known, fail to operate. One possible reason: their batteries may be dead or missing.

A detector with dead batteries is worthless. So is one whose batteries have been removed to avoid nuisance alarms. Therefore, you should pay special attention to features such as a low-battery signal and temporary-disabling mechanism that would encourage you to keep the smoke detector operating.

Detector types

The least expensive smoke detector is the ionization type. Photoelectronic detectors cost more to buy and may not be as easy to find. Smoke detectors that combine the two types are also available.

Ionization smoke detectors rely on a small amount of radioactive material to let the air between two electrodes conduct a constant electric current. When smoke particles disrupt the current, electronic circuitry sets off the detector's alarm.

Ionization detectors have had a reputation for responding quickly to the small, invisible smoke particles from fast-burning fires. The photoelectric models' forte was detecting the larger, clearly visible particles of slow-burning fires. A combination unit should combine the best of both designs.

How speedy a warning?

Consumers Union measured speed of response to smoke from four different burning materials. They simulated fast fires by igniting shredded paper, wood strips, and polystyrene. They simulated slow, smoldering fires (typically caused by a cigarette dropped on a mattress or upholstered furniture, these are one of the most common types of home fire) by placing wood strips on a controlled-temperature hot plate.

The most recent tests affirm that ionization detectors are the first to detect smoke from the test fires—but only by three or four seconds.

Manufacturers have apparently succeeded in making the photoelectrics more responsive to the smaller particles of smoke from a fast fire.

The tests using slow fires held no surprises. As in the past, the photoelectric models sounded off first by a substantial margin.

On every detector, the alarm is appropriately shrill and unpleasant, usually a high-pitched pulsating sound.

We tested to determine how feisty the alarms would sound when the batteries were weak. We tested each detector with batteries that were running low, then triggered the alarm and measured the sound level.

The alarms remain assertive with weak batteries, sufficiently loud to awaken most sleepers, if the alarm is near the bedroom. Heavy sleepers, hard-of-hearing individuals, and sleepers whose detector is positioned at some distance from the bedroom will want to choose a particularly loud model.

Nuisance alarms

Smoke detectors can't discriminate between the smoke from a dinner cooking or that from a house burning down. An alarm that sounds when there's no danger is a nuisance. (The problem of nuisance alarms may be exaggerated. According to a 1982 survey by Market Facts Inc., a survey research firm, more than twice as many nonowners as owners believe that smoke detectors go off too easily.)

Nuisance alarms can lead owners to disable their smoke detectors permanently, or to remove the batteries "temporarily" and then forget to replace them. In either case, the detectors provide no protection.

Some detectors may become cranky because they're in the wrong place. Avoid placing a detector near a kitchen or close to a fireplace or woodstove. Garages and furnace rooms are also poor locations, because car exhaust or a small back draft from your furnace can trigger a detector's alarm. Placing a detector near a bathroom can lead to trouble: droplets of moisture inside the chamber have the same effect as smoke particles.

Nuisance alarms may also sound because dirt has accumulated inside the detector's chambers. The air that constantly drifts through a smoke detector leaves behind dust and grime, which may eventually accumulate to the point where the unit triggers spontaneously or ceases to operate.

To keep your smoke detector operating properly, vacuum it yearly, cleaning with a vacuum wand. On a detector with a fixed cover, run the wand across the openings. If the cover of the detector is removable, gently vacuum the sensor chambers.

Another cause of nuisance alarms is insects. If your smoke detector is in their flight path, you'll know about it soon enough. Unfortunately, there's no fix you can make on an existing detector that wouldn't reduce the unit's sensitivity. As of March 1986, however, Underwriters Laboratories require all smoke detectors to have a bug screen or equivalent protection.

Batteries

The power source for the vast majority of detectors used in homes is a single 9-volt battery. After a year or so, the battery will need replacing. Replacements cost less than $2 and are simple to install.

To keep a smoke detector ready, you must know when the battery is failing. The detector should signal that fact with a gentle, periodic beep. How long a detector continues to beep a warning depends on the model. Some beep for at least seven days; others give you at least thirty days' warning that the battery is dying. A thirty-day warning is better—

just in case the battery starts failing while you're on vacation.

A number of smoke detectors give a second signal of battery distress. Those models have a pilot light that normally blinks slowly. When the battery is dying, dead, or removed, that pilot light goes out.

A few models utilize a second battery, which is used to power a small lamp. When the alarm sounds, the lamp goes on, illuminating a small area under the detector for at least four minutes. At night this could be a distinct advantage.

Test buttons

Any detector should have a test button to let you check if the unit is functioning. The button, for models made after November 1979, is a complete check. It indicates that the battery is live and the sounding device working,

and it simulates a fixed amount of smoke to check the unit's smoke-sensing capabilities. However, clogged air slots, which may not let real smoke enter, would not be discovered.

Use experience

A Consumers Union Annual Questionnaire included a number of questions about smoke detectors. More than one-third of the respondents said that they owned one or more detectors—a total of 90,000 units.

The detectors proved quite reliable: only 2.5 percent had required repair. The respondents had experienced a total of 1,879 fires, and their smoke detectors sounded an alarm in 96.6 percent of those fires. In 956 such incidents, the respondents believed, the alarms helped prevent or reduce injury or death; in 1,133 cases, they helped reduce or prevent property damage.

The 64 units that didn't respond to a

real fire may not all have been defective, in truth. In the absence of periodic cleaning, they may have gotten too begrimed to work. The batteries may have been faulty, or even missing. Weekly testing would quickly reveal such faults. The survey revealed that 67 percent of the units were checked less than once a month.

The respondents' worst problem by far appeared to be nuisance alarms. Almost one-third of the detectors had been tripped by innocuous smoke from the kitchen or fireplace (at that, though, the average was only three times a year).

What about radiation?

Most ionization smoke detectors contain a tiny amount of americium 241, a radioactive material. In sufficiently large doses, radiation from any source, including the sun, can do biological damage. Thus, the very presence of americium in the home may make some people nervous.

In the judgment of Consumers Union engineers, the hazards associated with americium in smoke detectors are very small. But questions regarding the acceptability of even a trivial risk from radioactive materials keep coming up.

In trying to answer these questions, Consumers Union's staff reviewed a number of scientific studies undertaken since 1977. They looked not only at the possible exposure of people to radiation from smoke detectors in the home, but also at exposure of workers to americium during the manufacture of the devices. They also considered possibilities of accidents or fire conditions that could release the material, and at long-term environmental consequences of the disposal of millions of the radioactive sources in discarded smoke detectors.

The available information led them to make the following estimates:

If there were no ionization smoke detectors, some 2,590,000 persons would die of cancer over the next seven years. Assuming every household in the country installed an ionization smoke detector, that number might grow to 2,590,001.

That added risk is equivalent to the additional radiation risk that would result from moving from your present home to one sixteen feet higher in elevation, and hence sixteen feet nearer the sun.

Where to put detectors

Ideally, smoke detectors should be on every level of a house, in every bedroom, in every hallway, and even in the basement. How many detectors you end up with will probably be determined by the size of your pocketbook. Other factors that determine how many are the size of your house, the number of rooms, and whether or not anyone in your family smokes.

If you can afford one smoke detector, make it a photoelectric unit. It may cost twice as much as an ionization model, but it's worth the extra price and shopping effort. The photoelectric models tested by Consumers Union responded to slow fires considerably faster than the ionization units.

The best location for a single detector is just outside the bedrooms.

If you install a multiple-detector system, choose one ionization detector and make the rest of the units photoelectric. Place one outside the bedrooms and one in the living area, but not too close to the kitchen or a fireplace. Additional detectors can go inside the bedroom of any smoker or in the basement (at some distance from the furnace).

All detectors can easily be mounted on a ceiling or a wall. A good spot is the middle of a ceiling. Nearly as good is someplace on the ceiling off to one side but no closer than six inches to a corner.

Recommendations

The best protection against fire is a smoke detector that responds quickly to all kinds of fires. Photoelectric and combination units do that best.

If you can, it's best to buy more than one detector. For a multiple-detector system, one detector should be an ionization or combination unit; the rest should be photoelectrics. An ionization unit's fast response to a fast-developing fire makes it good supplemental protection. Furthermore, since no two fires are alike, it's possible that a fire could produce a relatively transparent smoke that could greatly accentuate the difference in response times between the two types of detector.

Ratings of smoke detectors

(As published in an October 1984 report.)

Listed by groups in order of their response to slow fires. All responded reasonably well to fast fires. Within groups, listed in order of increasing price. Prices are suggested retail; + indicates shipping is extra.

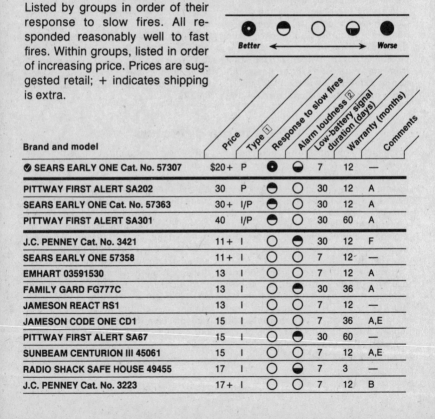

Brand and model	Price	Type ①	Response to slow fires	Alarm loudness ②	Low-battery signal duration (days)	Warranty (months)	Comments
✔ SEARS EARLY ONE Cat. No. 57307	$20+	P	◉	◒	7	12	—
PITTWAY FIRST ALERT SA202	30	P	◒	○	30	12	A
SEARS EARLY ONE Cat. No. 57363	30+	I/P	◒	○	30	12	A
PITTWAY FIRST ALERT SA301	40	I/P	◒	○	30	60	A
J.C. PENNEY Cat. No. 3421	11+	I	○	◒	30	12	F
SEARS EARLY ONE 57358	11+	I	○	○	7	12	—
EMHART 03591530	13	I	○	○	7	12	A
FAMILY GARD FG777C	13	I	○	◒	30	36	A
JAMESON REACT RS1	13	I	○	○	7	12	—
JAMESON CODE ONE CD1	15	I	○	○	7	36	A,E
PITTWAY FIRST ALERT SA67	15	I	○	◒	30	60	—
SUNBEAM CENTURION III 45061	15	I	○	○	7	12	A,E
RADIO SHACK SAFE HOUSE 49455	17	I	○	◒	7	3	—
J.C. PENNEY Cat. No. 3223	17+	I	○	○	7	12	B

Brand and model	Price	Type ①	Response to slow fires	Alarm loudness ②	Low-battery signal duration (days)	Warranty (months)	Comments
FYRNETICS LIFESAVER 0905	20	I	○	○	7	12	—
GENERAL ELECTRIC HOME SENTRY SMK6	20	I	○	○	30	12	C
PITTWAY FIRST ALERT SA76RC	20	I	○	◐	30	60	A
RADIO SHACK SAFE HOUSE 49456	20	I	○	◐	7	3	E
SEARS EARLY ONE Cat. No. 57313	25+	I	○	○	30	12	A,D
J.C. PENNEY Cat. No. 3102	30+	I	○	○	30	12	A,D
WARDS Cat. No. 61626	30+	I	○	○	30	60	A,D
PITTWAY FIRST ALERT SA120	40	I	○	○	30	12	A,D

① P = photoelectric; I = ionization; I/P = ionization/photoelectric.
② Measured after low-battery signal had finished.

SPECIFICATIONS AND FEATURES
All have: ● Test button. ● Low-battery beep signal. ● Instruction book judged good. *Except as noted, all:* ● Require one 9V battery. ● Have a pulsing, high-pitched alarm.

KEY TO COMMENTS
A – Flashing red light indicates working battery.
B – Has "hush" feature; alarm can be temporarily silenced in case of an accidental alarm.
C – Alarm is a warbling sound.
D – Requires 2 9V batteries; second is for emergency light.
E – Portable. Can be mounted on door.
F – Sample tested was identified as **BRK 79P**.

Meat thermometers

You can cook roasts and fowl by using your experience and intuitions. But you'll get more satisfactory results if you go by the food's internal temperature, the best indication of its degree of doneness. Hence the meat thermometer.

Most models we looked at are basically dials mounted on a metal stem. Within the stem is a twisted strip of two different metals, bonded together and connected to a pointer on the dial. When heated, each metal expands at a different rate, causing the dial's pointer to rotate.

A few other models are essentially glass tubes containing liquid that expands as it heats. The level of that liquid is read against a temperature scale.

Some dial thermometers are "professional," or chef's, models. Those need to be inserted only momentarily in meat to check its temperature. The others are left in place as long as the meat is in the oven.

Dial thermometers have certain advantages over tube models. You can insert a dial unit's pointed stem directly; glass-tube models generally can't be inserted into the meat without your first making a pilot hole, lest the tube break. (Most tube models provide a skewer.)

Furthermore, a number of dial models have a sliding marker that can be preset to a desired temperature. When the dial's pointer lines up with the marker, it's time to take the meat out of the oven.

Using a thermometer

Speed of response isn't important for a regular thermometer, which is designed to remain in place as a roast cooks. But you do want quick response, 30 seconds or less, in a professional model, since it is stuck into the meat only momentarily for a temperature check. The faster the response, the more likely you are to get a full, correct reading.

Dial models are generally easier to read than tube thermometers.

Tube thermometers, with their glass stems, are inherently more fragile than dial models. But if the tube is protected by a metal sheath, the problem diminishes.

The fluid in tube thermometers is an organic liquid that manufacturers say is harmless but unpleasant-tasting. Should a tube thermometer break in your roast, the manufacturers advise you to cut out a large enough section of the meat to remove any discoloration from the liquid, as well as any broken glass.

Several professional models can be recalibrated should they ever go out of adjustment. You dip the stem of the thermometer into boiling water and read the dial, noting how far the reading is from the boiling temperature (that's 212° at sea level, as little as 200° in high-altitude cities; your weather bureau can tell you what temperature to use). After the thermometer cools, you grip the nut on the stem with pliers and turn the dial by hand to compensate for the error.

Models that aren't adjustable can still be checked for accuracy in boiling water if their temperature range goes high enough. A number of dial models will give readings above 200°. Others have a test point above their usual scales, labeled "test, 212°" or merely "test."

A number of models carry reminders that this or that temperature corresponds to specific degrees of doneness for various meats. That saves you the bother of remembering whether 160° means rare for lamb and medium for beef, or vice versa.

How to use a thermometer

The trick with a meat thermometer is to insert its sensitive tip into the center of the thickest part of your roast or poultry, but well clear of any bones, masses of gristle or fat, or air spaces. Check roasting birds in the middle of their stuffing—especially if it's a raw pork-sausage stuffing.

Pork and some game meats demand special attention because of the possibility that they might contain *Trichinella spiralis,* the parasitic worm that causes trichinosis. Authorities used to recommend that fresh pork be cooked to 185°, to be sure that any parasites were dead. More recently, studies have demonstrated that an interior temperature of 170° is hot enough to make pork safe. Some thermometers still carry the old 185° recommendation; others recommend the lower temperature.

To use a professional thermometer, pull the meat out of the oven far enough so you can stick the instrument into the center of the thickest part. Leave the thermometer in place until its needle stops climbing (usually 15 to 30 seconds).

You really can't trust the accuracy of a glass-tube thermometer in a convection oven. The hot-air draft causes the thermometer to indicate a higher temperature than really exists in a test roast. Dial models seem less sensitive to hot air.

Recommendations

Most professional models are basically laboratory instruments with sharp points. Most of them are highly accurate—but also expensive.

Ratings of meat thermometers

(As published in a March 1985 report.)

Listed by types; within types, listed in order of estimated overall quality. Models of comparable quality are bracketed and listed alphabetically. Prices are suggested retail.

Better ←———————→ Worse

Brand and model	Price	Format [1]	Range (°F)	Accuracy	Readability	Comments
Regular models						
TEL-TRU BP40	$17.25	D	140 to 185	●	◒	E,G
TAYLOR 5939	8.49	D	120 to 200	◒	◒	A,E,G,J
ACU-RITE 00670	5.95	D	140 to 185	◒	○	A,E,G
TEL-TRU RM36	7.25	D	50 to 270	◒	◒	A,C,E,G
COOPER 323	7.99	D	130 to 190	○	◒	A,E,G
SPRINGFIELD 303	4.98	D	130 to 195	○	◒	G
TAYLOR 5937	6.98	T	105 to 185	○	○	C,D,G,J
OHIO 4631	2.80	T	140 to 185	○	○	G,K
ACU-RITE 00762	2.29	T	130 to 185	○	◒	G
SPRINGFIELD 613	2.29	T	140 to 190	◒	○	G
TRU-TEMP 290	3.70	T	130 to 190	◒	○	G
TRU-TEMP 121	4.16	D	120 to 190	●	◒	A,G

Professional models

Brand and model	Price	Format [1]	Range (°F)	Accuracy	Readability	Comments
COOPER MICRO-THERM 2236	12.95	D	0 to 220	◐	◓	A,B,F,G,H
TEL-TRU MT39R	18.50	D	0 to 220	◐	◓	A,B,H,I
CUISINART BA711	14.00	D	46 to 254	◐	◓	A,B
TAYLOR 6081 Professional	24.98	D	96 to 224	◐	◓	A,B,J
SPRINGFIELD 813	10.95	D	0 to 220	◐	○	A
H-B 21105	15.95	D	0 to 220	◐	◓	A,B,I
TAYLOR 5982	15.98	D	0 to 220	◐	◓	A,B,F,I,J
TRU-TEMP 341	16.00	D	100 to 400	●	○	A

[1] D = dial; T = liquid in glass tube.

KEY TO COMMENTS

A – Has test point at 212° F (boiling point of water at sea level), or has scale that includes that point.
B – Can be adjusted for accuracy.
C – Marked with yeast-rising temperature, judged an advantage.
D – Much of tube is encased in metal, making model more durable than others of its type.

E – Movable arrow at rim of dial can be pre-set to desired temperature.
F – Mfr. says thermometer can be left in meat during microwave cooking.
G – Labeled with meat-doneness reminders.
H – Mfr. says thermometer can be left in meat during cooking in conventional oven.
I – Has sheath with pocket clip. Sheath can serve as convenient handle when measuring temperature of hot liquids.
J – A later designation for all models of this brand is Sybron/Taylor.

Oven thermometers

You don't need a precision thermometer to verify the accuracy of an oven's thermostat. But you should expect reasonable accuracy—within, say, 20 degrees of actual temperature. You should also expect a thermometer to hold its reading long enough for you to read it when you open the oven door to check it. And, of course, its markings and divisions need to be clear and readable.

Recommendations

Many people use an oven thermometer only occasionally, to measure the average temperature of their oven. When you do that, place the thermometer as near the center of the oven as possible. Set the oven for the temperature you use most often and let it run for at least 15 minutes. Take a reading and record the temperature. Take several additional readings at 10- to 15-minute intervals, then average the results. Compare the average with your thermostat setting. If the oven temperature is incorrect, you can make a mark on the thermostat dial to remind yourself to set the oven higher or lower than its dial indicates. You may prefer calling a service technician to recalibrate the oven.

If you cook on both top and bottom shelves of the oven, you may want to check temperatures in both places. It's common to find a considerable difference in temperature between the two. If the two readings aren't the same, there isn't much you can do about it except change the length of cooking time to compensate for the difference.

If you leave your thermometer in the oven while cooking, it will probably get dirty and become more difficult to read. But a thermometer's divisions and markings aren't likely to be affected by occasional scrubbing with a nonabrasive cleanser; such a scrubbing should be adequate in most cases. Do not soak the thermometer, because the water could leak inside and destroy the vital parts.

Don't put the thermometer in the dishwasher or leave it in the oven during the self-cleaning cycle. Either action could destroy the thermometer, or at least ruin its accuracy.

Ratings of dial-type oven thermometers

(As published in a February 1985 report.)

Listed in order of estimated overall quality, based on accuracy, response time, and readability. Models of comparable quality are bracketed and listed alphabetically. Prices are suggested retail.

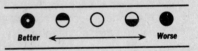

Better ←——————→ Worse

Brand and model	Price	Accuracy	Stability	Readability	Dial-division intervals (degrees)
⊘ SPRINGFIELD 103	$2.29	◑	◑	◑	10 F, 5 C
ACU-RITE 00820	1.98	◑	◒	◑	10 F, 10 C
COOPER 324	4.59	○	◑	◑	25 F, 10 C
TAYLOR 5931	4.98	○	◑	○	25 F, 25 C
TEL-TRU PS 600	5.00	○	◑	○	50 F
TRU-TEMP 2125	3.70	◒	◑	○	25 F

Mercury thermometers

When shopping for an oven thermometer, you'll find both dial and mercury-in-glass models on the shelves. Mercury thermometers generally offer very accurate readings. But the mercury can pose a health hazard if you break the thermometer.

Although a metal-encased mercury thermometer may be hard to break even when accidentally dropped, it poses a special hazard if left in the oven during the self-cleaning cycle. In such a hot oven, the thermometer is likely to burst and the mercury will vaporize. Mercury vapor is highly toxic. Even at room temperature, spilled mercury is a hazard.

Despite the potential hazard, mercury thermometers remain popular because of their accuracy. But, compared to dial-type thermometers, they are generally more difficult to read and less stable in the face of sudden temperature change.

You can slow the response time of a mercury thermometer by wrapping the bulb in several layers of aluminum foil. But a better choice would be to buy a good dial thermometer.

If you already own a mercury thermometer and don't want to get rid of it, use it with care. Should it ever break, sweep the mercury into a dustpan or pick it up with an eyedropper. Do not use a vacuum cleaner—it will vaporize the mercury and spread it through the room air. Place the collected mercury in a jar or bottle with a lid, and throw it away. Air the room thoroughly.

Refrigerator/freezer thermometers

A refrigerator or a freezer should be kept at the proper temperature. If it's too warm, your food will spoil too soon; if it's too cold, you'll be paying for electricity you don't really need. But most refrigerators and freezers don't have a useful temperature indicator. Usually, there's just a "colder-to-warmer" dial that has no temperature markings.

The solution: buy a refrigerator/freezer thermometer, an item that costs less than $5. Most models are accurate enough for home use.

Some are the familiar liquid-in-glass type: They show temperature changes by the movement of a colored liquid in a glass tube. Others are the dial type, in which the expansion and contraction of a metallic coil moves a pointer around a dial. Whatever the design, most models are marked in both Fahrenheit and Celsius scales, usually in two-degree or five-degree increments.

The best thermometers were accurate to within about one degree Fahrenheit. The next best were almost as good; they were off by only one to two degrees Fahrenheit.

Reading

The liquid-in-glass models are graduated in two-degree increments, and most dial models are marked off in five-degree increments. A lot of markings on a rather tiny scale can be difficult to read. More legible, if slightly less precise, are thermometers with bold markings every ten degrees or so.

All thermometers have markings representing what the manufacturers consider proper temperature ranges for food storage. Those markings are often too broad to be helpful. Use instead Consumers Union's recommendations of 37° for the refrigerator and 0° for the freezer.

Most thermometers are designed to stand at a convenient location on a shelf or to be hung from the shelf. If your refrigerator has glass shelves, you'll have to hang the thermometer from one of the brackets that support the shelf.

In a single-control refrigerator, put the thermometer near the center of the main space and away from any bulky foods. Let the temperature stabilize for several hours during a period when you know the door will not be opened—overnight, for example. Then, take a reading as soon as possible after you open the door. The faster you take the reading, the less the chance that warm room air will affect the thermometer reading. You should adjust the control until you get 37° consistently. If the freezer space then isn't reasonably close to 0°, you may have to readjust until you come close to a consistent 37°/0° combination.

Things may get trickier with a two-control refrigerator-freezer. The controls, though separate, don't work totally independently; a change of setting for one compartment may affect the temperature of the other. Getting a balance of 37° and 0° at the centers of the two compartments is apt to take more successive adjustments than with one-control models.

Recommendations

Legibility is an important consideration. Be sure the temperature markings on the one you choose are easy for you to read.

Ratings of refrigerator/freezer thermometers

(As published in an August 1984 report.)

Listed in groups in order of general average accuracy. Within groups, listed in order of estimated overall quality. Unless otherwise indicated, all are graduated in both Fahrenheit and Celsius scales, and can either be hung from a shelf or stood upright. Temperature ranges given are typical; with some models, not all samples may have the same range. Prices are suggested retail.

Brand and model	Price	Type	Temperature range (°F)	Comments
■ *Average error: approx. 1 degree Fahrenheit*				
OHIO 489	$3.85	Liquid	−30 to 70	A,C,E,F
TRU-TEMP 2126	3.50	Dial	−30 to 70	D,E
COOPER 335	4.59	Liquid	−40 to 80	—
SYBRON/TAYLOR 5925	4.98	Liquid	−40 to 80	—
■ *Average error: between 1 and 2 degrees Fahrenheit.*				
SYBRON/TAYLOR 5923	4.98	Dial	−30 to 70	D
SPRINGFIELD 203	2.29	Dial	−30 to 90	—
COOPER 325	4.59	Dial	−20 to 80	—
■ *Average error: between 2 and 3 degrees Fahrenheit.*				
TEL-TRU CR450	5.00	Dial	−25 to 85	A,D
ACU-RITE 01432	2.29	Liquid	−54 to 120	C,F
■ *Average error: between 3 and 4 degrees Fahrenheit.*				
ACU-RITE 00830	2.19	Dial	−40 to 140	B,C

Thermostats

A "setback" thermostat lowers (sets back) the house temperature at certain times of day. The object is to save energy in the winter, with your heating plant—or in summer, with an air-conditioning system. The tried-and-true setback thermostat is the electromechanical type. It uses a clock-timer to change from one temperature setting to another at the times you designate. You typically choose the high and low temperatures you want, then set the high and low cycles by positioning tiny pins on the clock-timer's face.

Electromechanical setback thermostats have been convenient and generally reliable over years of use. Unfortunately, that's not true of the newer breed of energy-saving thermostats—computerized models that can be programmed to control the heating system.

Consumers Union planned a report on computerized thermostats, but decided to forgo full testing after running into serious problems with them.

Computer bugs

A thermostat is essentially a switch in the heating system's low-voltage electrical control circuit. The thermostat is "off" when the house is warm enough; it's "on," telling the furnace to run, when the house needs heat.

Most of the computerized thermostats Consumers Union engineers looked at use the low voltage not only to control the furnace but also to ensure that their electronic clock maintains the correct time. Ironically, that special use of the control circuit often renders the thermostat's temperature-setback function useless.

In many heating systems, the voltage in the control circuit is interrupted during part of the furnace's cycle, cutting off power to the thermostat. When that happens to many computerized models, the clock fails to keep the correct time; it may lose or gain several minutes whenever the furnace cycles on or off. Consequently, the setback cycles you've so carefully programmed will go awry.

That problem seems more likely to occur in homes that have hot-water heat because such heating systems tend to have more complex controls than do forced-air systems. But the problem can arise with any brand of heating system. Similar problems may occur if you use a computerized thermostat to control central air conditioning.

And there's more. With an electromechanical setback thermostat that draws its power from a battery, it doesn't matter if the low voltage is on or off for part of the cycle. But a computerized thermostat that uses the control circuit for timing must have that low voltage at all times. You must buy the appropriate thermostat for your system, or hire a contractor to modify parts of the control circuit.

Unfortunately, there's no simple way a homeowner can tell if a given computerized thermostat will actually work with a given heating system. Even if you can determine that the thermostat you've chosen is compatible with the heating system's control box, some other part of the system may cause the clock to misbehave.

For all their supposed versatility— the ability to handle several setback cycles each day, or a different set of cycles for each day of the week—computerized thermostats lack flexibility in one key area. They may let the house warm up or cool down too much before turning the heat on or off.

In practice, a thermostat seldom shifts at precisely the temperature you've selected. But most conventional thermostats have what's known as an adjustable anticipator, a component that allows them to switch within, say, one-half degree above or below the temperature you want. A temperature swing that small is barely perceptible.

Computerized thermostats commonly have an electronic temperature sensor that's nonadjustable; it typically controls only to within two degrees above or below the chosen temperature. That four-degree "deadband" can lead to some unacceptably wide temperature swings.

Learning the program

A computerized thermostat can be quite complex to program.

On most computerized thermostats, changing the program means pushing the proper set of buttons in the proper order. The sequence of button-pressing often varies, depending on what you want the thermostat to do. Instructions printed on the thermostats themselves are often too skimpy to be useful.

Some computerized models are preprogrammed with a schedule of setback times and temperatures. That sounds convenient, but it's mostly a gimmick. If the preset schedule doesn't suit your needs, you'll have to change the program.

Conventional thermostats

Traditional setback thermostats have few of the shortcomings of their computerized cousins. They will work just fine with almost any oil- or gas-fired heating system (but not with an electric heating system or a heat pump, both of which need a special thermostat). They have an adjustable anticipator to reduce undesirably wide temperature swings. Most units are designed only for heating systems, but versions for central heating/cooling systems are available. Prices range from $50 to $168.

All the thermostats work, and no one brand works better or worse than another.

Setting the temperatures. On most, you move two levers to select the "normal" and the "setback" temperature you desire. Some work a little differently: You can set any normal temperature you want, but the setback temperature can be no more than 10 degrees below the normal setting. That's an undesirable limitation, though some people may not object to it.

Setting the cycles. Thermostats usually have a small clock wheel with slots around the rim for removable pins. You determine when the house will be warm or cool by shifting pins.

All let you set at least two setback cycles per day; most models allow you three.

Overriding the settings. You'd change to an alternate temperature setting in the middle of a cycle if, for example, you decide to stay up late one night and don't want to be rattling around in a chilly house. The override is also useful on weekends, when you may want to keep the house warmer for longer periods than on weekdays.

Installation. Putting a new thermostat in may well pose some problems for do-it-yourselfers. Some anticipators can handle only one ampere of current, which is less than some furnace controllers need. Also, if the furnace control circuit should short-circuit, the anticipator would act as a fuse and burn out. End of thermostat.

If you are considering a new thermostat, first check your existing thermostat or the furnace control box. Look for a label that tells you what the anticipator setting (in amps) is. If it's one amp or more, you'll have to use either a **Honeywell** or **Autostat** add-on.

Ratings of thermostats

(As published in an October 1985 report.)

Listed in order of estimated overall quality, based on suitability for do-it-yourself installation and ease of use. Prices are mfr. suggested retail; + indicates shipping is extra.

Brand and model	Price	No. cycles per day	Minimum setting interval	Comments
EMERSON 1F76-353	$60	Up to 3	12 min.	C
SEARS CAT. NO. 9101	50+	Up to 3	12	A,C,H
ROBERTSHAW T32-1042	60	1 or 2	12	A,B,C,E,G
ROBERTSHAW T33-1042	60	1 or 2	12	A,B,C,E
HONEYWELL T8085A	140	Up to 3	30	D
HONEYWELL T8082A	168	Up to 3	30	F

SPECIFICATIONS AND FEATURES

All: ● Have adjustable anticipator. ● Control same program each day. ● Control 24-volt circuits.

Except as noted, all: ● Have cycle-change switch. ● Have independent high and low temperature settings. ● Use replaceable alkaline battery. ● Have half-cycle capability. ● Are for heating systems only; essentially similar versions are available for heating and cooling.

KEY TO COMMENTS

A – Has no cycle-change switch.
B – Only high-temperature setting is fully adjustable; setback temperature can be no

more than 10 degrees below high setting.
C – Current capacity limited to one amp. Unit is vulnerable to destruction if control circuit shorts.
D – Has no battery. Requires third wire to furnace in order to power clock. Clock will lose time during power outages.
E – Has no half-cycle capacity.
F – Uses built-in nickel-cadmium battery, which has unpredictable service life and can't be easily replaced by owner.
G – Uses replaceable nickel-cadmium battery.
H – Heating/cooling version.

Recommendations

You don't need a setback thermostat to save energy. The one you have now will do just fine, provided you diligently turn it down each day and provided you don't mind a chilly house in the morning.

A setback thermostat can be set to raise the temperature before you wake up, sparing you some discomfort. A setback thermostat is also consistent and dependable, keeping the house at the temperatures you set day after day.

If you want the advantages that a setback unit offers but don't want to replace the thermostat you now have, consider the $50 **First Alert Autostat** discussed on pages 274–75. It offers much of the flexibility promised by computerized thermostats, but none of their problems.

A thermostat add-on

You don't need to buy a new thermostat to gain the convenience and dependability of a setback model. Instead, you can hang the $50 **First Alert Autostat** on the wall next to the existing thermostat.

The **Autostat** is essentially a motor-driven hand, controlled by a computer chip, that reaches over to adjust the lever on the existing thermostat. The **Autostat AS555** is for rectangular thermostats; the **AS550** is for round thermostats.

The **Autostat** offers the same kind of flexible programming that other computerized thermostats do—up to four temperature changes per day, separate programs for weekends, and half-cycle capability. It's also easier to program than many other computerized thermostats. Nonetheless, setting up the unit so the thermostat will produce the temperatures you want involves some trial and error.

Because of play in the lever-moving mechanism, the **Autostat** must be adjusted carefully so it doesn't overshoot the mark.

The **Autostat's** biggest drawback is noise: its little motor whirrs for a few seconds whenever it changes the temperature setting. That may be disconcerting to people who expect a thermostat to sit silently on the wall.

Overall, the **Autostat** is a worthwhile alternative if you want the advantages of a computerized setback thermostat but none of its timing or compatibility problems.

Timers

A kitchen timer keeps track of work in progress. It tells you when to take a cake out of the oven or when to release a child from piano practice. A cycling timer *does* things for you. It can turn a light on and off, start the morning coffee, or switch on the air conditioner a half hour before you get home from work. A whole-house timer does the work of a number of cycling timers and permits remote control of lights and appliances programmed to it.

The various types are quite different in construction and function.

Cycling timers

Most cycling timers are, in effect, a 24-hour electric clock, with graduations marking 12 daytime and 12 nighttime hours. On first use, you set tabs or pins to the times you want a lamp or appliance to go on and off. Then you set the clock, rotating its dial to the correct time of day. Plug the appliance you want cycled into the timer, and the timer takes over. The timer can control plant lights, turn on the coffee percolator, or light up the house for your return after dark. Indeed, the thrust of promotion for these timers seems to be for home protection. The idea is that lights or radios going on and off in an empty house may deter would-be intruders. Lights that turn on at dusk also provide you with the security of coming home to a lit house rather than a dark one.

Cordless models plug directly into a wall outlet. They're reasonably compact but are none too convenient if plugged into an outlet that's near the floor or blocked by furniture.

Table models, with cords about four feet long, don't have that drawback. But then, they take up table or counter space you might have better use for.

Accuracy

Electric clockwork is inherently accurate—the mechanisms are pegged to the cycling of household alternating current and usually vary little. If set to switch at the same time each day, the switching action shouldn't vary by more than a minute of the same time each day (unless the unit has a built in management to vary its on/off times, mimicking human caprice in a way meant to deter burglars).

Most timers have either 15-minute or one-hour setting increments. A few have half-hour increments. The smaller the increment, the more exact the setting.

Features

Some cycling timers are easier to live with than others:

Setting. It's nice to have an easy-to-read digital display and push buttons atop the unit.

It's also convenient to have on and off tabs that you move to position on the timer's clock face, or a ring of pins that you merely press opposite the on and off times you want.

It's inconvenient if you have to use a pen or other pointed tool to move the pins, if you have to remove and replace loose pins, or take off the dial to remove and reset the pins.

Cycles. A number of timers can be set for only one on/off cycle per day. But that's enough if you just want light turned on over plants or before you arrive home from work. Others can provide multiple cycles. With those, you can have a light in a bathroom or kitchen turn on and off several times in an evening, to give the impression someone is home.

A few timers can be set for only a half cycle. They'll just shut off that laundry-room light that everyone forgets, or turn on an entryway light that you shut off when you get home.

Cycle override. If you want to use a timer-controlled device—a lamp, say—at other than the programmed time, a "one-time" override switch allows you to turn a lamp on and off earlier than the scheduled time without interfering with subsequent cycles. Say a timer blacks out a light while you're still reading. You then merely work the timer's override switch as you would ordinarily work the lamp switch: Switch on to turn the lamp back on; switch off to turn the lamp off and get the cycle back on schedule. Should you forget to turn the switch off, most timers will do it for you at the next programmed off time.

A "permanent" override switch, a feature on some models, completely bypasses the preset cycling mode while still maintaining the correct time on the timer clock. It allows you to turn a lamp on or off manually at any time. But to get the cycling mode working again, you have to remember to turn off the permanent override at some later time.

Either kind of override switch, then, has advantages and drawbacks. The handiest arrangement is to have both kinds of override available, as a few timers do.

Electrical considerations. Most appliances today have polarized plugs: One prong is wider than the other, to fit into modern polarized outlets in household circuits. Polarized plugs and outlets reduce the risk of electrical shock; the plug can be inserted into an outlet only one way. However, that plug design prevents you from reversing the

Ratings of cycling timers

(As published in an August 1983 report.)

Listed in order of estimated overall quality. Differences between closely ranked models were slight. Bracketed models were judged equal in overall quality. In columns, ✓ means yes and ✗ means no. Prices are suggested retail; + indicates shipping is extra.

Legend: Better ○ ◐ ● Worse

Brand and model	Price	Ease of reading			Cycle override		Setting interval (min.)	Multiple cycles	Variable cycles	Comments
		Present time	On/off	One-time	permanent					
⊘ RADIO SHACK MICRONTA DIGITAL 63886	$30	◐	○	✗	✓		1	✗	✗	A,B,C,H,I
INGRAHAM TIMEMASTER 24-HR. 12047B	10	○	○	✗	✓		60	✓	✗	I
INGRAHAM TIMEMASTER MULTIPLE PROGRAM 12050	15	○	○	✗	✓		60	✓	✗	A,I
AMF/PARAGON SUPER TIME COMMAND 48 C296	30	○	○	✗	✓		15	✓	✗	A,E
INTERMATIC TIME-ALL EB11	19	◐	●	✓	✓		30	✓	✗	A,F,G,M
RADIO SHACK MICRONTA MULTI-PROGRAM 63864	12	◐	○	✓	✓		60	✓	✗	—
SEARS PROGRAMMABLE Cat. No. 6666	12+	◐	○	✓	✓		60	✓	✗	—
INTERMATIC SUPERCOP D711	13	○	◐	✓	✗		15	✗	✓	—
INTERMATIC MASTER CONTROL D811	15	◐	◐	✓	✓		60	✓	✗	A,G
RADIO SHACK MICRONTA MULTI-PROGRAM 63865	16	◐	◐	✓	✓		60	✓	✗	A,G
INTERMATIC MASTER CONTROL II ED811	18	◐	◐	✓	✓		60	✓	✗	A

Brand and model	Price	Ease of reading — Present time On/off	Ease of reading — On/off	Cycle override — One-time	Cycle override — permanent	Cycle override — Setting interval (min.)	Multiple cycles	Variable cycles	Comments
TIMEX VARIABLE 8511	13	◐	○	X	✓	15	X	✓	K
RADIO SHACK MICRONTA 63862	9	○	○	✓	✓	15	X	X	—
INTERMATIC TIME ALL D111	11	○	○	✓	✓	15	X	X	—
AMF/PARAGON TIME COMMAND 12 C124	15	◐	◐	✓	X	30	✓	X	F,G
SEARS VARIABLE Cat. No. 6640	10+	○	○	✓	X	15	X	✓	—
WESTCLOX MULTI PROGRAM 5201	11	◐	◐	①	✓	30	✓	X	D,F,J
TIMEX 8501	11	○	○	X	✓	15	X	X	K
AMF/PARAGON TIME COMMAND C102	11	●	◐	✓	X	30	X	X	F,G,L
INGRAHAM DO-ALL 12010	8	◐	○	X	✓	15	X	X	I

① Override can put on/off cycles out of phase.

SPECIFICATIONS AND FEATURES

All: ● Have fixed on/off times. ● Handle up to 1,875 watts of appliance load.

Except as noted, all: ● Are cordless plug-in models. ● Are electromechanical. ● Have polarized plugs and polarized appliance receptacle. ● Can accept only 2-wire appliances. ● Have captive pins or tabs for setting on/off times. ● Handle up to 1,250 watts of incandescent bulbs.

KEY TO COMMENTS

A – Table-top model.
B – Electronic.
C – Has battery backup, an advantage.
D – Has unpolarized plug and receptacle.
E – Has 3-wire plug; will accept 3-wire appliances, including 120-volt air conditioners.
F – Has loose pins for setting on/off times, a disadvantage.
G – Can be set for half cycles.
H – On/off times set electronically; does not use pins, an advantage.
I – Handles only 600 watts of incandescent bulbs.
J – Handles only 840 watts of incandescent bulbs.
K – Handles only 875 watts of incandescent bulbs.
L – Handles only 1,000 watts of incandescent bulbs.
M – Handles 1,500 watts of incandescent bulbs.

orientation of most of the cordless timers to clear a piece of furniture, a corner, or other obstruction.

Timers also differ in the amount of electricity they can handle. Most are rated to cope with up to 1,875 watts of "appliance" load (the kind imposed by a coffeepot or toaster), but only 1,250 watts of the load imposed by incandescent bulbs (which draw more than their rated wattage when first turned on).

Recommendations

If you're in the market for a number of timers, check the discussion of a central system, on page 279. A central system offers some special conveniences—at rather specialized prices.

With a central system, you can reset all your timed gadgets from the command post. When daylight saving time ends, you don't have to make the rounds of the house to reset individual timers.

Timing for the whole house

If you want to control multiple devices, all over the house, you can buy individual timers for each room, or you could consider a central system that gives you independent, remote control. For $130, for instance, the **BSR System X10** gives you a plug-in timer, with digital display, that you can program to send on/off signals over your house circuits to two individual modules. The lamps or appliances you want to control merely plug into the modules. Those modules, in turn, plug into your wall outlet. (Others are also available in versions that wire directly into household circuitry.) Working a lamp's switch lets you override its controlling module—to turn a light on, say, when the module says "off."

To get independent control of several lamps, you set the corresponding module to a unit code that matches one on the timer. There's no limit to the modules you can use on any one unit code.

The **BSR** timer controller can handle eight unit codes. (If you need 16 unit codes, you buy a second timer.) The timer will also let you turn modules on and off by hand, but not easily. You can use the timer, by hand or automatically, to "sweep" the house, before you go to bed, of lights left on accidentally. If you don't want a timer but do want to turn distant lights on and off by hand easily, or dim them, you buy a basic controller without the timer function (that basic box handles 16 unit codes). If you want timing plus remote manual control, you buy a timer and a basic controller.

The system is accurate in turning lights on and off at preset times. (You can also set it for variable cycles.) It can provide two on/off cycles for each module in a 24-hour period. It has a back-up battery so that a power failure won't throw its timing off.

Each component of the **BSR** system can be bought separately. A basic controller, **014301**, lists at $50; timer **TC262**, $90. There's a plug-in module with a dimmer function for incandescent lamps (**0145011**, $23) and one without a dimmer for appliances and fluorescent lights (**AM286**, $23). There's a wire-in wall-switch module (**014711**, $23). And there are many more, for specialized uses such as controlling an air conditioner, a thermostat "fooler," or the like. There are also special controllers to handle phone or computer commands, and so on.

The **BSR** system is also compatible with one marketed by **Leviton**, so mod-ules of that brand can be used, too. One such is a wire-in wall socket module (**Leviton 6227**, $42), with one controlled and one uncontrolled outlet.

Note: Installing a wire-in switch or outlet module is no more complex than replacing a conventional switch or outlet. But it's not a job for the casual do-it-yourselfer. Many jurisdictions require that such work be done, or at least inspected, by a licensed electrician.

Kitchen timers

You don't need split-second accuracy for many cooking chores. But some folks are finicky about their three-minute eggs; and a good timer shouldn't be off by more than 10 seconds in a three-minute span. Foods that need longer cooking are less sensitive. For something that takes 45 minutes to cook, a half-minute's error is acceptable.

Accuracy, however, is more than just a matter of how a timer's mechanism runs. You also have to be able to read the dial exactly and to set the dial with some precision to the amount of time you want.

If you often time long cooking periods, consider the span you can set, too. Mechanical timers are usually limited to a maximum of 60 minutes. To use them when cooking a roast, you'd have to reset the timer after the first hour. An electronic model is more versatile.

Even an accurate timer will allow dinner to burn if you can't hear it signal. Be sure the signal is loud enough to be heard from an adjoining room.

The electronic **Triple Timer** will beep for up to a minute. An electronic model that beeps until you turn it off has an advantage: it's not that loud, and might otherwise be missed in a noisy home. But it can also be a problem. If it's battery-powered and goes off when nobody's home, it will run its batteries down. And the insistent beeping could be annoying if you're on the telephone, say, and want to ignore the signal for a while.

Convenience

Even an accurate timer won't be satisfactory if it isn't handy to use:

Setting. A timer's dial, if it has one, should turn smoothly and allow a reasonably good grip. It's also nice if the timer can be set with one hand without slipping on a smooth tabletop.

Sometimes it's possible to bear down on a mechanical timer, or wedge it against another countertop object, to help you set it one-handed. But in general, setting a mechanical timer is best done with two hands.

An electronic timer is, in contrast, generally easy to set with just one hand.

Ratings of kitchen timers

(As published in an August 1983 report.)

Listed in order of estimated overall quality; differences between closely ranked models were slight. Bracketed models were judged equal in overall quality and are listed in order of increasing price. Except as noted, all are wind-up timers with maximum time setting of 60 min. Prices are suggested retail rounded to nearest dollar; discounts may be available.

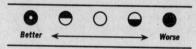

Better ←→ Worse

	Price	Accuracy	Audibility	Ease of setting — 1 hand	Ease of setting — 2 hands	Comments
PRESTO 04200, A Best Buy	$14	◉	◖	○	◖	—
AMERICAN FAMILY TWO TIMER 1400	30	◉	○	◖	◖	A,C,D,G
WEST BEND KITCHEN CONTROLLER 40003	80	◉	○	◖	◖	E
MIRRO QUARTZ M0304	22	◖	○	◖	◖	B
ROBERTSHAW LUX MINUTE MINDER 2428	11	◖	○	◖	◖	—
WEST BEND TRIPLE TIMER 40000	34	◉	○	◗	◗	A,C,E
INTERMATIC TIME KEEPER R10	14	◖	◖	●	◖	—
MIRRO SINGLE RING M0302	12	○	○	●	◖	—
MARK TIME BELL TIMER	9	●	◖	◗	◖	—
MEMATIC MINUTE TIMER	16	●	○	○	◖	—
AMERICAN FAMILY BIG TIMER 1000	11	◖	○	◖	◖	F,G

KEY TO COMMENTS
A – Electronic; battery-powered.
B – Mechanical; battery-powered, quartz-timed.
C – Will time up to 10 hr.
D – Will time up, like a stopwatch, as well as down.
E – Contains 3 timers; will time up to 3 separate foods.
F – Judged much less durable than most.
G – Can be wall-mounted.

Toasters

The pop-up toaster hasn't changed much over the years. Toasters are commonly available in two- or four-slice versions. Some four-slice models have the slots arranged side by side, with separate controls for each pair of slots, while others have two pairs of slots placed end to end.

With few exceptions, toasters toast quite well, but their price tags do vary considerably.

Toasting

Consumers Union's toaster testers toasted hundreds of slices of white bread three different ways: with a single slice; with a full capacity of either two or four slices; and with six batches, one after another. All the toasters turned out a good range of light to dark toast. But on the single-slice test, four-slice toasters with slots placed end to end didn't toast the slices quite as well as the other types of toasters.

Consecutive batches were tested because once you have set a toaster's controls, it should be able to produce batch after batch of toast to the same doneness. The toasters did that quite well, too.

As for toasting time, two-slice models average less than a minute and a half to toast both slices. Four-slice toasters take about two minutes to toast a full load.

Whole-wheat bread and English muffins take longer to toast than ordinary white bread. Both require a darker color setting. Depending on freshness, you might have to send an English muffin through a second cycle to get it toasted.

Many toasters do much the same job, so you may want to select on the basis of price. Of course, there are various controls, handles, and appearances. But the importance of these things is easy to judge for yourself.

For example, the controls of most toasters are mounted on the end. A few have the controls on the side. Which position is best for you depends on where you put your toaster.

Most toasters have a sliding lever to let you choose light or dark toast, but a rotary knob can be easier to adjust. Some sliding levers can be so stiff that it takes two hands to make an adjustment—one hand to hold the toaster and the other to move the lever. Like other small appliances that generate heat, toasters draw a fairly large amount of current. Most two-slice models use $6\frac{1}{2}$ to 9 amps. Any of these can be used on the usual 15-amp branch circuit with other low-wattage items, such as lights. Four-slice models draw $11\frac{1}{2}$ to 14 amps; they should not be used at the same time as heating appliances plugged into the same 15-amp circuit.

Safety

What toasts the toast is electric current passing through electrically live wires. Modern toasters all use a double-pole electric switch, which ensures that the heating elements are not live when the toaster is off but still plugged in. Nevertheless, you still should unplug a turned-off toaster before poking anything but bread into it and before cleaning crumbs out of the bottom, where there is additional electric wiring.

Recommendations

If all you want from your toaster is a couple of good pieces of toast, just about any two-slice model with features and appearance that appeal to you should do the job just fine. Shop for price.

For cooking larger quantities of toast, a four-slice model with the slots side by side is the toaster of choice. It's more flexible than the other kind of four-slice toaster because it has separate controls for each pair of toast slots. Here, too, it's all right to shop for price.

Before deciding definitely on buying an ordinary toaster, consider that a toaster oven (below) offers a great deal more. In addition to toasting oversize breads, it heats frozen foods, bakes potatoes and meat loaf, warms rolls, and grills cheese sandwiches. For still more money—and more utility—you should think about a toaster oven-broiler (page 284).

Toaster ovens

A toaster oven is a lot more versatile than a toaster. But it doesn't work as a broiler. Another appliance, the toaster oven–broiler (see page 286), has all three functions. Although the toaster oven is less expensive, it's also more limited—though it can heat some frozen foods, bake potatoes and meat loaf, warm rolls, melt cheese on sandwiches, and toast bread.

While toaster ovens are a large part of the toasting-machine market, there aren't many to choose from. All work well as small countertop ovens, but perhaps not as well as bread toasters.

Cooking

You can heat a variety of frozen items in the toaster ovens—pizzas and pot pies, turnovers and TV dinners—with good results. The times required to cook these foods properly can vary from the instructions on the package, but that happens in many full-sized ovens, too.

Controls. The controls on toaster ovens are as good as or better than those on toasters. With some larger toaster ovens, however, it is possible to have both toaster and oven modes on simultaneously, since there are separate controls for each function. The disadvantage is that if you stick in a slice of bread to make toast and forget the oven's on, you could end up with burned toast.

Recommendations

A toaster oven is more versatile than a plain toaster. (It is also much larger, something to consider if you're short on counter space.) Yet a toaster oven is not as versatile as a toaster oven–broiler, which costs slightly more.

Toaster oven–broilers

(Toaster oven–broilers and broiler ovens)

Every portable cooking appliance costs money and takes up space. It makes sense, then, to look for one that fills as many needs as possible. If you are interested only in toasting bread, a plain toaster should fill the bill (page 282). If you want to heat frozen foods, bake potatoes and meat loaf, warm rolls, and grill cheese sandwiches too, a toaster oven is probably the appliance to head for (page 283).

How about a unit that can work as a broiler as well? A toaster oven–broiler may offer the versatility you need. Perhaps you merely need a roomy auxiliary oven, and flexibility is secondary. The answer then may be a broiler oven. In short, you need to set some priorities before you start shopping.

Cooking

Here's what Consumers Union testers found when they tried these appliances at cooking chores:

Broiling. In each model, and in the oven of a conventional electric range, they broiled six quarter-pound ground-round patties. The range oven came closest to their expectations, turning out hamburgers that were crisp and well done on the outside, rare to medium rare inside. The best toaster oven–broilers did almost as well, but the remainder broiled unevenly.

Cooking times with toaster oven–broilers ranged from about 19½ minutes for the slow models to 13 minutes for fast ones.

CU testers also broiled one-inch-thick steaks in several of the ovens. The steaks emerged well done both inside and out. Thicker steaks came out better: well done and crisp outside, rare inside.

Baking. All the ovens matched the regular range in baking four half-pound potatoes, which came out with crisp skins and well-cooked interiors.

Cake layers were a different story when baked in each oven that could take an eight-inch cake pan. In general, the broiler ovens did better with layers of white-cake mix than the toaster oven–broilers, with the largest broiler ovens giving the best results of all. When the testers followed the temperature recommendations in the instructions for the toaster oven–broilers, the layers often came out too light or too dark, or were cracked or striated. So they experimented wtih higher or lower settings. Results still were not top-notch. At best, the toaster oven–broilers were only fair to good at baking cake. The majority of the broiler ovens were judged good at the chore.

Toasting. Toaster oven–broilers have toast-color controls that can be set to produce toast in a range from barely browned to almost burned. By and large, the toasting models gave four slices of toast of the same color again and again.

Nevertheless, if toasting bread is your main planned use for one of these appliances, you'd do better with a regular toaster; a good two- or four-slot model should certainly cost considerably less than a toasting oven. The heating elements of toaster oven–broilers (and ordinary toaster ovens, for that matter) are set at some distance from the bread; they tend to dry out the bread a bit before the surface reaches the proper color. On the other hand, the little ovens are handier for items such as English muffins or rolls, which are often too plump to fit a standard toaster.

Rotisserie. A few broiler ovens offer a special feature, a motor-driven skewer that can be used for rotisserie cooking. You first impale the bird or roast on the skewer and lock it in place with thumb-screw-tightened pronged clamps at either end. You then may balance the loaded skewer on a two-tined cradle and position the skewer in a pair of notches inside the oven. The cradle makes removing the cooked food safe and convenient. In another design you merely hold the skewer by its wooden handle, insert its pointed end into a receptacle on one side of the oven's interior, and push the other side of the skewer into position through a large fore-and-aft gap in the side wall, leaving the handle protruding from the cabinet.

Both arrangements are easy to use. In about an hour you'll have cooked a 2½- to 3-pound broiling chicken. The main limitation with such models is the size of the item you can cook: no bigger than about six to eight inches in diameter. But if you're really interested in countertop rotisserie cooking, there's an even better choice—an "open-hearth" broiler, discussed on page 38.

Slow cooking. In regular operation, these appliances cycle on and off. Set for, say, 250°F, they will heat to a bit hotter than that, shut off for a while, then reheat. The result, if the thermostat is reasonably accurate, will be an average temperature in the neighborhood of 250°. Some recipes requiring liquid and long-term cooking will, however, do better with steady, low heat. To handle that problem, there may be a special slow-cook setting that lets the elements run continuously at a low level.

While you may be able to get pretty good results in an oven without the slow-cook feature, it's not a particularly good idea to try long-term cooking in such an oven. The thermostat can be tricky to adjust and the oven temperature may cycle in a way that could leave food for appreciable periods at temperatures below 140°, thus spurring bacterial growth.

Other factors

Set side by side on a store's shelf, these boxy appliances often look much alike. Looks are deceptive, however. Even within type, some ovens are easier to live with than others.

Size and capacity. If your kitchen is small or your counter space limited, you may find that the size of these appliances will limit their usefulness. Even the smallest is about 15 inches wide and 10 inches deep with the door closed. The larger ones take up about 21 by 14 inches of space. Check the dimensions of the unit you want to buy to be sure you have room for it.

It's also worth weighing internal size against the cooking uses you have in mind. A relatively roomy appliance, of course, can bake or broil more food at a time than a smaller one. While the plain broiler ovens tend to be the roomier units, internal accommodations can vary substantially.

Most of the very smallest-capacity

models are toaster oven–broilers. Those models can hold six hamburgers, but they won't take items thicker than four inches or a single eight-inch cake pan. You can, however, warm even a large (10¾-by-7-by-1-inch) TV dinner in them.

Models of midsize or larger can hold a cake pan or eight hamburgers.

Broiling pans and shelves. Shelves are retained in grooves or on projections along the inside walls. All models have a broiling pan to hold the fats and juices rendered when you broil meat. In most, the food rests on a rack or a slotted metal plate that fits over the drip pan.

Most toaster oven–broilers come with one shelf, in one position, but in a few you can raise or lower the shelf a notch. Or there may be an extra shelf and shelf position to allow heating of an extra shallow dish, such as a TV dinner. Most toaster oven–broilers have a guard on the rear of their broiler tray that keeps food from being dumped off at the back.

Broiler ovens typically offer a choice of two positions for their single shelf. Sometimes there's a removable handle that clips onto the shelf or pan, making it easier to take the food out. Otherwise, you'll need a kitchen mitt to protect against burns.

Controls. Typically, you activate the heating elements—the bottom one for baking, the top one for broiling—with a dial or a set of push buttons. That broil/bake selector is often combined with a "cooking guide," a chart on the unit that gives recommended temperature settings for cooking. On most models, the selector includes an off position; otherwise, there's a separate on/off switch. The thermostat is usually calibrated in twenty-five- or fifty-degree intervals and easy to use.

Sometimes there's a timer. The best ones provide automatic shutoff for baking and broiling, but don't force you to use the shutoff. And they can also be used by themselves as a kitchen timer. Toasting models, of course, shut off when the toast is done, and they give you an audible signal of the fact as well.

Cleaning. Typically, there's a "continuous-clean" interior that's supposed to rid itself of grease and grime at normal cooking temperatures. Consumers Union's testers noticed no such effect during the course of their performance tests. Because of its dull, usually mottled appearance, a continuous-clean finish may look cleaner longer than an ordinary finish does. But it's no substitute for an occasional scrub-down by hand. In the long run, such a finish may prove something of a disadvantage, since its rough, soft surface eventually makes cleaning very difficult.

Getting the crumbs out of the toasting models can be more of a problem too. The least handy in this respect lack a removable crumb tray; you have to sponge crumbs out or upend the entire appliance to dump them. Some others have a handier setup, a tray that opens from underneath. Better yet are trays that slide out from the front.

Door. Broiler ovens are made with glass doors that come off for cleaning. But only a few toaster oven–broilers have a detachable door. On most of the broiler ovens, the door fits poorly and is a nuisance to open and close because the hook-hinges fit so imprecisely. Most of the toaster oven–broilers and all the broiler ovens have a detent or enough friction to allow the door to be left ajar during broiling. (The slightly open door lets hot air escape; that makes for better radiant cooking.) A number of toaster oven–broilers offer special conveniences: some ovens turn themselves off when you open the door, and others move their shelf forward when the door opens.

Ratings of toaster oven–broilers and broiler ovens

(As published in a February 1982 report.)

Listed by types; within types, listed in order of estimated overall quality. Bracketed models were judged about equal in overall quality and are listed alphabetically. Prices are list; + indicates shipping is extra.

◉	◓	○	◒	●
Smaller ←				→ **Larger**

Brand and model	Price	External dimensions (HxWxD) (in.)	Relative capacity	Advantages	Disadvantages
Toaster oven–broilers					
GENERAL ELECTRIC CT02000	$110	9½x16¾x13	○	A,B,D,E,I,J,K,M,N	a
SUNBEAM 20560	64	8¼x16x9½	●	A,C,G,J,K,L,N	b,c,d,e,g
GENERAL ELECTRIC T26	69	7½x15½x10	●	A,G,L	f,g,q
GENERAL ELECTRIC T131	80	7½x15½x10	●	A,G,L	f,g,q
PROCTOR-SILEX 0220AL	45	7½x15¾x10¼	◒	C,F,L	c,f
PROCTOR-SILEX 0221AL	50	7½x15¾x10¼	◒	C,F,L	c,f
PROCTOR-SILEX 0235W	70	7½x15¾x10¼	◒	C,F,I,L,N	c,f
NORELCO TO4500	74	8¼x15¾x11½	◒	A,N,O	a,h,i,p
TOASTMASTER 340	88	8½x17¼x11½	◒	E,K,N	a,b,c,h, k,m,n
TOASTMASTER 320	67	8½x15¼x11½	◒	E,N	a,c,h,k, m,n
Broiler ovens					
MUNSEY BB3	72	8¾x19½x11¼	◒	H	b,c,e
MUNSEY 7650	82	11x21¼x12¼	○	H	b
SEARS Cat. No. 6921	57	9¾x20¼x13	◓	—	b,f,j,l
TOASTMASTER 5242	82	9¼x20¾x13¼	◓	—	f,j,l
TOASTMASTER 7009	125	12¾x21½x13¾	◉	H,I,N	h,l,o

SPECIFICATIONS AND FEATURES
All have: • Door with glass window that can become hot enough to burn. • Removable drip pan for broiling.

Except as noted all: • Broiled satisfactorily. • Have a signal light. • Have a "continuous-clean" interior. • Have handles at the sides (all handles stayed relatively cool). • Have a

thermostat marked in degrees F. ● Have a partly open door position for broiling. ● Have 1 shelf. ● Have a tray or rack that fits into drip pan. ● Have a 3- to 4-ft. power cord.
All toaster oven–broilers: ● Have a toast switch and a toast-color control. ● Toasted satisfactorily, but did not bake cake layers as well as most broiler ovens. ● Chime or click audibly to signal that toast is done.
Except as noted, all toaster oven–broilers: ● Have a nonremovable door, hinged at bottom. ● Have 1 shelf position. ● Can toast 4 slices of bread.
All broiler ovens: ● Have a removable door.
Except as noted, all broiler ovens: ● Baked cake layers better than the toaster oven–broilers, but not as well as a conventional oven. ● Have 2 shelf positions.

KEY TO ADVANTAGES

A – Elements turn off when door is opened.
B – Back and interior side panels, as well as door, are removable, making interior very accessible for cleaning.
C – Shelf advances when door opens.
D – Has timer with alarm; if used, will turn off unit after baking and broiling; can be used by itself as a kitchen timer.
E – Exterior stayed relatively cool.
F – Crumb tray slides out easily from front.
G – Crumb tray opens from underneath.
H – Convenient door.
 I – Fastest and most energy-efficient model when baking potatoes.
J – Toasted with better consistency and uniformity than most.
K – Broils hamburgers better than most.
L – Backguard on broiler tray keeps food from being pushed off at back.
M – Can toast 6 slices at once.
N – Has more shelf positions than others of its type.
O – Door opens partially after toasting.

KEY TO DISADVANTAGES

a – Lacks removable crumb tray.
b – Lacks signal light.
c – Some samples may have sharp edges.
d – Awkwardly long (5½-ft.) cord.
e – Has rotary toast switch, slightly less convenient than lever type of most others.
f – Thermostat tended to be inaccurate at some bake-temperature settings.
g – Would not hold an 8-in. cake pan.
h – Lacks carrying handles.
 i – Use of on/off push button slides unit unless it's up against a wall.
j – Door is ill-fitting, may come off hinges when opened.
k – Among slowest models when baking potatoes.
 l – Less energy-efficient than most when baking potatoes.
m – Toasted top of bread slices uniformly, but left underside darkly striated.
n – Slightly slower than most when toasting.
o – Broiled hamburgers very unevenly.
p – Toast got lighter with successive loads.
q – Door does not stay in partly open position for broiling.

Energy use, safety

You might think that a small, countertop oven would offer a bit of operating economy in such chores as baking potatoes. It may or may not. Assuming an electric rate of 7.75 cents per kilowatt-hour, our range's oven used about 7½ cents' worth of electricity to bake four potatoes—though of course you could cook lots of other things in there at the same time. Most of the broiler ovens used about 6 to 9½ cents' worth of electricity to bake the four potatoes. A large broiler oven could prove rather expensive to run. The toasting ovens, however, cost less, with the most efficient model costing only about 4 cents.

A microwave oven used 2 cents of electricity to bake four spuds (but microwave baking may not be entirely pleasing to you—see pages 209–210).

Cost differences in broiling were smaller. The toaster oven–broilers grilled hamburgers for an average of 1¾ cents in electricity; the broiler ovens, for 2⅓ cents. You'd pay more to broil six hamburgers in the oven of a regular electric range. But the range oven would be cheaper with large quantities.

The glass doors of all the ovens can get hot enough to burn you; so can some metal parts on the outside of the cabinet. Most, however, have plastic sides with molded-in handles; those models can be moved safely even when hot.

A few safety tips: Don't use an extension cord unless absolutely necessary. If you do, use a short cord rated to handle at least 15 amperes. Don't use any of these models outdoors. Beware of sharp edges on the sheet metal—a number of our samples had razor-sharp edges on their crumb trays, drip pan, door, or the cabinet itself. Look over any countertop oven you buy and file smooth any sharp spots you find.

Recommendations

Don't expect these ovens to bake as well as a range, or to handle large orders of food with as much finesse. Still, as an auxiliary oven, one can come in very handy. Unless you need the maximum in broiling capacity, almost any toaster oven–broiler would be a better choice than a broiler oven.

Trash compactors

A compactor will reduce the volume of household trash to a fraction of its unmashed volume. That means that 20 pounds of trash, instead of occupying, say, three 20-gallon cans, will be reduced to a disposable bag of about 1½ cubic feet (about 11.2 gallons), still weighing 20 pounds. A trash compactor is not the total answer to home waste disposal. Some manufacturers recommend against placing any food waste in a compactor; others, against compacting wet garbage. You shouldn't compact flammable material, toxic chemicals, and insecticides.

People living in communities that have voluntary or mandatory separation of wastes for recycling will have other items on their keep-out list. And, with the newspapers, cans, and bottles to be recycled eliminated from the garbage, the need for compacting the remaining wastes is minimized (as is the burden on already overloaded town incineration facilities).

While a compactor may help you fit more into your old garbage containers, you won't be able to throw those containers away. Bags full of even mashed trash do need a container for protection; otherwise, they are easy prey to the jaws and claws of animals.

With most compactors, you dispose of trash by pulling out a drawer, placing the trash in a bag-lined bin, and closing the drawer. To compact, you switch the machine on.

If you're going to use a compactor as a substitute for your kitchen trash can, a model that's inconvenient to load would quickly dramatize for you the number of times a day you discard something. If, for reasons of space or aesthetics, you're going to locate the compactor in the garage or basement, you'll be taking accumulated trash to the compactor in loads; in that event, cycle time (the speed of compacting each load) and the amount of uncompacted trash you can put in at one time become important convenience factors. Whatever the location, it should be easy to get the compacted trash out of the machine and it should be easy to clean the interior of the compactor.

Compacting

A motor-driven ram, fitted with a mashing faceplate, begins a slow descent into the bin and squeezes the trash until the resistance of the compacted mass slows the motor (or, if there's very little trash in the bin, until the ram reaches a factory-set lower limit). Then the motor reverses, returning the ram to its original position, and the motor shuts off. If there is not much trash in the bin, there will be little or no compaction.

As the bin fills, you hear unmistakable sounds of compaction, especially when a bottle or glass jar is crushed. The machine has an interlock that pre-vents the motor from starting unless the drawer or door is closed. And there's a key switch to be turned on before you can operate the control to start the motor.

The amount of trash a compactor can take before the bag has to be removed depends upon the clearance factors of bin size and degree of compaction. If you cannot readily put a half-gallon milk container into the bin, for example, you may be inclined to remove the bag before it's entirely full.

There's another limit to capacity, however—the difficulty of closing the bag when it's completely full.

Nothing but trash

If you use a compactor as your kitchen trash basket, you may be opening and closing it as many as 20 times a day. Therefore, that should be easy to do. In particular, the drawer should be easy to pull out. In addition, it should be possible to line the bin with a bag. The bag should be easy to remove and to carry when it is full.

Odors and cleanliness

Unless you wash your trash, some smelly things are bound to get compacted. To combat odors, there may be an activated-charcoal filter and a fan for circulating air inside the compactor and through the filter. Or there may be a provision for automatically spraying deodorant onto the compacted waste. You can do that by hand, of course, with any compactor.

In the end, however, odor control depends on what you compact, how long it stays in the bag, and how hot the kitchen gets.

Even if a kitchen trash compactor poses no significant health hazards related to contamination with harmful organisms, still, for aesthetic reasons as well as to cut down on odors, the compactor should be kept as clean as the kitchen garbage pail. There should be good access for removing or scrubbing the ram's faceplate. Here, though, be extremely cautious. Never tackle the faceplate with your bare hands; it may be covered with minute glass splinters. For that matter, you really should do all compactor-cleaning wearing gloves.

The environment

A large proportion of the nation's collectible (municipal) solid waste is still placed in open dumps where it may be burned or left to accumulate. There, compacted refuse in a plastic-lined or all-plastic bag will not decompose or degrade as readily as wastes left to the elements.

Most of the incinerators used to dispose of a small fraction of solid waste are old-fashioned and inefficient—so inefficient that they contribute heavily to air pollution. Some of those incinerators may not operate at high enough temperatures to burn compacted refuse thoroughly, leaving a less than sanitary residue for eventual dumping. More modern incinerators operate at higher temperatures and also eliminate noxious fumes and fly ash; the more expensive systems can utilize the heat to generate electrical power. And in installations that shred all solid waste before incineration, the compacted plastic bags would present no burning problem.

Another small fraction of our solid waste goes into sanitary landfill operations. Bulldozers spread the waste over a specific area, compact it into the ground, then cover the waste with two or three feet of compacted dirt. Compacted trash and its plastic bag are chewed up and dispersed in this process.

Pyrolysis—incineration without oxygen—is an initially more expensive solution that few municipalities can yet afford to adopt, though it is nonpolluting and has attractive side benefits. Incineration by pyrolisis is totally contained; nothing escapes into the air. The end results are gases, liquids, and solids. The gas generates enough heat to run the plant, with a surplus for other uses. The solid residue, or char, can be screened to recover metals for recycling, and the remainder is processed into briquette fuel. Compacted refuse presents no problem if the mass is shredded, classified, and separated for recyclables for incineration, sanitary landfill, or pyrolysis.

Recommendations

The trash compactor doesn't appear to be any ecological boon. If your town's disposal system can handle the solid mass, you may save the sanitation department a little time; it should take slightly fewer man-hours to load compacted bags than to load an equivalent weight of uncompacted refuse. And compacting your trash may extend the lifetime of landfill sites, in towns that use that disposal method.

But coping with compacted trash successfully by other methods requires waste-disposal facilities that your community may not have. Indeed, some communities may not accept compacted trash because of fears that their incinerators may not be able to cope with it satisfactorily. Compactors require individual householders to lay out hundreds of dollars for a piece of equipment that seems to offer, as its main advantage, the opportunity of cutting down the number of trips to one's garbage containers. Too, compactors encourage the use of electrical energy to perform a function most people get along very nicely without.

Who, then, needs an expensive appliance that does nothing but reduce the volume of household trash? There could be an advantage for the few who have to haul their own trash for some distance. For others, the advantages seem limited—especially where local recycling programs would mandate your salvaging reusable materials from your trash before you compact it.

Electric typewriters

The old portable electric typewriter—the motorized version of a manual machine, complete with clacking metal type bars—may be going the way of the icebox and the 78-rpm record player.

In fact, the new portable electric typewriters aren't just electric—they're *electronic.* Gone are all those levers and shafts topped with little bars of type. In their place are electronic circuits etched onto chips and a printing element like that found in a computer printer.

These modern typewriters are products of the word-processing revolution now spreading from the office into the home. Some of the technology used in word-processing printers has been transferred to the typewriter. Some electronic typewriters even come in versions that can be used as printers for home-computer systems.

At first glance an electronic typewriter doesn't look very different from other electric typewriters. Its printing element, however, is obviously different.

Instead of type bars, an electronic machine prints with a daisy wheel—a hub with "petals" or spokes, radiating outward. Each spoke carries a character: a letter, number, punctuation mark, or symbol. As you type, the daisy wheel spins quickly, stopping for an instant as the proper spoke flicks against the ribbon to type a character on the paper. Electronic models are much smoother and quieter than the traditional typebar models because only the lightweight daisy wheel moves down the line as you type. And there's nothing to tangle and jam.

Daisy wheels can be changed fairly easily, so you are able to switch from one typeface to another. More impor-

tant, the wheel can turn very quickly, so this typewriter should be able to print at a faster speed than more familiar kinds of printing mechanisms.

The electronic typewriter was presaged by another kind of typewriter, one that uses the golf-ball–shaped printing element introduced by **IBM** in 1961. That design eliminated some of the typewriter's levers and first made it possible to switch typefaces easily. Both **IBM** and **Avanti** have incorporated that design into a portable model.

More primitive electric typewriters—the electrified versions of manual typewriters—are still around. They're known as "typebar" machines because the bars that hold the type do the printing. Most have all the typewriter's motions—printing, carriage return, space bar, backspace, and so forth—powered by electricity instead of by hand. But a few still have some manual functions. Although typewriters that make you return the carriage by hand are still sold, they don't capture the spirit of electrification at all.

Electric typewriters can also be categorized by their correction-making system, an increasingly important consideration.

The electronic typewriters have a memory—yet another advantage. You can keep typing even while the daisy wheel is returning to the left margin. The machine stores the characters in its memory during the return and then prints them out in a rush.

An electronic portable provides other features that can reduce many typing chores—setting margins and tabs and making corrections, to name a few—to one or two keystrokes. Some machines can also be used as a computer printer as well as a typewriter. Prices range from about $350 to $700 (a computer-compatible machine is generally the more expensive). Discounts bring the actual cost down to about $250 to $500.

Old-style type-bar machines, priced from about $225 to $500, are still around, and several are included in the Ratings for comparison purposes.

Typing

Fancy electronics and easy correction methods don't mean much if a manufacturer has neglected another aspect of typewriter design—how the typewriter feels to use.

The feel of a particular typewriter is determined by such things as the slope of the keyboard, the shape of the key tops, the distance between keys, and the effort it takes to trigger them. Additional aspects of a typewriter's feel can be transmitted through your fingers: whether the keys feel shaky or smooth, the overall steadiness of the machine, the immediacy of the keys' response. Other motions of a typewriter—the speed of the carriage return, the action of the repeating keys and the tabs, how the typewriter moves during shifting for uppercase letters—also affect the rhythm of your typing.

Keyboard feel. This is a highly subjective factor. There were some differences of opinion among the testers, but a large number of them agreed on one point: among the electronic models, the **Olivetti Praxis 20** was the least comfortable machine to use. Fully half the testers said that the machine slowed them down. Some blamed sluggish keyboard response. Others said that the keys didn't feel "normal."

Five electronic models, noted in the Ratings, are singled out for their superior keyboard feel. Two others are better than average. Other models are downrated because of the size, shape, angle, or spacing of their keys or for their rough key action.

Portable electric typewriters have long been more convenient to use than manual machines. In some areas, the new electronic models have borrowed features from home computers to make typing even easier and more automated.

The return. An automatic-return feature can do away with the need to touch the return key at the end of each line. You engage the system by pressing the proper key or keys just once. The daisy wheel then automatically returns to the left margin whenever you press the space bar or the hyphen key while you're typing in the "hot zone"—usually the six spaces before the right margin stop. (Word-processing programs for computers work in a somewhat similar fashion.) The automatic return was very popular with our panelists.

The quiet return on electronic machines and on the **IBM** and the **Avanti** "standard" units makes those models very pleasant to use. The daisy wheel or ball element glided smoothly to the left margin, while the heavy carriage return of the type-bar models is noisy and harsh. The type-bar models may shift on the table as their carriage slams back, unless you cushion-mount them to minimize the nuisance.

Three electronic models—the **Olivetti-Underwood** and both **Olivetti Praxis** models—have a disappointingly slow return. Fast typists could find that a problem.

Repeating backspace. Only the **SCM Sterling Cartridge** and **Electra XT**

type-bar models lack this feature. You depress the backspace key lightly for a single backspace, hold it down for repeat backspacing. (On a few electronic models, both the backspace and the repeat key must be depressed.)

Ideally, repeat backspacing should be smooth and swift. But the type-bar electrics and some electronic models backspaced in a series of annoying jerks.

An "express" key is useful for backspacing. It moves the daisy wheel rapidly back to the left margin of the line you're typing.

Repeat spacing. All the machines tested let you hold down the space bar or a separate repeat spacer to move the carriage (or printing element) quickly along the line. Repeat spacing was smooth and swift on many of the electronic models, not quite so smooth on the electrics and on a few electronic models. It was especially rough and noisy in the three **SCM** type-bar models. Some testers judged that the **Olivetti Praxis 20** had a sluggish space bar.

Repeating keys. All the machines have them. Typically, the keys that repeat are the X (capital and lower case), underline/hyphen, and period. Most of the electronic machines have a separate repeat key that can make any letter or function repeat. The **IBM** has only a repeating underline/hyphen.

Printing capitals. The electronic models could print a string of capital letters nearly effortlessly. All you need to do is tap the shift-lock key to put the daisy wheel in the all-capitals position.

Since the shift-lock key doesn't stay down on an electronic typewriter, you can't tell just by looking at the key whether the lock is engaged. All the machines have an indicator light that lets you know.

You may dislike the noise and vibration that accompany the shifting and shift-locking processes in the type-bar machines. Three type-bar models—the **SCM Sterling Cartridge,** the **Brother,** and the **SCM Electra XT**—can be especially annoying. On those machines, the type-bar "basket" doesn't shift downward to print capitals; instead, the carriage moves up.

The **Olivetti Praxis 20** electronic model drew some criticism from the panelists because of its small right shift key.

Setting tabs. Several machines let you set as many tabs as you like, anywhere you like. Other models provide a dozen or more adjustable tabs—certainly sufficient, in our judgment. But the **Adler Satellite II** and the **Royal Alpha 2001** provide only five adjustable tabs.

Most of the type-bar models have permanently preset tabs every ten spaces. Preset tabs are much less useful than adjustable ones.

Setting the tabs on the electronic models is very easy—all it takes is a light press of a key. CU's testers weren't as pleased with the tab-set and the tab-clear buttons on the two type-bar models that have adjustable tabs. Also, they found using the tab button on those two models produced a good deal of noise and vibration.

Several electronic machines have preset tab stops built into their memory. You can easily clear those settings to add your own. On most of those machines, the tabs you've set aren't retained when you turn off the machine. Only the check-rated **Royal** and **Sears** remember your last tab settings.

You may like the idea of a machine that sounds an audible beep when its tabs are set or cleared. The **Brother Compactronic 60,** the **Sears Communicator 3,** and the **SCM Ultrasonic III Messenger** and **Memory Correct III Messenger** had a reverse tab feature, allowing you to move from tab to tab going right to left. That could be a time saver if you're preparing a big table of figures.

Setting margins. Most of the electronic models provide you with preset margins when you turn them on. Generally, the settings are reasonable for typing on paper that's inserted so its left edge is at zero on the margin scale. But the settings on the two **Silver Reed** models, the **Olivetti-Underwood,** and the **Olivetti Praxis 41** are far off to the right. The **Royal Alpha 2015** and the **Sears Communicator 2** remember margin settings even when you turn off the machine.

Each machine chimes or chirps when you're near the right margin. Some of the signals may not be loud enough, and some sound only five or six spaces before the margin. With so little warning, you can expect to make frequent use of the margin-release key.

Special effects. Some electronic machines, noted in the Ratings, can automatically arrange the material you're typing in special ways: each line centered between margins, individual lines centered on a tab stop, lines spaced so the right margin is even, and so on. It would be impractical to try to manipulate copy in those ways on an ordinary electric typewriter.

Paper handling. Some electronic models have a powered paper injector that automatically raises the paper bail and turns the platen to roll in the paper. Inserting paper is very easy with those machines.

The **Royal Alpha 2001** and the essentially similar **Adler** tend to skew the paper as it is inserted. The only machine that misaligned paper while rolling it up was the **Olivetti Praxis 20;** a raised edge behind the platen sometimes caught the bottom of the sheet. Some models, noted in the Ratings, tend to snag and bend or tear the corners of paper, envelopes, or index cards.

The platen knobs on the **Brother Correct-O-Riter** and the **SCM Electra XT** are small, and neither machine has a paper guide. The small knobs on the **SCM Sterling Cartridge** also earn demerits. The knobs on the **Olivetti-Underwood** and **Praxis 41** don't protrude far enough for easy gripping. And on the **Olivetti Praxis 20,** they don't protrude at all; only the top of each knob is accessible.

Most of the machines CU tested have a variable platen. You can release the platen from its ratcheting mechanism and change the position of the typing line. That's especially handy when you need to fill out a form or go back to a page you've already typed and line up new copy with old. (On the other models you can also release the platen from its ratcheting mechanism, but you can't change the position of the line.)

How the print looks

The quality of an electric typewriter's type depends largely on the kind of ribbon it uses. There are two main kinds. Plastic film is coated with ink on one side and is mounted in an easily inserted (or removed) cassette or cartridge. It produces a very black, crisp letter, but you can use a film ribbon only once. Nylon fabric, which makes a slightly fuzzier, grayer letter, gets reused and so it lasts longer.

All the electronic models and the ball-element **IBM** use a film ribbon. The electronic machines and the **IBM** produced excellent print—black and crisp—on typing paper.

The **SCM Coronet XL** and the **Sterling Cartridge** type-bar models can also use a film ribbon. Print from those machines isn't quite as crisp, but it is just as black; we judge it very good. The rest of the type-bar models and the ball-element **Avanti** use fabric ribbon. They produce markedly inferior print.

But even some of the machines that produce outstanding printwork when typing straight text don't do quite as well with other normal typing tasks.

Return addresses. Not all typewriters can type a return address clearly at the very top of the upper left-hand corner of an envelope. In the Ratings, those models that can are judged excellent for that feature. The others produced smudges, slight scratches, and incompletely formed characters. A machine may do better if you start a line or two lower. If a typewriter can't produce a clear return address on an envelope it's not likely to do much better with index cards.

Underlines. Most typewriters produce excellent underlines, straight and dark. But some, particularly the type-bar models, may produce underlines that lack uniformity.

Corrections

All the machines tested include some means of correcting mistakes. The electronic typewriters and the **IBM** use the "lift-off" method (originated by **IBM**), in which a separate correction ribbon temporarily displaces the printing ribbon. When you make a correction, the correction ribbon lifts the ink off the paper. Lift-off corrections are neat and nearly invisible, as a rule.

With electronic models that are the easiest to use for corrections, all you do is press lightly on the correction key; the machine automatically moves the correction ribbon into place and strikes over the offending character. The memory in those machines also lets them handle long corrections with aplomb. Simply hold down the correction key and the machine "types" backward, lifting off characters. Some machines can erase up to ten characters at once; others can erase an entire line at a time. The Ratings list includes each machine's memory capacity for corrections.

CU found the remaining electronic models a notch less convenient. Either the correction key is difficult to depress or you must press two or three keys to get multiple corrections.

Most electronic machines also have a "relocate" key that's handy if you're about to finish typing a line and spot a mistake near the beginning of the line. Once you've made the correction, the relocate key quickly skips the printing element back to where you left off.

The **Avanti** and the type-bar electrics use the "cover-up" method for correcting errors. They conceal the mistyped character by striking over it with white ink. Typically, the upper half of the ribbon has black ink for printing, and the lower half has white ink for corrections. Cover-up corrections are usually quite obvious, and they work well only on white paper. (Lift-off corrections work on colored paper as well as white.)

The electric machines don't have a memory and thus don't provide for automated corrections. You must press a key or change the ribbon selector to the "correct" setting, backspace the proper number of spaces, and cover up each character manually by typing over it.

Corrections are even less convenient with the **SCM Coronet XL** and **Sterling Cartridge** type-bar models. First, you must substitute a correction cartridge for the ribbon cartridge. After making the corrections, you must switch cartridges again. Removing and snapping in the cartridges isn't difficult, but is, in the testers' opinions, "a waste of time" and "annoying."

Each time you make a cover-up correction, the white ink is depleted from a spot on the ribbon. On a typewriter that uses fabric ribbon, the white ink is used up randomly from various spots on the ribbon. After the ribbon has undergone a fair amount of use, you may have to correct a character two or three times to get adequate results.

Cost of ribbons

Film ribbons can be used only once, whereas fabric ribbons can be used as long as the ink in the cloth holds out. And a film-ribbon cassette tends to be more expensive than a spool of fabric ribbon.

Among the electronic typewriters, CU estimates that film ribbons marketed by the manufacturers for use with their machines, will last for an average of 46 pages of normal typing (that's 250 words to a page, double-spaced). The **Silver Reed** models, however, can turn out only about 25 pages on one ribbon; the **Olivetti** models, only about 30. The **Adler** and the **Royal Alpha 2001** will last for about 73 pages.

Of all the machines, both electronic and electric, the **IBM** is the true ribbon champ. It can produce more than 170 pages on one ribbon.

Most lift-off correction ribbons last for 1,200 to 1,600 individual strikeovers. Here, too, the range is telling. Inexperienced typists should know that the correction ribbons for the **SCM** models yield about 2,800 strikeovers, while the **Olivettis** give only about 1,100 corrections per ribbon.

If you are an economy-minded typist, the **IBM** may be a good choice. The cost of printing and correction ribbons for that machine works out to less than $7 per 100 pages of typing. (We've assumed that a film ribbon costs $5 and a correction ribbon $3.) By contrast, ribbons for the **Olivettis** and the **Silver Reeds** run more than $20 per 100 pages of typing. On average, the film printing and correction ribbons will cost about $16 per 100 pages.

Ratings of electric typewriters

(As published in a November 1984 report.)

Listed by groups in order of overall preference of CU's panel of typists. Differences between closely ranked models were slight. Models judged equal are bracketed and listed alphabetically. Weights include carrying case or lid. Prices are suggested retail; + indicates shipping is extra.

Excellent	Very good	Good	Fair	Poor
●	◒	○	◒	●

Brand and model	Price	Dimensions (HxWxD), in.	Type	Print element [1]	Weight, lb.
✅ ROYAL ALPHA 2015	$700	$5^3/_4$x$19^1/_4$x$15^1/_4$	Electronic	D	26
✅ SEARS COMMUNICATOR 2, Cat. No. 5303	445 +	$5^3/_4$x$19^1/_4$x$14^3/_4$	Electronic	D	26
OLYMPIA ELECTRONIC COMPACT 2 [3]	599	$5^3/_4$x$19^1/_4$x$15^1/_4$	Electronic	D	21
BROTHER COMPACTRONIC 60	600	$5^3/_4$x$18^1/_4$x$15^1/_2$	Electronic	D	19
SEARS COMMUNICATOR 3, Cat. No. 5307	545 +	$5^3/_4$x$18^1/_4$x$15^1/_2$	Electronic	D	19
OLYMPIA REPORT ELECTRONIC	499	$5^3/_4$x$19^1/_4$x$15^1/_4$	Electronic	D	25
BROTHER CORRECTRONIC 50	530	$5^3/_4$x$18^1/_4$x$15^1/_2$	Electronic	D	19
SCM ULTRASONIC III MESSENGER	635	6x$18^1/_2$x$15^1/_4$	Electronic	D	28
SCM MEMORY CORRECT III MESSENGER	499	$5^3/_4$x$18^1/_2$x$15^1/_4$	Electronic	D	27
BROTHER CORRECTRONIC 40	480	$5^1/_2$x$18^1/_4$x$15^1/_2$	Electronic	D	19
ADLER SATELLITE II	475	$5^1/_2$x$18^3/_4$x14	Electronic	D	25
ROYAL ALPHA 2001	440	$5^1/_2$x$18^3/_4$x14	Electronic	D	24
SILVER REED EX42	395	$5^3/_4$x18x14	Electronic	D	19
SILVER REED EX43	459	$5^3/_4$x18x14	Electronic	D	19
OLIVETTI-UNDERWOOD 3500	545	5x17x$13^3/_4$	Electronic	D	20
OLIVETTI PRAXIS 41 [3]	645	$6^1/_4$x19x$15^3/_4$	Electronic	D	21
IBM PERSONAL [3]	695	$7^1/_2$x$17^1/_2$x$15^1/_2$	Electric	B	32
OLIVETTI PRAXIS 20	395	$4^1/_2$x15x$12^3/_4$	Electronic	D	14
AVANTI 1400 [3]	495	5x$19^1/_4$x$14^3/_4$	Electric	B	21
SCM CORONET XL	289	$5^3/_4$x18x$13^1/_4$	Electric	T	24
SEARS THE SCHOLAR Cat. No. 5377	245 +	$5^1/_2$x$15^1/_2$x14	Electric	T	20
OLYMPIA E-R12	275	$5^1/_2$x$15^1/_2$x14	Electric	T	20
ROYAL ACADEMY	290	$5^1/_2$x$15^1/_2$x$13^1/_2$	Electric	T	21
SCM STERLING CARTRIDGE	279	$5^1/_4$x16x$12^1/_2$	Electric	T	21
BROTHER CORRECT-O-RITER 3800	300	$5^3/_4$x$16^1/_4$x14	Electric	T	20
SCM ELECTRA XT	225	$4^1/_2$x16x$12^1/_2$	Electric	T	20

Panel preference	Keyboard feel	Print quality — Paper	Print quality — Envelopes	Print quality — Cards	Functions — Return	Functions — Repeat	Functions — backspace	Corrections — Ease	Corrections — Quality	Memory capacity (characters)	No. of characters provided	Advantages	Disadvantages	Comments
●	●	●	◐	●	●	●	●	●	●	[2]	102	A,E,F,G,H,J,K,L,M,N,O,Q,R,S	d,n,v	A,J
●	●	●	◐	●	●	●	●	●	◐	[2]	102	A,E,F,G,H,I,J,K,L,M,N,O,Q,R,S	d,n,v	—
◐	●	●	●	◐	●	●	●	●	◐	46	100	E,F,G,H,I,K,L,M,N	d,n,v	A,E
◐	○	●	●	●	●	●	◐	●	●	[2]	98	A,E,F,G,H,L,M,N,O,Q,R,S,T,V	m,s	A,C
◐	○	●	●	●	●	●	◐	●	●	[2]	98	A,E,F,G,H,L,M,N,O,Q,R,S,T,V	m,s	A,C
◐	○	●	◐	●	◐	●	●	●	◐	46	100	E,F,G,H,I,K	d,n,q,v,y	—
◐	○	●	●	◐	●	●	◐	●	●	[2]	98	E,F,G,L,M,N,R,S,T	m,s	A,C
○	◐	●	●	◐	●	●	●	●	●	[2]	90	A,I,K,M,O,P,Q,S,T,U,V	e,l,n,q	A,F,I
○	◐	●	●	○	●	●	●	◐	●	[2]	90	A,I,K,M,O,P,Q,V	e,l,n,q	A,I
○	○	●	◐	●	●	●	◐	●	◐	16	98	E,F,G,M,N	m,n,q	C,I
○	●	●	●	◐	●	●	◐	●	●	20	102	E,F,G,I,K,M,S,W	d,f,j,q,v	D,I
○	●	●	●	◐	●	●	◐	●	●	20	102	E,F,G,I,K,M,S,W	d,f,j,q,v	D
○	◐	●	◐	●	●	◐	●	◐	●	16	88	B,F,M,N	o,x,cc	C,D
○	◐	●	◐	◐	●	◐	●	◐	●	16	88	B,F,M,N	o,x,cc	A,C,D
○	○	●	◐	●	○	◐	◐	◐	◐	10	102	B,D,F,M,N	i,o,t,u,x,cc	—
○	○	●	◐	●	◐	◐	◐	◐	◐	10	102	D,F,M,N	d,i,o,t,u,x,cc	AE
○	◐	●	◐	●	●	●	○	◐	●	1	88	B,W	d,r,t,bb	B,E
◐	○	●	●	●	●	◐	●	◐	◐	19	102	B,D,F	a,c,i,o,p,q,t,u,w,aa,cc	C
◐	◐	○	◐	●	●	◐	●	◐	◐	1	88	—	i,r	E,H
●	○	◐	●	●	●	◐	●	●	◐	1	88	B,C	b,h,k,t,dd	G
●	○	●	●	◐	●	●	◐	●	◐	1	88	—	h,n,r,t,dd	H
●	○	●	●	◐	●	●	◐	●	◐	1	88	—	g,h,n,r,t,aa,dd	H
●	○	●	●	●	●	●	◐	●	◐	1	88	—	g,h,n,r,t,aa,dd	H
●	○	◐	○	○	●		[4]	●	◐	1	84	C,N	b,g,h,k,n,t,u,z,dd	G
●	○	○	◐	●	●	◐	◐	◐	◐	1	88	—	b,g,h,r,u,z,aa,dd	C,H
●	○	○	○	○	●		[4]	◐	◐	1	84	C	b,g,h,n,t,u,z,dd	H

1. D = Daisy wheel; B = Ball; T = Type bars.
2. Memory holds 1 typed line.
3. Not supplied with case and not, strictly speaking, portable.
4. Backspacer doesn't repeat.

SPECIFICATIONS AND FEATURES

Except as noted, all have: ● Easily changed daisy-wheel or ball printing element. ● Film ribbon cassette or cartridge judged easy to change. ● Lift-off correction ribbon on spools, judged difficult to change. ● Ability to make multiple corrections at the press of a single key. ● Platen that can accept paper 11 in. wide. ● Line-spacing capability of 1, 1½, or 2 lines. ● At least 3 repeat keys, plus repeat return or index key. ● At least 12 adjustable tabs. ● Pica type, with elite and micron type (15 characters/in.) also available. (On daisy-wheel models, type size can be changed; not so with ball-element or type-bar models.) ● Rigid carrying case.

All ball-element and type-bar models: ● Are much noisier than electronic models. ● Can correct only 1 character at a time.

KEY TO ADVANTAGES

A – Has automatic return capability.
B – Margin signal sounds 8 or more spaces before right margin stop.
C – Type can be switched on at least one key, increasing versatility.
D – Has characters for typing in French, Spanish, German, and Italian.
E – Has characters for technical use.
F – Repeat key can repeat any character.
G – Daisy wheel judged very easy to change.
H – Has powered paper-injector lever.
I – Beeps when new margins or tabs are set.
J – Retains margin and tab settings when machine is turned off.
K – Repeat spacing judged extremely smooth and quick.
L – Has reverse index key, which can automatically roll paper back to top of page.
M – Has "relocate" key, useful when making correction in already-typed line.
N – Has half-space key.
O – Can automatically underline every character and space.
P – Can automatically underline every character.
Q – Can automatically center copy between margin stops.
R – Can automatically indent every line.
S – Can automatically keep decimal points lined up in columns of figures.
T – Can automatically type copy with uniform ("justified") right margin.
U – Can automatically center copy around selected tab stop.
V – Has reverse (right-to-left) tab feature.
W – Cost of ribbons relatively low (less than $10 per 100 pg.).

KEY TO DISADVANTAGES

a – Paper shifted when rolled up and down.
b – Uncomfortable carrying handle.
c – Daisy wheel a bit difficult to replace.
d – Bent or ripped corners of some paper, envelopes, or cards.
e – Correction key difficult to depress.
f – Only 5 adjustable tabs.
g – Has only preset tab stops.
h – Lacks repeat return or index key.
i – Printing-element return slower than most.
j – Return key inconveniently placed, according to CU's panelists.
k – Separate cartridge must be inserted in place of printing cartridge when making corrections.
l – Noisier than other electronic models. (**Avanti** was noisiest of all models tested.)
m – Margin warning signal judged not loud enough.
n – Margin warning signal sounds only 5 or 6 spaces before right margin.
o – Multiple corrections require depressing 2 or 3 keys.
p – Printer lagged noticeably with fast typists, breaking their rhythm.
q – Unlike most electronic models, lacks express backspace key.
r – Lacks "on" light.
s – "On" light judged not visible enough.
t – Produced relatively uneven repeat underline.
u – Platen knobs judged difficult to turn.
v – Has only one platen knob.
w – Right shift key judged too small.
x – Preset margins were too far to the right with edge of paper at zero on scale.
y – Backspace and repeat key must be depressed for repeat backspacing.
z – When typing capitals, carriage moves up; CU's panelists said that was disconcerting.
aa – Lacks variable platen.
bb – Only the hyphen/underline key repeats.
cc – Cost of ribbons relatively high (more than $20 per 100 pg.).
dd – Shift lock required hard push.

KEY TO COMMENTS

A – Computer-compatible.
B – Available only with elite type.

C – Has lid cover instead of case.
D – No provision for micron type.
E – Comes with dust cover.
F – Comes with elite-type daisy wheel; pica wheel available for $10.
G – Comes with fabric ribbon cartridge, but can use film ribbon cartridge.
H – Uses fabric ribbon and "cover up" correction ribbon on spools (in cassette for **Avanti**); ribbon on spools judged difficult to change.

I – According to the company, this model had been discontinued by November 1984. The test information has been retained here, however, for its use as a guide to buying.
J – According to the company, this model had been discontinued by November 1984.

Versatility

The ball-element machines and most of the type-bar electrics have 44 keys and can print 88 characters. The **SCM Sterling Cartridge** and the **SCM Electra XT** have only 42 keys; they lack the 1/! and +/= keys.

Most electronics, on the other hand, can print from 90 to 102 characters. The extra keys on the three **Olivetti** models, for example, provide accent and other marks needed to type in Spanish, French, German, and Italian. Other electronic models offer mathematical and technical symbols, brackets, and paragraph and section symbols.

Two **Brother** models—the **Compactronic 60** and the **Correctronic 50**—and the **Sears Communicator 3** show an additional 23 characters on the keyboard. But some of those extra charac-

ters are available only with the "international" daisy wheel, a $20 extra.

The three **SCM** type-bar models offer a lesser amount of character flexibility. The **Sterling Cartridge** and the **Electra XT** allow you to switch characters on one type bar; the **Coronet XL** allows switching on two type bars. SCM offers a wide variety of symbols for those type bars, along with overlays for the keys.

CU tried to test all typewriters with pica type, a size that yields 10 characters to the inch. The **SCM Ultrasonic III Messenger** came with elite type, which produces 12 characters to the inch; a pica daisy wheel costs $9.95 extra. The **IBM** is available only with elite type.

Changing a daisy wheel is easy in most models. Access to the wheel is limited, however, in the **Olivetti Praxis 20.**

Portability

Most of the typewriters weigh about 20 to 30 pounds ready to travel. At 14 pounds, the **Olivetti Praxis 20** is a real lightweight. But then, its platen is the shortest in this group.

Most of the machines come with a rigid case. Eight come with just a lid cover. When you prepare to type, it's much easier to remove a lid than to lift the machine from a case. But a case should

give better protection in transit.

Four machines, noted in the Ratings, do not come with a case or lid and, strictly speaking, are not portable machines. Of the four, the **IBM** is the heaviest (32 pounds). The others weigh 21 pounds.

Most of the case handles were fine, but the Ratings single out four uncomfortable ones.

Computer interface

Nine of the electronic typewriters can be connected to a computer to serve as printers. The **Royal 2015** and the **Olympia Electronic Compact 2** have a computer interface built in. All they need is a connecting cable (about $50). The other models require an interface ($150 to $175) as well as cable.

Many computers use what's called a "Centronix-type parallel" interface, and all nine typewriters are set up for that kind of connection. The **Brother, Olivetti,** and **SCM** interface units also have "RS232C serial" connections, which demand absolute electronic compatibility between computer and printer. If you want to use one of those typewriters with a computer that has a serial interface, be sure the dealer can make the hardware work properly.

Most of the typewriters accurately printed out text sent via a parallel interface in the **IBM PC** computer CU used in the tests. But the **SCM Ultrasonic III** and **Memory Correct III** made several errors.

The same errors cropped up with another interface. It's hard to determine whether the two **SCMs** would experience the same problem with a different computer.

As daisy-wheel computer printers go, the machines are relatively slow. Five—the **Brother Compatronic 60,** the **Sears Communicator 3,** the **Brother Correctronic 50,** and both **SCM** models—can print about 10 characters per second. The **Olympia** prints about 9 characters per second, while the **Royal,** the **Silver Reed EX43,** and the **Olivetti Praxis 41** manage only about 7.

None of these typewriters would be suitable as the only printer for a word-processing computer. The typewriters are too slow, and they may not be built to withstand heavy use. But they could serve well as a backup printer, called upon to produce occasional documents and correspondence that must look crisp and professional, and earning their keep as a typewriter the rest of the time.

Recommendations

Once you get used to an electronic portable, you probably won't want a type-bar model ever again. CU's testers strongly preferred even the lowest-rated electronic models to any of the type-bar models. They also preferred most of the electronic portables to the **IBM Personal Typewriter** with its ball element.

The testers had no problem choosing the best of the electronic machines. The clear choices were the $700 **Royal Alpha 2015** and the $445 **Sears Communicator 2,** both check-rated. They scored well in all important tests, and they're loaded with convenience fea-

tures. The **Royal** is also computer-compatible.

The six models in the next Ratings group also performed well and are worth considering.

The type-bar machines are much lower in price than the electronic models, but they are much noisier, rougher, and less convenient than the electronics. And they lack the modern special features of the electronic machines.

Whichever typewriter you decide on, make sure you try it before you buy. The feel of a typewriter is so subjective that only you should be the final judge.

Dot-matrix typing printers

For all their versatility, none of the electronic typewriters covered in the accompanying report will make a suitable traveling companion. They're rather large and they can't run on batteries. This is where the small printer (sometimes called the "personal printer") enters— a lightweight, compact typing machine.

CU tested four small printers—the $449 **Brother EP44,** the $250 **Canon Typestar 5,** the $350 **Sharp PA1000,** and the $350 **Silver Reed EXD15.** The **Silver Reed** weighs 13 pounds and requires an electric outlet. The others weigh 6 or 8 pounds and can run on batteries.

One reason these printers are so small is their printing element: a "dot matrix" head that forms characters by pushing an array of tiny pins or rods against the platen. In the **Brother,** the **Canon,** and the **Sharp,** the platen is just long enough to take a sheet of typing paper. You can't type envelopes, cards, or carbon copies on any of these printers.

But in their own miniaturized way, these machines are as versatile as larger electronic machines.

Each printer has a tiny display screen above the keyboard to show the characters you've typed. The **Sharp**'s display shows 70 characters—a full line of type; the others show 15 characters.

In their "direct print" mode, three machines print the characters as they are typed. The **Silver Reed** prints one word at a time. If you're typing in direct print, making corrections is impossible.

Fortunately, all the printers give you some way to make corrections. The **Brother** and the **Silver Reed** have a "correction print" mode: the printer will store a certain number of characters before printing them; if you notice a mistake in the display, you can correct it and proceed with your typing. The **Canon** and the **Sharp** have a "line print" mode that allows you to key in a full line (and make any corrections) before it's printed. The **Brother** also has line-print.

The **Sharp** and the **Brother** both have a memory: up to 2,200 characters for the **Sharp,** up to 3,700 for the **Brother.** The **Brother** can also be connected to a computer to print out data or to transmit data to another printer. However, the **Brother** uses an RS232C serial interface, a computer connection that our engineers have found to be extremely finicky. Before you buy a **Brother** to use with a computer, be sure that your computer dealer can make the two machines work together.

You can set up all the printers for automatic carriage return. But that feature works best on the **Sharp** and the **Silver Reed**. They have a "word wrap" feature that carries a long word at the end of a line down to the beginning of the next line.

All the printers can automatically underline words. All but the **Brother** can expand letters to twice their normal width, and the **Silver Reed** and the **Canon** can print in either pica or elite type.

The keyboard on these printers is more complex than the keyboard on a standard typewriter. And, although CU's testers found the keyboards generally easy to use, there are some problems. The **Brother**'s keys—like those on pocket calculators—are too flat, too small, and too far apart. The **Brother**'s shift lock is too small, and the **Silver Reed**'s tab key is in an awkward place.

Poor print quality is often a problem with dot-matrix printers. In an earlier test on the first **Brother** printer, the **EP20**, its poor print was the biggest drawback found. The later machines generally produced good-looking print, though not as crisp and clear as the print produced by an electronic typewriter with a daisy wheel.

These printers must use photocopying paper, "thermal" paper (a heat-sensitive type that doesn't hold up well), or a special smooth typing paper available from Brother. The testers got the best results with the smooth paper ($7.95 for 500 sheets).

The cost of ribbons for these units (at $5 per ribbon) is about $15 per 100 pages for the **Silver Reed,** to $25 per 100 pages for the **Canon.**

If you need a small, truly portable typewriter to handle correspondence while you travel, the **Sharp** or the **Canon** would be a good choice. The **Canon** is the cheapest; CU paid $225 for theirs. A carrying case is $23. The **Brother** is the machine to choose if you can't be separated from a computer (but many of CU's testers disliked its keyboard).

Manual typewriters

If you need a truly portable typewriter, you should consider a manual model. Manuals are light—usually 10 to 15 pounds—and can go anywhere. "Portable" electric typewriters usually weigh about 20 to 30 pounds and, of course, they can't stray from an outlet. Electrics are expensive, too.

But if you must have a typewriter to produce truly good-looking copy, you'll have to buy an electric model. With manual typewriters, the appearance of the copy depends to some extent on the touch. A heavy-handed typist may produce dark copy, a light-handed typist, light copy. Someone who types unevenly or in rapid spurts may produce uneven-looking copy.

A typewriter that doesn't feel right will obviously be unpleasant to use for any length of time. Many things can affect keyboard feel. The typewriter's touch may be too heavy or too light. The tops of the keys may be too large, too small, too flat, too curved, or too slippery. The keys may be too far apart or too close together. Or they may feel shaky or rough when you strike them.

But judgments of keyboard comfort are highly subjective; so the person who's going to use the machine should try several in the store before buying. Touch controls, which vary the amount of pressure needed to work the keys, are generally useful.

Typing

A typewriter's keyboard may feel perfectly comfortable, but if you have a difficulty with the carriage return, the space bar, the shift key, or the backspace key, you still will not be at ease with the typewriter.

Since the space bar gets so much use, an inconvenient one can be a particular nuisance. Such things as a space bar that's too thin or inadequate spacing between the frame and the space bar can be especially irritating.

The shift keys get a lot of use, too. On the more expensive typewriters the type basket drops down when you shift to type capital letters. On the lower-priced models, the carriage moves up as you shift. Some people find a moving carriage annoying.

All typewriters have a shift key on both sides of the keyboard and a shift lock on the left side. On some models, only the shift key on the left side will

release the lock; that's a minor disadvantage.

The carriage-return lever should be easy to use, not too short, and not too stiff.

The backspace key should be easy to reach and easy to push, and the typewriter should sit firmly on its rubber feet during use. Some lightweight models don't.

A typewriter should produce good-looking copy no matter how heavy or light the typist's touch. The letters should be distinct, fully formed, and uniformly dark. Underlines should be even, and punctuation marks shouldn't produce holes.

Bond paper. Typists with a light touch shouldn't have problems, but if your touch is heavy, check to be sure the machine you select doesn't produce shadows—a second impression to the right of the original letter.

Your machine should print under-lines without leaving occasional gaps or an uneven look.

Cards and envelopes. The top edge of a card or an envelope must be held firmly against the platen if the typing is to look good. Unfortunately, many typewriters don't provide a sufficiently firm hold. As a result, typing in the upper left-hand corner of cards and envelopes isn't completely free of un-wanted ink marks or smears.

On a few models, the metal pieces used to hold material against the platen tend to snag the top corner of cards and envelopes. On some machines with a short platen, business envelopes will prove to be a tight fit.

A few typewriters won't type close to the left edge of a large envelope—a drawback that detracts from the ap-pearance of a typed return address.

Other factors

There are a number of useful control and convenience features.

Adjustable tab stops. Most machines have fully adjustable tabs, for indenting paragraphs or to type charts and tables. Models with tab stops permanently set every ten spaces present a much less convenient arrangement, but still are better than no tabs at all.

Some machines have a paragraph-indent feature that doesn't require the use of the tab. By holding down the margin-release key while you return the carriage, you'll start a new line five spaces in from the left margin.

Horizontal spacing. Most machines can move the carriage forward half a space at a time—a provision that comes in handy if you want to squeeze in a let-ter to correct a word.

Several typewriters have a repeat-spacer in addition to the normal space bar. You hold down the repeat-spacer and the typewriter spaces along auto-matically, just as an electric typewriter does.

Vertical spacing. Just about every machine can type single-space, space-and-a-half, or double-space.

Many machines have a variable line spacer, which disengages the line-spac-ing mechanism and allows you to roll the platen freely. It's a useful feature if you have to type out a form on which the lines don't match the typewriter's line spacing. It's also handy for lining up copy on an already-typed sheet.

Paper-handling aids. A good type-writer should have several features de-signed to make it easy to handle paper in the machine. These include: a paper guide to line up the left edge of the paper; a support behind the carriage to keep paper from flopping down behind the machine; an erasure plate to sup-port the paper when you correct mis-takes; a paper bail (the metal bar that holds paper against platen) that lifts up when you use the paper-release lever; and rollers on the paper bail.

Margin bell. All machines have a bell that rings several spaces before you reach the right margin. The trouble is that you may not hear the bell on many of these because the ring is too faint. Typically, the bell rings six or seven spaces before the margin; that ought to be room enough to finish a word before moving on to the next line.

Page-end indicator. This handy de-vice warns you when you're getting close to the bottom of the page. Typi-cally, the indicator is a scale on the plat-en or the paper support. A few models merely have a "window" in the metal below the platen, which doesn't give you much warning; when you see the bottom of the sheet in the window, you'd best make the line you're typing the last.

Ratings of portable manual typewriters

(As published in a November 1979 report.)

Listed in order of estimated overall quality, based on panel tests and engineering judgments. Differences between closely ranked models was judged slight. Dimensions are for typewriter ready to use; weight is for typewriter plus carrying case. Prices are list, rounded to nearest dollar; + indicates additional shipping charge.

⊙	◓	○	◒	●
Better	←		→	Worse

Brand and model	Price	Height x width x depth (in.)	Weight (lb.)	Platen width (in.)	Print quality: Bond paper	Print quality: Envelopes	Print quality: Cards	Advantages	Disadvantages	Comments
HERMES 3000	$278	$7\frac{3}{4}$x$14\frac{1}{2}$x$13\frac{3}{4}$	$16\frac{3}{4}$	$9\frac{3}{4}$	●	●	●	A,B,F,G,I,J	I,v,cc,ee	—
OLYMPIA SM9-13′′	235	$6\frac{1}{2}$x$17\frac{1}{4}$x13	24	$13\frac{1}{4}$	◓	◓	◓	A,F,G,J,K	f,y	*D*
ROYAL CHAMPION L	235	$6\frac{1}{4}$x$14\frac{1}{4}$x$13\frac{1}{2}$	$17\frac{3}{4}$	$9\frac{1}{2}$	●	◓	●	A,E,F,G,J	f,g,cc	—
OLYMPIA B12	150	7x$17\frac{1}{4}$x$13\frac{1}{2}$	$15\frac{1}{2}$	$12\frac{1}{2}$	◓	◒	◒	A,B,C,J	bb	—
OLYMPIA SM9-$9\frac{1}{2}$′′	215	$6\frac{1}{2}$x$13\frac{1}{4}$x13	$20\frac{3}{4}$	$9\frac{1}{2}$	◓	◓	◓	A,F,G,J	f,y	—
SMITH-CORONA CLASSIC 12	190	$5\frac{3}{4}$x$16\frac{3}{4}$x$12\frac{3}{4}$	$19\frac{1}{4}$	$11\frac{3}{4}$	○	●	●	A,B,C,F,G,I,J,K	b,i,bb,cc,dd	*F*
SMITH-CORONA GALAXIE TWELVE	185	6x$16\frac{1}{4}$x$12\frac{3}{4}$	$19\frac{1}{2}$	$11\frac{3}{4}$	○	●	●	A,B,C,F,G,I,J,K	b,i,bb,cc,dd,ee	*C,F*
FACIT 1620	149	6x$14\frac{3}{4}$x12	$16\frac{1}{2}$	10	◓	◓	○	A,E,G,J	b,g,i,q	*D*
HERMES 3000S	148	$6\frac{1}{2}$x$13\frac{1}{2}$x13	$16\frac{3}{4}$	10	◓	◓	○	A,E,G,J	b,g,i,q,cc	—
BROTHER ACCORD 12	170	$5\frac{3}{4}$x15x13	$14\frac{1}{2}$	12	◓	◒	◒	C,H	f,ff,gg	—

Brand and model	Price	Height x width x depth (in.)	Weight (lb.)	Platen width (in.)	Print quality Bond paper	Print quality Envelopes	Print quality Cards	Advantages	Disadvantages	Comments
BROTHER ACCORD 10	140	5³/₄x13¹/₄x13	10¹/₂	9¹/₄	◑	◑	◑	C,H	f,o,w,ff,gg	B
OLIVETTI STUDIO 46	170	6³/₄x17x14¹/₄	21	12	○	◑	◑	C,G,H,K	g,j,u,aa,bb,cc	D
OLIVETTI LETTERA 35	140	5¹/₂x13³/₄x14	16	9³/₄	○	●	●	G,H	g,j,t,u,aa,cc,gg	—
ROYAL CUSTOM III	150	6³/₄x15³/₄x12³/₄	16¹/₂	9¹/₄	◑	◑	◑	D,F,G	c,e,j,r,t,bb,cc,ee,hh	—
ROYAL SABRE	140	6¹/₄x15¹/₂x12³/₄	16¹/₄	9¹/₄	◑	◑	◑	D,G	c,e,j,r,t,bb,cc,ee,hh	A
ROYAL SAFARI	100	6¹/₄x15¹/₂x13¹/₄	10¹/₄	9³/₄	◑	◑	●	B,J	d,j,o,r,s,u,v,w,ff,gg,ii	—
OLIVETTI LETTERA 25	110	5¹/₄x13³/₄x14¹/₄	12¹/₄	9³/₄	○	◑	●	H	g,j,n,t,u,aa,cc,ff,gg,hh,ii	—
OLYMPIA OLYMPIETTE 3	103	5¹/₄x13³/₄x12¹/₄	11	9¹/₂	○	○	—	—	e,g,h,o,s,w,bb,ee,ff,gg	B,E
OLYMPIA OLYMPIETTE 2	98	5¹/₄x13³/₄x12¹/₄	11	9¹/₂	○	○	—	—	e,g,h,n,s,w,bb,ee,ff,gg	B,E
BROTHER CHARGER 12	130	5x14³/₄x11¹/₄	13	12	○	○	○	I	c,m,n,s,u,z,ff,gg,hh,ii	B
BROTHER CHARGER 11	100	4³/₄x12¹/₄x11¹/₂	10¹/₄	9¹/₄	○	○	○	I	c,m,n,s,u,w,z,ff,gg,hh,ii	B
ROYAL CARAVAN	108	4¹/₄x12¹/₄x11³/₄	11¹/₄	9¹/₄	●	◑	○	F	a,b,c,h,p,v,y,cc,ff	E
SMITH-CORONA COURIER	110	5³/₄x15x13¹/₄	11¹/₂	11¹/₂	○	○	○	F,I,J	a,c,m,o,r,s,t,v,w,z,gg	—
OLIVETTI LETTERA 32	130	4³/₄x12³/₄x13	11¹/₂	9³/₄	●	●	●	G,H	a,c,g,j,k,m,t,u,x,aa,gg,ii	—
HERMES ROCKET	99	4¹/₄x13¹/₄x11³/₄	9	9¹/₄	◑	○	○	F,I	a,c,g,j,k,l,n,r,s,u,v,w,x,z,cc,ee,gg	—

SPECIFICATIONS AND FEATURES

Except as noted, all have: ● Satisfactory touch pressure. ● Adjustable tab stops. ● Horizontal half-space. ● Variable line spacer. ● Paper guide and paper support. ● Erasure plate. ● Type basket that shifts to type capitals. ● Audible margin bell that sounds about 7 spaces from right margin stop. ● Pica type, with elite type also available. ● 44 keys. ● Settings for 1, 1½, and 2 vertical spaces.

KEY TO ADVANTAGES

A – Judged better than most in overall quality by panel.
B – Made darker carbon copies than most.
C – Has repeat-space feature.
D – Ribbon can be changed without soiling hands.
E – Margin bell sounds 9 spaces before right margin stop.
F – Has page-end indicator.
G – Has 2 carriage releases.
H – Has paragraph indentation feature.
I – Has key to clear jammed type bars.
J – Paper release raises paper bail.
K – Carbon pack easy to insert.

KEY TO DISADVANTAGES

a – Judged worse than most in overall quality by panel.
b – Tended to produce shadows on copy when used by typists with heavy touch.
c – Touch pressure judged unsatisfactory.
d – Carbon pack difficult to insert.
e – Inserting large envelope more difficult than with most.
f – Tended to bend cards or envelopes.
g – Top sheet of carbon pack shifted when platen was rolled up and down.
h – Can type no closer than about 4 spaces to left edge of large envelope.
i – Carbon tended to punch through top sheet of carbon pack.
j – Judged worse than most in quality of underline.
k – Space bar judged awkward to use.
l – Backspace judged difficult to depress.
m – Carriage-return judged hard to use.
n – Lacks tabs.
o – Has preset tabs.
p – Has only 5 hand-adjustable tabs.
q – Tab controls judged inconvenient.
r – Paper release judged inconvenient.
s – Carriage shifts to print capital letters.
t – Margin bell sounds less than 6 spaces from right margin stop.

u – Margin bell judged too faint.
v – Has only 1 shift-lock release.
w – Paper bail lacks rollers.
x – Tended to move about more than most.
y – Changing ribbon judged more difficult than with most.
z – Has only 42 keys.
aa – Has only 43 keys.
bb – Lacks carriage lock for storage.
cc – Case handle judged uncomfortable.
dd – Extremely high force needed to release shift lock with right-hand shift key.
ee – Lacks horizontal half-space feature.
ff – Lacks variable line spacer.
gg – Lacks paper guide.
hh – Lacks paper support.
ii – Lacks erasure plate.

KEY TO COMMENTS

A – Lacks paper bail.
B – Elite type not available.
C – Has settings for 1, 2, and 3 vertical spaces.
D – Comes with dust cover.
E – Manufacturer recommends use of plastic part to hold type bars for storage.
F – Has 2 type bars that can accept changeable type slugs.

Recommendations

If portability and price are not your prime concerns, and if consistently high-quality printwork *is* your concern, choose a portable electric typewriter.

If you're buying a typewriter for yourself, type on a few models before you buy, to find the one that feels right for you. If you're buying it as a gift, make sure to get return privileges.

Vacuum cleaners

The upright was the first kind of vacuum cleaner to gain wide distribution, and it's still the most popular. An upright rolls out of the closet ready to go. The beating brushes in its nose make it a champ at cleaning carpets. But put that upright on bare linoleum, and it flounders.

You can rig up a hose and various attachments to suck up crumbs from the kitchen floor or dust the lampshades, but the arrangement isn't a happy one. You must do some fiddling to make the conversion, and then you have the awkward, lumbering body to tug along. And where uprights really fall down—quite often literally—is on stairs.

The canister vacuum cleaner, with a squat or streamlined little body on wheels, was developed to do those things an upright can't do well. A canister is easier to carry than an upright; it trails along behind you, albeit not as obediently as you'd like; it does somewhat better at standing on stairs; and it usually has more suction. A canister's small swiveling rug and floor tools are more adept at poking under low chairs and beds. But the canister that relies solely on suction rarely does a good job on carpets. Without the action of beating brushes, it does not loosen and pick up embedded dirt.

A canister with a "power nozzle" sacrifices convenience in the attempt to combine the best features of the two other designs.

There is yet another variety, the lightweight upright—sometimes called an electric broom (see page 318).

Full-sized uprights are descendants of old-fashioned manual carpet sweepers. Much of their cleaning power derives from rotating beaters and brushes that loosen and sweep up dirt lodged in carpet pile. Their suction serves mainly to transport the loosened soil into the dirt bag.

As a class, uprights are at their best on broad expanses of carpet. In tight quarters, they behave awkwarkly. And although many of them can be fitted with a hose and tools, low suction and the inconvenience of the conversion usually make an upright ill-suited to barefloor and above-the-floor cleaning.

The plain canister uses suction alone and a set of specialized tools to pick up dirt from bare floors, windowsills, upholstery, and the like. Low-slung and compact, a canister is readily portable. But its suction, though more powerful than an upright's, doesn't do as well with deeply embedded carpet dirt. On a carpet, a suction-only canister is helpful mainly for tidying away surface litter.

Fitted with a power nozzle (a motorized attachment with rotating beaters and brushes) a canister vacuum can clean carpets about as well as an upright. And the superior canister suction remains available for work with other attachments.

However, the power nozzle itself isn't, strictly speaking, a handy attachment. Though somewhat smaller than a proper upright, the nozzle must still trail its canister along over long stretches of carpet. And, when it's not on the carpet, the power nozzle is one more item in a canister's cluttered inventory of wands and tools that have to be assembled, disassembled, and stored.

Further, if a power-nozzle canister is designed to perform like two machines in one, it's also priced high, mostly over $200.

Using the machine

Given the imperfect state of the art, the object of your shopping for a new vacuum cleaner is to find one that makes vacuuming as painless as possible. Your choice depends not only on what you then want to clean (mostly carpet? mostly hardwood or tile?), but on which shortcomings bother you least. Is the machine hard to push and pull? Will it be too heavy to carry from room to room or up and down stairs? Is it too noisy? Can it be easily emptied?

Here are some particular convenience details to note:

Switch. A canister's on/off switch should be plainly visible from standing height and easy to operate by foot. Most canisters have a pedal switch on the hose side of the tank.

The on/off switch of most uprights is either a slide or toggle on the handle or a step-on button at the base.

Noise. At their worst, vacuums can be rather raucous appliances; but some models let you switch the vacuum's speed down a notch or more, a feature that should let you reduce noise. But don't count on it: not many multispeed models offer the advantage of an appreciably quieter operation at a lower speed.

Nozzle-height adjustment. For deep carpet cleaning, the nozzles of a number of models are supposed to be raised or lowered to suit the particular carpet you're working on. Too high a setting and the rotating beater/brush won't agitate the nap enough to loosen dirt, too low and the whole nozzle digs in so deep that it becomes hard to push.

Setting the nozzle may prove a small nuisance. With some models, the adjustment is inconvenient, or the position reached is only vaguely indicated. Even those models with clearly marked settings, say the manufacturers, are best set by ear—by listening to the vacuum's pitch as the nozzle is lowered on the carpet, and by making the adjustment at the point where the whine changes to a buzz.

Note, though, that some models clean carpets well without making you adjust the brush height by hand. A few uprights and most power-nozzle canister units don't require adjustment in normal operation. (Either they have no adjustment mechanism or the mechanism is to be adjusted only for carpeting of unusually high pile.)

Suction reducers

When you're cleaning curtains, drapes, or other light materials, a vacuum cleaner's suction might well inhale the material right into the nozzle. So most canisters provide a ring, slide, knob, or lever that lets you bleed air into the hose to reduce suction.

An upright doesn't need a suction reducer: when used with attachments, its suction is already quite low.

Ease of pushing. Upright models with power-assisted wheels are easy to push, though it takes a bit of a knack to keep them under control. (With the power-drive switched off, such machines are unwieldy.)

Most power-nozzle canisters are easier to maneuver than the bulky uprights. But a loose and wobbly joint between the power nozzle and wands of some models provokes a sense of uneasy control. It can require constant correction to keep them from wandering.

In hunting for surface litter you will find few suction-only canisters that move smoothly over carpet. Most tug and grab and stub. One area that's hard to clean properly is under furniture. Unpowered rug and floor tools can get into the smallest spaces. Power nozzles require another inch or so of headroom. And uprights, with their fat noses, need the most.

Bag changing. Most models work only with disposable dust bags. Unfortunately, disposable bags aren't necessarily neat and easy to change. Some spills are inevitable if a bag fills through an opening in the bottom. A number of canisters use bags with a cardboard collar that requires tricky folding and fitting in a cramped space. And a mess is a sure thing if a too-full bag becomes trapped inside a machine and has to be ripped out.

Bag indicators. A few canisters attempt to show, usually by way of a flag rising in a window, how much dirt is in the bag. Others signal with a light or a pop-up button when the machine feels the bag needs changing. However, all the indicators are just rough guides, not a substitute for regular checks on the bag itself.

Cord storage. Storing a power cord can be rather a nuisance with a movable appliance. The most convenient cord storage takes the form of a reel with a button or pedal release that sends the cord zipping back into the machine. Next best is a reel that releases with a yank of the cord.

If there isn't a reel, the cord has to be wrapped by hand—with varying degrees of difficulty and security—on projecting hooks or fittings.

Hoses. Some canisters have a thin plastic hose that tends to kink rather than curve when going around bends. The reinforced hose of most others is more substantial and more tractable.

Wands. It's good to have wands made of metal and with some provision—usually a button popping through a hole—for locking wands and tools in place. A few canisters have locking plastic wands. Friction-fit plastic wands are difficult; once firmly fitted, they're hard to separate.

Tools. Typically, a canister comes with a power nozzle or a passive floor/rug nozzle, a wall/floor brush, an upholstery nozzle, a dusting brush, and a crevice tool. Attachments are often optional for uprights.

Use as blowers. A vacuum cleaner that can be converted into a blower can be handy for hard-to-reach areas, such as the coils of a refrigerator. Most power-nozzle canisters and a few suction-only ones can be used that way.

Filters. A canister's motor pulls air through the dust bag, so it needs protection against dirt that may evade the bag and against the possibility of a broken bag. In most canisters, there's a filter in the airflow behind the bag and before the motor.

Most uprights are of a design that requires no motor protection. In only a very few models is dirt apt to be drawn past the bag and toward the motor, and in those, the motor is protected against such an occurrence.

Current draw. Uprights and suction-only canisters draw moderate electric current, from 3 to 9 amps. Power-nozzle canisters usually draw more—as much as 12 amps for the most power-hungry models. So, with power nozzles in particular, it's wise to turn off as many electrical devices as possible on the circuit powering the vacuum before beginning to clean. Otherwise, you may pop a circuit breaker or blow a fuse. Avoid using light-duty extension cords with vacuum cleaners, too.

Safety. Manufacturers' warnings against the use of vacuum cleaners outdoors or on wet surfaces should be heeded.

Vacuum cleaners are involved in many home accidents, mostly injuries to fingers and toes from the revolving brushes in uprights and power nozzles. Such injuries are largely preventable with a little common sense. First of all, wear shoes when cleaning house. Don't try to free something caught in the roller without first turning off the cleaner. Don't convert uprights to accessory use without unplugging the machine. And don't leave children alone with an upright or power nozzle.

Maintenance

Except for changing bags (and it's best to use replacement bags purchased by mail from the manufacturer—or from an authorized dealer), there's little to do. But note that the brush roller in power nozzles and uprights is driven by a belt. The belts can and do break. In general, belts in uprights are easier to replace than the ones in power nozzles. It's a good idea to keep a spare belt on hand to save the frustration of shopping for one in the middle of a house cleaning.

Recommendations

If your home has expanses of carpeting, an upright is a wise choice. If your decor tends toward bare floors, loose mats, and a busy layout, you will do better with a suction-only canister. If, however, your home doesn't lend itself to such distinct typecasting, you might need a power-nozzle canister, a two-in-one machine—or even two or three machines, an upright, a suction-only canister, and possibly a lightweight upright (see page 318). Your needs will obviously be dictated by whether you want the convenience of an upstairs and a downstairs cleaner, whether you live were dust accumulates rapidly, and so on.

Ratings of vacuum cleaners

(As published in an August 1982 report.)

Listed by types; within types, listed, except as noted. In order of estimated overall quality. Bracketed models, judged approximately equal, are listed alphabetically. Weights are to nearest 1/2 lb. and include stored tools, hose, and nozzle for canisters. Prices are suggested retail and, unless otherwise indicated, include attachments only for canisters; + indicates shipping is extra.

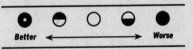

Better ← → Worse

Uprights

Brand and model	Price	Carrying weight (lb.)	Deep carpet cleaning	Bag capacity	Relative quiet	Switch convenience	Bag-changing convenience	Cord-storage convenience
HOOVER U3101-900	$340 [1]	21	◉	◉	◓	◓	◓	◉
EUREKA 5060	290	18	◓	◉	◓	○	◓	○
ELECTROLUX DELUXE 1451E	353 [1]	24	○	◓	◓	◓	●	○
PANASONIC MC663	220	14½	○	◓	○	◓	◓	○
PANASONIC MC671	145	13	○	◓	○	◓	◓	○
EUREKA 1425	100	11½	○	◉	○	○	○	○
EUREKA 1428AT	150 [1]	11½	○	◉	◓	○	◓	○
HOOVER U4159	160	17	○	◉	◓	○	◓	○
J.C. PENNEY Cat. No. 6304	165	17	○	◉	◓	○	◓	○
SINGER U60	80	12½	○	◉	◓	◓	○	○
SINGER U68	125	12	○	◉	◓	◓	○	○
SINGER UHD4	220	14	○	◉	◓	◓	◓	○
HOOVER U4119	95	15½	○	◓	◓	◓	◓	○
J.C. PENNEY Cat. No. 1016	78+	15	○	◓	○	○	○	○
ROYAL 801	290	14	◓	◉	◓	○	○	○
ROYAL 153	190	13	◓	◉	◓	○	—	◓

Power-nozzle canisters

Brand and model	Price	Carrying weight (lb.)	Deep carpet cleaning	Suction	Bag capacity	Relative quiet	Switch convenience	Bag-changing convenience	Cord-storage convenience
HOOVER S3199	$380	30	◑	◑	◑	○	◑	○	⊙
HOOVER S3193	240	24½	◑	◑	◑	◑	◑	○	○
ELECTROLUX DELUXE 1453	300	23	○	○	●	○	○	⊙	◑
EUREKA 1745	220	24½	○	◑	○	◑	○	◑	○
EUREKA 1773	300	27½	○	◑	○	◑	○	◑	◑
HOOVER S3189	160	21	◑	○	○	◑	○	◑	○
PANASONIC MC881	220	21	○	○	○	○	⊙	◑	⊙
PANASONIC MC883	270	24	○	○	○	○	⊙	◑	⊙
SINGER C17	230	24½	○	○	○	○	○	◑	◑
DOUGLAS C6675	164	17	◑	○	◑	●	●	◑	○

Suction-only canisters

Brand and model	Price	Carrying weight (lb.)	Deep carpet cleaning	Suction	Bag capacity	Relative quiet	Switch convenience	Bag-changing convenience	Cord-storage convenience
HOOVER S3187	160	22	—	○	○	○	◑	○	⊙
PANASONIC MC771	140	15½	—	○	○	◑	⊙	◑	⊙
HOOVER S3183	90	16	—	○	○	○	○	○	○
SINGER C16	135	19½	—	○	◑	◑	○	◑	◑
EUREKA 3710	140	19	—	○	◑	○	◑	◑	◑
DOUGLAS B6674	75	10½	—	○	◑	◑	●	◑	○
EUREKA 3320	90	12½	—	◑	◑	◑	◑	○	◑

1 Price includes hose adapter and above-floor cleaning attachments.

SPECIFICATIONS AND FEATURES
Except as noted, all: ● Have 1-speed motor.
● Cannot be used as blower. ● Use disposable dust bags. ● Have 15- to 20-ft. cord.

All uprights: ● Were unstable on stairs.

Hand-held cordless vacuum cleaners

Indoors it's a nuisance to get out your upright or canister vacuum cleaner for small spills and light accumulations of debris. A good minivac can suck up the dirt with much less fuss. Outdoors, a small, cordless appliance seems perfect for cleaning a car's interior, especially if there's no electrical outlet within easy reach.

Like many other cordless devices, the minivacuum owes its existence to the rechargeable battery. Aside from its charging stand, a cordless vacuum typically consists of two elements: a motor/handle element that holds the battery and a nozzle/collector element that sucks up and stores the dirt. To empty most models of their dirt, you simply push a release button, separate the two elements, and shake out the collector.

Cleaners on the market are alike in many ways. They have a sturdy, shock-resistant plastic body about 14 or 15 inches long and about 3 or 4 inches high, and they weigh less than 2 pounds. When not in use, the cleaner is stored on a combined battery charger/holder, which can be mounted on a wall. Plugged into a wall outlet, a charger draws less than 5 watts.

Before you use a new cleaner, be sure to give it a 16-hour charge, or whatever the manufacturer recommends. Next, to condition the batteries and break in the motor, completely discharge and recharge the battery at least 3 times. (A vacuum cleaner will perform better if its batteries are completely discharged and recharged now and again.)

Suction and capacity

Good cleaning depends partly on the strength of a cleaner's suction—the ease with which it sucks up dirt. It also depends on how long a vacuum cleaner can maintain suction without significant drop-off.

No cordless vac will provide prolonged cleaning power: The hardiest models are likely to give you 10 to 15 minutes of sustained effort.

A cordless vacuum will do its job best when completely free from the debris of an earlier cleaning mission, so empty it before putting it back on the charger.

Cleaning car interiors

Most cordless cleaners are effective for regular cleanups of a reasonably tidy car interior. However, a full-sized vacuum cleaner will do the job far better if the car is really dirty.

Car cleaning was easiest with the **Dustbuster Plus,** the **Norelco 2,** and the **Douglas 6.0,** mainly because of their superior staying power. A brush and a tapered nozzle extension, or crevice tool are helpful attachments when working inside a car.

Gravel, hair, and dried grass lodged in a car will often be too much for these units. And some models won't be able to cope with sand.

Use care when cleaning up sand: the cleaner may eject fine particles through the air-exhaust ports with surprising force—an unpleasant surprise if the vacuum is close to your face, as it may well be in a confined space such as a car's interior. Protecting your eyes is a sensible precaution.

Cleaning household debris

Potting soil. Most cleaners can pick up spilled potting soil from a hard surface with no more than about two passes. On a rug, however, pickup is more difficult, often needing more back-and-forth passes.

Hair. A cleaner that can pick up hair from a hard surface is a good choice for parents who give their toddler a haircut in the bathroom or kitchen.

Cigarette butts. Any unit should be able to pick up the occasional cigarette butt, but a full ashtray dumped on a rug is best left to full-sized cleaners.

Convenience

Be sure the cleaner you buy fits your hand and that you are strong enough to maneuver the appliance. Check on the ease of disassembling the unit for cleaning.

Most manufacturers recommend using a brush to clean the vacuum. An old two-inch paint brush should serve nicely.

Noise. This is a minor concern.

Storage. Several models hang nozzle-down from the charger; they can drop debris from the nozzle when you switch the cleaner off. You'll have to get into the habit of pointing the nozzle upward before shutoff. That way, the vac will inhale any loose material from the nozzle.

Features. Cordless vacuum cleaners don't have the sort of dust bag found in a full-sized vacuum; most store dust and debris in the collector. All have a replaceable filter to protect the motor from dust.

Most models let you wind excess charger cord neatly behind the holder. That's a nice convenience, as is a rack for extra tools.

Recommendations

Don't use a cordless vacuum on wet material or in a damp area such as a patio after rain; moisture could damage the machine.

Lightweight vacuum cleaners

It's a chore to drag out a full-sized vacuum cleaner (see page 314) to pick up the crumbs from last night's snack or the sand tracked in from this morning's beach outing. You can use a carpet sweeper to touch up rugs, and a broom to clean bare floors. Or you can use a lightweight, upright vacuum cleaner on rugs as well as bare floors.

These vacuums have several advantages. They're light (five to nine pounds). They store in small places. And they're a lot cheaper than conventional full-sized vacuums.

Apartment dwellers short on space might rely on a lightweight machine for all vacuuming chores, but these little vacuums aren't intended for heavy-duty cleaning. They have relatively low suction, and most have limited brush action. Therefore they don't provide the deep carpet cleaning of a full-sized upright vacuum. A typical lightweight has a nozzle brush, but the brush just floats freely over carpets.

Some lightweights come with attachments to do above-the-floor cleaning. But, because of their low suction, they can't handle that job as well as a full-sized canister vacuum.

Cleaning

A lightweight cleaner picks up, but not as efficiently as you might hope:

Surface cleaning. These machines do much better at removing dust and litter from bare floors than from carpeting. Most will leave a dusty carpet looking somewhat linty.

Edge cleaning. On bare wood, most lightweights will draw in dust adjacent to walls. But most aren't able to completely pick up heavier dirt, such as beach sand.

On carpets, a lightweight vacuum will commonly leave 1½ inches of border area virtually untouched.

Convenience

There's more to convenience than light weight and small size. There's ease of maneuvering, frequency and ease of emptying the bag, and noise.

Maneuvering. Pushing any vacuum cleaner over a hard floor is easy. Pushing one over carpeting may not be. Since a lightweight is so slow to pick up soil, it can take a lot of extra pushes and pulls to clean a section of carpet.

Most models have little wheels or rollers built into the nozzle, making it reasonably easy to maneuver the machines over medium-pile carpet.

Some lightweights don't have wheels. Since the nozzle bears down directly on the carpet, they are not very easy to maneuver on carpet. Check to be sure the handle is long enough. If you have to stoop while pushing, vacuuming becomes exhausting, or even impossible.

A second handle or grip down low, near the motor housing, makes the machines easy to maneuver when vacuuming stairs.

Exhaust deflection. Air is drawn into a vacuum cleaner through its nozzle, filtered through the machine, and ex-

Ratings of lightweight vacuum cleaners

(As published in a July 1983 report.)

Listed in order of estimated overall quality. Essentially similar models are bracketed and listed alphabetically. Prices are suggested retail; + indicates that shipping is extra.

Legend: Excellent ● Very good ◕ Good ○ Fair ◑ Poor ●

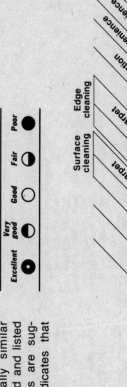

Brand and model	Price	Weight (lb.)	Surface cleaning — Bare floor	Surface cleaning — Medium-pile carpet	Edge cleaning — Bare floor	Edge cleaning — Medium-pile carpet	Maneuvering on carpet	Exhaust deflection	Emptying convenience	Switch convenience	Initial suction	Noise	Capacity (oz.)	Comments
⊘ REGINA ELECTRIKBROOM POWER TEAM HB6910	$72	9											1/2	A,J,K
BISSELL DYNA-CLEAN 3045-7	58	6 1/4											6	E
HOOVER QUIK-BROOM S2039	75	8											4	E,J
J.C. PENNEY Cat. No. 0539	65+	7											6	E,G
BISSELL DYNA-CLEAN 3004-4	52	6 1/4											6	—
BISSELL MULTI-VAC DYNA-CLEAN 3030-9	52	6											4 1/2	I,J
REGINA ELECTRIKBROOM B4316	42	5 1/4											1/2	G,K
SEARS KWIK-SWEEP Cat. No. 62362	35+	5 1/4											1/2	H
HOOVER QUIK-BROOM S2041	70	8											4	J

Brand and model	Price	Weight (lb.)	Surface cleaning — Bare floor	Surface cleaning — Medium-pile carpet	Edge cleaning — Bare floor	Edge cleaning — Medium-pile carpet	Maneuvering on carpet	Exhaust deflection	Emptying convenience	Switch convenience	Initial suction	Noise	Capacity (oz.)	Comments
HOOVER QUIK-BROOM S2015	50	6¼	○	●	●	◐	◐	○	◐	◐	◐	●	3	B
REGINA ELECTRIKBROOM HB7439	52	6	●	●	◐	●	◐	●	◐	◐	◐	◐	½	F,J,K
SEARS KWIK-SWEEP Cat. No 62382	55+	6	●	◐	●	●	●	●	●	◐	●	◐	½	F,H
REGINA ELECTRIKBROOM B6228	48	5½	●	●	●	●	●	●	◐	●	●	◐	½	E,K
SEARS KWIK-SWEEP Cat. No. 62372	45+	5½	●	●	●	●	●	●	●	●	●	◐	½	E,H
EUREKA 163 Type A	70	5¼	○	●	●	◐	◐	○	◐	●	○	◐	1½	C,D
HOOVER QUIK-BROOM S2057-900	70	6¾	●	●	◐	●	●	●	◐	●	●	◐	4	I,J

SPECIFICATIONS AND FEATURES

All have: 15-20 ft. power cord, which must be manually wound up for storage. *Except as noted, all have:* ● Suction-only operation, making them unsuited for deep carpet cleaning. ● Single motor speed. ● Nozzle brush that locks in position for hard-floor cleaning, floats for carpet cleaning. ● Optional attachments, including upholstery nozzle, brush, and crevice tool, available for approx. $15 to $25.

KEY TO COMMENTS

A – Only model tested with power nozzle. Judged fair at deep carpet cleaning.
B – Relatively short handle; requires stooping to push.
C – Modified hand vacuum is poorly balanced. Weight is concentrated in motor housing near handle; nozzle tends to float off floor and wander.
D – Blows exhaust air on user.
E – Two motor speeds.
F – Three motor speeds.
G – Nozzle brush is adjustable.
H – Optional attachments not available from Sears, but Regina attachments can be used.
I – Attachments included.
J – Has auxiliary handle near motor housing; useful when vacuuming stairs.
K – Essentially similar versions of **Regina** models as follows: **HB6910** as **HB6010** and **HB6710; B4316** as **B6518** and **B4516; HB7439** as **HB7739** and **HB7839, B6228** as **HB4528** and **B6628.**

pelled back into the room. The air coming back into the room should be diffused so that it doesn't disturb loose dust, particularly on hard floors.

Emptying. Lightweight machines often don't have much dirt-holding ability. The smaller the capacity, the more often the vacuum cleaner has to be emptied.

The dust cup should be easy to empty. Or if there's a disposable paper bag it should be easy to remove and empty without spilling.

Switches, cords. An on/off switch is most convenient when it is located within reach of the handle. If the switch is located on the motor housing you have to use one hand to grip the vacuum's handle, the other to work the switch.

Initial suction. At best, a lightweight vacuum has only about half the suction strength of a full-sized machine. Any machine with low initial suction is likely to be too feeble to work well with optional cleaning attachments.

Extra speeds. Several models feature two or three motor speeds. In all, the highest speed is "normal." The optional speeds are slower. Since performance at normal speed is barely adequate, you probably won't have any use for the slower speeds.

Recommendations

Most lightweights do a good job of picking up dust on hard floors. That job alone makes one convenient to have handy for quick cleanups.

Steam vaporizers

A vaporizer can't cure a cold, but it can make you feel better. Medical authorities agree that the only purpose of a vaporizer is to add moisture to the air, soothing your parched nose and throat. It makes no difference whether the moisture added is warm or cool, medicated or unmedicated.

But cool mist from an ultrasonic machine may not give you the pampered feeling you get with steam. In a steam vaporizer an electric current passes through a solution of water and minerals, heating it to generate steam.

Moisturizing

How much moisture you'll need from a vaporizer depends on the size of the sickroom and the humidity level you want. You won't be able to control humidity levels precisely, since the typical steam vaporizer doesn't have a humidistat. But you can judge the size of the room and plan for the worst. A vaporizer with an output of about one-half pint per hour should suffice for a small bedroom. A large bedroom would probably require moisture output of around three-quarters of a pint per hour, perhaps more if the room is not well sealed.

Steam vaporizers typically cook out a pint or more per hour if the manufacturer's instructions are followed; if that's too much for the room, you can reduce the steaming rate by using a smaller amount of minerals in the solution.

Overvaporizing probably wouldn't hurt the patient, but moisture can condense on windows or walls, and the sickroom could become uncomfortably clammy. The added moisture could conceivably damage outside walls if they do not contain an adequate vapor barrier.

An important factor is running time between refills. The water capacity quoted by manufacturers of vaporizers won't necessarily give you an exact idea of how long the units can run without a fill-up. That's because a unit's output can vary considerably, and because all vaporizers leave a bit of water in the bottom of the reservoir when they have supposedly run dry. So, consider manufacturer ratings as being on the shy side. It makes sense to pick a vaporizer that will run for at least twelve hours, so the unit can run all night without requiring a refill.

Other factors

It's worth considering vaporizers for characteristics other than their all-important moisture-generating ability.

Noise. The gentle bubbling of a steam vaporizer shouldn't bother anyone.

Filling. The easiest way to fill a large vaporizer is with a pitcher at the place of intended use, even though that may mean several trips with the pitcher. The alternative is to carry as much as twenty-one pounds of water ($2\frac{1}{2}$ gallons) sloshing about in a large, flat container. What's more, you might need a faucet with a long spout (such as on a bathtub or kitchen sink) to fill some of the larger models.

The fill level is clearly indicated on most models, generally by a label or mark on the inside of the tank. On many you can see the water level through a translucent tank, a window, or a transparent cover.

Some units may have the fill level marked on the outside of a fairly opaque tank, which makes fill-up something of a guessing game. With steam models, a tank filled to the brim can overflow when you reinsert the boiler containing the electrodes.

For safety's sake, a steamer should be unplugged during refills. If there is an overflow, wipe the unit dry before plugging it in; if you don't, the overflow could be electrically live.

Cleaning

Steam-vaporizer reservoirs need only a rinse after each use, but the boiling chambers require regular cleaning. Over the long term, the electrodes themselves are likely to require cleaning, too, particularly if hard water has been used. Otherwise, the output will dwindle or become erratic.

The most direct method of cleaning is to disassemble the boiling chamber and sand or scrape away the deposits on the exposed electrodes. This can be difficult to do on some models because only the bottom of the boiling chamber comes off, requiring you to clean by poking up through the bottom.

Some makers suggest that stubborn deposits can be softened by soaking electrodes overnight in vinegar before cleaning. You may be able to buy cleaning tablets to dissolve the electrode deposits, but the tablets won't do the whole job. Scale and debris can flake off the electrodes and collect in the bottom of the boiling chamber.

It's a good idea to avoid running a steamer with very low water levels because the solution becomes concentrated and more active chemically, which increases the likelihood that the electrodes will corrode. A few models have carbon electrodes, which do not corrode but do collect mineral deposits. You'll have to scrape gingerly to avoid gouging the relatively soft carbon.

Steamers take a few minutes to reach their full output. Vapor production begins to tail off when the water level drops enough to leave a major portion of the electrodes above the water level. Furthermore, the rate of steam production depends heavily on how well the water conducts electricity. Distilled water is essentially not conductive, while hard water is highly conductive. Consequently, people who live in soft-water areas generally have to add mineral salts to adjust the water's conductivity for a satisfactory steaming rate.

Manufacturers variously suggest adding borax, baking soda, or common table salt. Borax and baking soda are less conductive than salt, so you have to add more of them. With salt, a pinch or two is all you need, even with very soft water. Because only a little bit of salt is needed, the exact amount is critical; a pinch too much can make the vaporizer start spitting hot water along with the steam—and not necessarily at once but possibly a few hours later. Very hard water can be too conductive, so you may have to blend tap water with distilled water or rainwater in order to slow down the vaporizer.

The water in a steam vaporizer is electrically live whenever the unit is plugged in.

Modern vaporizers made of sturdy plastic in low, wide designs minimize the likelihood of a vaporizer's accidentally breaking or tipping over. These designs also minimize the risk of being burned by the hot water or getting a shock by coming in contact with the water while it's electrically live.

Steam vaporizers present a second potential problem if they are not cleaned regularly. If scale or debris is allowed to build up at the bottom of the boiler chamber, it can bridge the electrodes and lead to a short circuit.

To prevent this kind of problem, buy a vaporizer with ample capacity for overnight duty, so that the unit won't boil dry; always add the same amount of salts, so that the steaming rate will be predictable. It's also important to clean the boiler chamber regularly.

Before attempting any service on a vaporizer, you should, of course, be sure that it is unplugged.

Recommendations

Vaporizers are for sickroom use; they are not whole-house humidifiers. Although they can't cure anything, they can ease the miseries of a severe cold or flu.

The best time to buy a vaporizer, of course, is before someone in your family becomes ill and requires one. In an emergency, you might have to settle for whatever is available without regard to performance or price.

If you relish the warmth of steam and are soothed by the cozy sound of percolation, you won't go far wrong with just about any steam vaporizer. Performance doesn't vary much from model to model and depends mainly on the mineral content of the water used.

Unless your storage space is very limited, it makes sense to buy a large model for the convenience of extra running time between refills.

Waffle makers

Homemade waffles used to be a meal staple, served with butter and syrup, jam, creamed fish and fowl, or just about anything else that struck the cook's imagination.

There are still people who want their waffles made fresh, not thawed from the freezer. Some are traditionalists who favor a waffle with smallish indentations. Others favor the Belgian waffle, with its larger, deeper indentations. Appliance manufacturers cater to those preferences; so did we in choosing the eighteen waffle makers we tested. Four make round waffles; the others make them rectangular or square.

Most traditional waffle makers have multipurpose grids. Some grids can be flipped to provide a flat surface for grilling sandwiches or hamburgers; some have a round, patterned surface on the flip side, used for baking a kind of wafer cookie called a pizzelle.

The absence of a grill or a pizzelle maker isn't really important. No one who owns a range and frying pan needs a grill. And most people probably don't feel the urge to make pizzelles very often.

The perfect waffle?

The CU testers' goal was to produce six consecutive batches of medium-brown waffles in each appliance. They aimed for waffles that were evenly browned, crisp on the outside, soft and light on the inside.

As a group, traditional waffle makers did the best job. The best of the Belgian-waffle makers also made very good waffles.

Time and temperature

Most waffle makers need about half an hour, from preheat to finish, to make six medium-brown waffles. That's why waffles are a Sunday kind of breakfast.

Many waffle makers have a browning control that lets you choose how dark you want your waffles, plus a signal light. But the lights are generally hard to see, and they aren't all that helpful. In most models, the light indicates only that the waffle maker has come up to temperature, not necessarily that the waffles are cooked. You'll do better to time the baking yourself. CU's testers didn't find most browning controls to be very useful. They cooked light or medium-brown waffles quickly by turning the control to its darkest setting and then adjusting the baking time as needed.

Grilling

Waffle makers that double as grills can be used two ways: open, flat, with the lid and the base used as grilling surfaces; or closed, as both sides of the food are cooked at the same time. Open, they can obviously grill twice as much food as they can closed, but they take twice as long because you have to flip the food over to cook both sides. Quarter-pound, ground-chuck patties cook on the open grill in four minutes instead of two; grilled-cheese sandwiches, about six minutes instead of three. Both jobs take about the same time in a frying pan as on the open grills.

The first side of burgers cooked on the open grills comes out nicely browned, but the second side hardly browns at all. Inside, the meat isn't very evenly cooked. Hamburgers cooked in the closed grills come out much better—nicely browned outside and rare or medium-rare inside.

With either cooking method, it's important to preheat the grill; otherwise, the hamburgers won't brown. Preheating takes from $7\frac{1}{2}$ to $11\frac{1}{2}$ minutes.

Grilled-cheese sandwiches come out better when you cook them on the open grills. Closing the lid compresses the sandwiches so the cheese oozes out, creating sandwiches that aren't very pretty and grills that are a mess to clean.

Safety and other concerns

The plastic handle used to open and close a waffle maker's lid during cooking should be large enough to provide a good grip, so that your fingers won't touch a surface that can reach nearly 300°F. The handle should also be designed to shield fingers from escaping steam.

Most waffle makers have carrying handles on opposite sides of the appliance. The handles should be generous-sized and convenient, not so narrow that you could burn yourself if you tried to move the appliance too soon after cooking.

Nearly all waffle makers draw between 500 and 1,100 watts. If you're using a waffle maker and another high-wattage appliance (such as a toaster oven) on the same 15-amp circuit, you could blow a fuse or trip a circuit breaker. A few lights alone on the same circuit shouldn't be a problem.

Convenience

Most waffle makers we tested are steady on their feet or stands. Rubber inserts in the feet make the appliance particularly stable. Round waffle makers have three feet, and tip back too easily when the lid is raised. That's annoying if you're trying to pry loose a waffle that's clinging to the lid.

Most models convert easily from waffle maker to grill.

Cooking grids have a nonstick coating. As long as you oil them lightly before each fresh use, you shouldn't encounter serious sticking problems. Once in a while, you may have to oil the grids between batches. The little bits that may cling to the grids are easily dusted away with a pastry brush when the grids are still slightly warm from cooking. When you want to wash away

Ratings of waffle makers

(As published in a July 1985 report.)

Listed by types; within types, listed in order of estimated overall quality. Bracketed models, judged approximately equal in quality, are listed in order of increasing price. Prices are mfr. suggested retail; + indicates shipping is extra.

Rating key: ● Excellent ◐ Very good ○ Good ◒ Fair ● Poor

Traditional waffle makers

Brand and model	Price	Size of waffle grills, in.	Performance [1]	Extra functions [2]	Advantages	Disadvantages	Comments
GENERAL ELECTRIC G48T	$53	8x8	Excellent	G	A,B,C,H,I	q	B,F
TOASTMASTER 269	67	9x9	Excellent	G	B,C,G,L,N	n,q	B,E,F
BROIL KING 785	70	9¼x5½	Excellent	G,P,S	B,C,D,F,K	a,d,h	A,B,G,I
BROIL KING 736	60	9x9	Very good	G	B,C,N,Q	a,n	B,D,F
RIVAL 9705	60	9x5¼	Very good	P	A,D,Q	l	A,C,F
TOASTMASTER W252	34	7 dia.	Excellent	—	A,G	i,m,o,p,q	B,F
MUNSEY WC-1	33	8x8	Fair	—	A,Q	c,g,k,o,p,q	C,H

Not acceptable

■ *The following models were judged Not Acceptable because lid handles do not protect user adequately from burns.*

Brand and model	Price	Size of waffle grills, in.	Performance [1]	Extra functions [2]	Advantages	Disadvantages	Comments
TOASTMASTER 270	54	8x8	Very good	G	B,C,H,I	e,j	B,F
SEARS Cat. No. 48238	44+	8x8	Very good	P	E,H,I	e,j,p	B,F
TOASTMASTER 290	58	8x8	Very good	P	E,H,I	e,j,p	B,F
SEARS Cat. No. 64778	42+	8x8	Fair	G	B,C,H,I	e,j	B,G

Belgian-waffle makers

Brand and model	Price	Size of waffle grills, in.	Performance [1]	Extra functions [2]	Advantages	Disadvantages	Comments
SEARS Cat. No. 64861	25+	7¼ dia.	Very good	—	—	b,i,m,o,p,q	B,G
J.C. PENNEY Cat. No. 4616	35+	7¼ dia.	Very good	—	—	b,i,m,o,p,q	B,G
TOASTMASTER 250	37	7¼ dia.	Very good	—	—	b,i,m,o,p,q	B,G
BROIL KING 780	50	9x4¾	Good	—	A,J,Q	o,q	C,G
RIVAL 9710	60	9¼x5½	Good	—	P,Q	f,l	A,C,F,J
OSTER 712-06	46	8¼x4	Fair	—	A,H,M,O,Q	f,o,p	A,C
MUNSEY BW-2	26	5¾x5¾	Fair [3]	—	Q	b,c,f,g,k,o,p,q	C,H

1 *Based on 6 consecutive batches of medium-brown waffles, which usually took 30 min. for preheating and baking.*

2 *G = Grill with range of temperature settings; P = Pizzelles; S = Sandwich maker.*

3 *Performance improved to ◯ when making dark brown, crisp waffles.*

SPECIFICATIONS AND FEATURES

All: ● Have plastic feet or stands. ● Have nonstick-coated grids.

Except as noted, all: ● Weigh 3 to 6½ lb. ● Use 500 to 1,100 watts. ● Have grids that form 4 individual waffles and that can be reversed for cooking other foods. ● Have browning control and light (which is hard to see) to indicate when unit is hot enough to begin cooking. ● Have overflow rim. ● Open flat for cleaning. ● Have plastic lid handle that provides adequate protection from burns. ● Have plastic carrying handles that should be used only when unit has cooled down. ● Are stable when lid is raised. ● Have crevices near handles or control that are somewhat difficult to clean. ● Can be stored on end.

Unless otherwise indicated, none: ● Have on/off control or light that signals when waffles are cooked.

KEY TO ADVANTAGES

A – Faster than most.
B – Hamburgers cooked with grill closed judged excellent or very good; those cooked on open grill judged good.
C – Grilled-cheese sandwiches cooked on open grill judged excellent; those cooked on closed grill judged good.
D – Pizzelles judged very good.
E – Pizzelles judged good.
F – Makes 2 filled, divided sandwiches per batch; judged very good.
G – Has on/off control.
H – Easy-to-see light indicates when unit is hot enough to begin cooking.
I – Light indicates when waffles are cooked to specified degree.
J – Has large, effective overflow rim.
K – Long plastic lid handle provides excellent protection against burns.
L – Plastic lid handle provides very good protection against burns.
M – Luggage-like lid handle provides very good protection against burns and safe, easy way to carry unit after cooking.
N – Plastic carrying handles judged very safe for carrying unit, even when it's still warm.

O – Rubber inserts in feet for extra stability.
P – Only Belgian-waffle maker with removable grids for easy cleaning.
Q – Housing judged fairly easy to clean.

KEY TO DISADVANTAGES

a – Slower than most.
b – Waffle grid isn't sectioned for serving-sized portions: unless cut by hand, entire waffle is a large serving.
c – Lacks signal light to indicate when unit is hot enough for cooking.
d – Controls are relatively inaccessible.
e – Small overflow rim.
f – Lacks overflow rim.
g – Lid handle provides less protection from burns than most.
h – Lid handle doubles as carrying handle and, as such, is awkward to use.
i – Plastic carrying handles become hot during cooking: must be allowed to cool before carrying unit.
j – Fairly narrow carrying handles provide little protection against burns.
k – Very narrow carrying handles provide minimal protection against burns.
l – No carrying handles.
m – Easily tipped when lid is raised.
n – Grids somewhat difficult to remove and replace.
o – Grids cannot be removed for cleaning.
p – Lid does not open flat for cleaning.
q – Unit does not stand on end for storage.

KEY TO COMMENTS

A – Grid sectioned to make 2 waffles.
B – Control can be set for light, medium, and dark waffles.
C – Lacks control for brownness of waffles.
D – Heavier than most: 8½ lb.
E – Heaviest, 9½ lb. Draws 1,400 watts.
F – Chrome-plated housing.
G – Painted metal housing.
H – Painted metal lid, plastic base.
I – One sample did not heat up and cook properly. Another sample's lid fell backward when it was raised.
J – One sample's plastic stand was not smoothly finished and scratched countertop.

Waffle recipes: Beyond the basics

Consumers Union waffle lovers felt encouraged to branch out from the basic **Bisquick** waffle batter. Herewith, two of their favorite recipes.

Gingerbread waffles

The recipe makes two nine-inch square or four seven-inch round waffles. One section of waffle with a dollop of whipped-cream topping (see below) or ice cream is a nice-sized serving.

1/4 cup butter or margarine
1/2 cup dark brown sugar
1/2 cup light molasses
2 eggs, separated
1 cup milk
2 cups flour
1 1/2 teaspoons baking powder
1 teaspoon cinnamon
1 teaspoon ginger
1/4 teaspoon powdered cloves
1/4 teaspoon salt

Cream together the butter and sugar. Beat in the molasses, the egg yolks, and milk. Sift together the dry ingredients and mix them into the batter. Beat the two egg whites until soft peaks form, then fold them into the batter. Bake in preheated waffle maker. Serve with sweetened whipped cream (1 cup heavy cream whipped with 2 teaspoons of sugar). For a fancier topping, sprinkle two tablespoons of chopped crystallized ginger into the whipped cream.

Sourdough waffles

Those who like sourdough bread will enjoy this recipe. It makes tangy waffles that are crisp on the outside, slightly chewy on the inside. You will need a sourdough starter, however—either a packaged product or homemade. The home-brew starter can take about a week to ferment, so don't count on sourdough waffles for tomorrow's breakfast.

Sourdough starter
1 cup water
Approx. 1 3/4 cups flour
1 egg
2 tablespoons vegetable oil
1/2 cup nonfat dry milk (optional)
2 tablespoons sugar
1 teaspoon baking soda
1/4 teaspoon salt

Prepare sourdough starter. When it has fermented, mix 1/2 cup of the starter with the water and enough flour to make a moderately stiff batter; don't worry about lumps. Cover the bowl and let it sit overnight in a warm place. Next day, remove a half-cup of the batter and put it back with the starter. To the remaining batter add the egg, oil, and, if you wish, the nonfat dry milk. Combine the remaining dry ingredients; stir them into the batter just before you pour it onto the grill. Bake at the waffle maker's darkest setting, if it has a browning control. Makes three seven-inch round or two eight-inch-square waffles.

excess oil, the removable grids have a decided advantage: They can be dunked in a sinkful of warm, sudsy water.

The flat grids for grilling usually require thorough cleaning—sometimes soaking—to remove hamburger grease or globs of sticky cheese. Most manufacturers recommend washing by hand, not in the dishwasher.

Most waffle makers have an overflow rim to catch excess batter. The grills have a channel around the edge to direct grease toward a spout in one corner and then into a dish you place under the grill.

The waffle makers with a chrome-plated metal housing show fingerprints easily. Those with a painted metal housing can scratch or discolor from heat.

Most of the time, cleaning the outside of these appliances requires no more than a wipe with a damp cloth. Occasionally, you may need to use a mild liquid cleaner to remove stubborn spots, but abrasive cleansers should never be used. Metal polish may help shine up the chrome-bodied machines. Batter overflow can present a more serious cleaning job.

Recommendations

A waffle maker isn't a necessity. There are always pancakes or frozen waffles. But if you consider pancakes a second-rate alternative, if you would rather go without than eat a frozen waffle, or if you think Sundays are meant for lavish breakfasts, a waffle maker can be one of life's small luxuries.

Traditional waffles, with their small indentations, seem more substantial than the airier Belgian variety. But both taste good as the base for strawberries and whipped cream or for creamed main-dish foods. For breakfast, though, it's easier to smear butter over traditional waffles.

Anothere thing to consider is that a waffle maker that also functions as a cookie maker might be more useful than a waffle maker that doubles as a grill.

Washing machines

The latest crop of washing machines is very versatile—you can have a great deal of control over the way the laundry is handled.

There are a variety of top-loading, front-loading, regular-size, large-capacity, compact, portable, or stackable machines. Then there's a multitude of features and combinations of features, cycles, temperatures, speeds, and water levels, and a further choice between dial controls, push buttons, or electronic touch controls. Knowing a few basics will help you narrow the choices.

Type. Washers can be either top- or front-loading. Top-loaders are by far the most popular in this country, while the European favorite is the front-loader.

Aside from the differences implicit in the names, the two types of washers differ in the way they agitate clothes. Top-

loaders circulate clothes up and down
and back and forth; front-loaders tumble clothes up and around, much the
same way that a dryer does.

An advantage of loading from the
front is that the top of the washer can be
used as a work area all the time.

Size. The size you choose depends on
how much laundry you do and how
much room you have. The most versatile machine is one that offers large
capacity but adjusts to small loads. The
most generous sizes of the large-capacity models are top-loaders, some of
which take as much as 16 pounds of
mixed, dry laundry. In general, front-
loaders hold less than top-loaders.

"Stackables" are compact washers
that are available separately or as part
of a unit that includes a dryer. Some are
small enough to fit inside a closet.

Water level. You can adjust the
amount of water to the size of your
wash. On most large-tub models the
minimum fill takes about half as much
water as the maximum fill. The difference can amount to 20 or 30 gallons of
water in a single wash.

The fill level is regulated by a pressure switch rather than a timer. So, even
if you live in an area where your water
pressure is low or irregular, your machine will continue to fill until the water

reaches the level you've selected.

Adjusting the water level to the load
will help you save water, energy, and
detergent. But the most economical
way to use your machine is to accumulate enough laundry to run the machine
only with a full load.

Water temperature. You can control
the temperature of the wash water and
the rinse water separately. These are
the standard options for wash and rinse:
hot/warm, hot/cold, warm/warm,
warm/cold, and cold/cold.

The less hot water you use, the
cheaper the machine will be to run. The
only time you need a hot wash is when
you have a load of very dirty clothes.
Otherwise, a warm or cold wash will do.
And always use a cold rinse because
warm water won't rinse any better, and
it may help wrinkle permanent-press
fabrics.

Speed control. Separate or built into
the cycle selector, these differing
speeds are part of the typical top-loader. This feature allows for gentler handling of delicate fabrics such as silk and
for fewer wrinkles when spinning out
permanent-press fabrics. Front-loaders
do not offer multiple speeds. They run
at a slow speed for wash and a higher
speed for spin.

Performance

There are a number of characteristics
to consider.

In absolute terms, front-loading machines use the least water. Fully filled, a
front-loader uses 25 to 30 gallons; top-
loaders anywhere from 40 to 57 gallons.
But top-loaders can hold as much as 50
percent more laundry than a front-loader. So one has to compare the gallons
used with the laundry done to arrive at
an efficiency judgment. As it turns out,
a front-loader is still the most efficient,

using less water per unit of laundry than
even the most efficient top-loader.

Hot-water efficiency is a measure of
how costly a machine is to operate.
That's because the greatest cost of running a washing machine is the cost of
providing the hot and warm water it
uses (the cost of electricity to run the
motor is negligible).

Again, a front-loading machine is the
most efficient. For example, in a typical
front loader, heating the water for what

the U.S. Department of Energy considers a typical year's worth of laundry would cost about $40 in electricity or $15 in gas (assuming a rate of 7.75 cents per kilowatt-hour and 62 cents per therm).

Other machines can cost considerably more. To do the same amount of laundry in an efficient top-loader would cost $75 in electricity or $30 in gas. With a relatively inefficient top-loader it would cost about $110 in electricity or $45 in gas.

If you're concerned about cutting down on water consumption, you might consider buying a "suds-saver" model that allows you to store the wash water and reuse it later on. Some people are skeptical about reusing dirty, cooled wash water. But many consumers seem to be interested in this feature.

Suds-savers can cost $20 to $40, but not all manufacturers promote suds-savers.

Permanent-press handling. If you dry your clothes on a line, it's important to have a machine that cools the clothes effectively before the spin-dry phase. That will help keep wrinkles from setting in permanent-press fabric. But if you use a clothes dryer, you don't really have to worry about how your washer handles permanent-press items; the dryer's permanent-press cycle should remove most wrinkles.

How the washer handles permanent press is important mainly for those who line-dry their laundry. A washer with at least two speeds minimizes compaction of the clothes by using a slow spin. And a machine with a permanent-press cycle cools down the wash with sprays of water or with rinses before it goes into the final spin. The extra-rinse method on some models is even more effective.

Unbalanced load. A washing machine will operate best when its load is spread evenly around the tub. With an unbalanced load, some machines show no signs of stress; others stop automatically when the load is even moderately unbalanced; still others don't spin properly. The best solution to the unbalance problem is to redistribute the wash load manually as soon as the problem occurs.

Water extraction. The amount of water left in your clothes after they're spun out determines drying time, whether in the dryer or on the clothesline. The best models leave about five pounds of water in a load of mixed fabrics, while others leave about six pounds.

Linting. The amount of lint left in your wash can depend on what you wash, the degree of agitation, and the machine's method of lint disposal. Front-loaders and the most effective top-loaders will leave only slight deposits of lint.

Lint filters vary somewhat in convenience for emptying, but not enough to be significant. A few washers flush the lint out with the wash water, which may pose a problem for drains that are already partially blocked.

Most models have a bleach dispenser. Its job: to channel the bleach into the outer tub instead of onto the clothes in the inner tub; the bleach is diluted before the water reaches the clothes. There's no dispenser on front-loaders.

Sand disposal. Most machines do a good job of getting rid of any sand that's collected in beach towels and clothing.

Oversudsing. Some machines, especially front-loaders, tend to froth when you use a high-sudsing detergent, especially if the water is soft. You can get around it by using a low-sudsing detergent.

Ratings of washing machines

(As published in a June 1985 report.)

Listed by types; within types, models, listed in order of estimated quality. See also features table beginning on page 336. Prices are average and range for white models, as quoted to CU shoppers in a 15-city survey.

Better ● ◕ ○ ◑ ● Worse

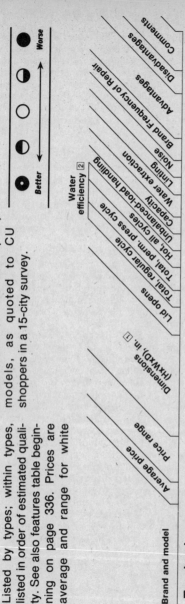

Top-loaders

Brand and model	Average price	Price range	Dimensions (HxWxD), in. [1]	Lid opens	Total, regular cycle	Hot, perm. press cycle	Total, all cycles	Unbalanced-load handling	Capacity	Water extraction	Linting	Noise	Water efficiency [2]	Brand Frequency of Repair	Advantages	Disadvantages	Comments
MAYTAG A712	$586	$519-655	43³/₄(51¹/₄)x25¹/₂x27¹/₄ [3]	Back											B,D,L,P,R	f	A,C
WHIRLPOOL LA9800XM	547	439-669	43¹/₂(50)x29x26	Back											M,R	n	A,G,I,J,L
SEARS 23921	574	490-669	43(52³/₄)x29x26	Left											J,M,O,R	d,j	A,F,I
GIBSON WA28D7WP	450	389-540	43(54¹/₄)x28¹/₄x27¹/₄	Left								[5]			F,H,L,P	d	—
SPEED QUEEN HA7001	516	459-600	43¹/₄(51)x25³/₄x28 [3]	Back											D,G,L,P,R	b	I
KELVINATOR AW800A	448	369-539	43¹/₂(54¹/₄)x28¹/₄x27¹/₄	Left								[5]			F,L,P	d	B
GENERAL ELECTRIC WWA8480B	493	420-600	43(50¹/₂)x27x25	Back											A,D,N	e,m	C,G,K
HOTPOINT WLW5700B	476	388-589	43³/₄(50¹/₄)x27x25	Back											D,N	m	C,G,I
WARDS 6841	567	479-609	44(55¹/₂)x27x27¹/₄	Left											C,E,I,M,P	d,h,l	B,D,E,K

Model					Ratings			
NORGE LWA9120S	461	400-515	43½(55½)x27x27	Left	(rating symbols)	[5]	E,M,P	d,h B,D,K
ADMIRAL W2OB8	458	396-520	43¼(55½)x27x27¼	Left	(rating symbols)	[5]	E,M,P	c,d,h B,D,K
FRIGIDAIRE WCIM	481	390-598	44¼(54)x27x26¼	Left	(rating symbols)		F,L,M	o B,D
WHITE-WESTINGHOUSE LA800E	470	420-520	43¾(53½)x27x27	Left	(rating symbols)		F,L,M	l,o B

Front-loaders

Model					Ratings			
WHITE-WESTINGHOUSE LT800E	577	470-699	43¾x27x27(39¼) [3]	Down	(rating symbols)	[5]	C,F,K, M,P,Q	a,e,g, C,H,I k,l
WARDS 6514 [4]	602	490-690	36¼x27x26¼(39¼) [3]	Down	(rating symbols)	[5]	C,F,P,Q	a,e,g, A,C,H,I i,k

[1] Parenthetical figure is height or depth with lid open. Leveling legs can add up to 1½ in. Additional depth may be needed for hoses.

[2] Measured at maximum fill for cycles indicated.

[3] Can be flush with wall; others need up to 4-in. clearance.

[4] Stackable model; requires painted top (approx. $20 extra) to be free-standing.

[5] Insufficient data.

SPECIFICATIONS AND FEATURES

All have: ● Presettable water-level control. *Except as noted, all:* ● Were judged average in handling permanent-press clothes for line-drying ● Were judged average in sand disposal. ● Have 40-45 min. regular cycle. ● Have 35-45 min. permanent-press cycle. ● Have no special prewash setting. ● Have safety switch on lid that stops spin action only. ● Have timer that can't be rotated when machine is on. ● Have top, lid, and tub finish of porcelain enamel; rest of cabinet, baked enamel. ● Have lint filter that must be cleaned periodically by user.

KEY TO ADVANTAGES

A – Can be set to use less water than others for small loads.
B – Uses less electricity to run motor than others.
C – Safety switch on lid or door stops all action.
D – Safety switch on lid stops agitation and spin.
E – Safety switch stopped tub faster than most during spin.
F – Lid or door locks during spin.
G – Stainless-steel tub.
H – Plastic tub.
I – Has top-mounted detergent dispenser.
J – Has top-mounted dispensers for detergent, bleach, and softener; convenient but side-by-side position could cause mix-ups.
K – Has top-mounted bleach and softener dispenser.
L – Washer speeds, temperature combinations can be set independently of chosen cycle.
M – Has prewash setting.
N – One cycle has automatic second rinse.
O – Water-temperature selection manual or automatic.
P – Allows front access for repairs.
Q – Better than average in sand disposal.
R – Judged better than average in handling permanent-press clothes for line-drying.

KEY TO DISADVANTAGES

a – Range of fill controls very narrow.
b – Range of fill controls fairly narrow.
c – Controls poorly marked and difficult to set.

d – Timer can be rotated forward with machine on, causing wear and tear on components.
e – Low-sudsing detergent may be required by **GE**, likely to be required by front-loading **White-Westinghouse** and **Wards.**
f – Tub took longer to stop than others when lid was opened during spin.
g – Noisy when changing action.
h – Makes loud clank at end of spin cycle when tub brakes.
i – Front-mounted controls may require stooping to set.
j – Lid does not lie flat, so access from left is hampered.
k – Top has painted finish, easily scratched.
l – Top has intrusive projections.
m – Changing cycle speed while machine is running may damage machine.
n – Can't be set for slow agitation, fast spin.
o – Worse than average in sand disposal.

KEY TO COMMENTS

A – Regular cycle shorter than most.
B – Regular cycle longer than most.
C – Permanent-press cycle shorter than most.
D – Permanent-press cycle longer than most.
E – Has "pre-spot" feature, a minor convenience.
F – Agitator's top and bottom sections move differently.
G – Mfr. offers manual or toll-free number for do-it-yourself repair.
H – No choice of speed; same speed for all washing, higher speed for extraction.
I – "Self-cleaning" lint filter convenient, but lint discharged into partially blocked drain may promote blockages.
J – Has electronic touch-panel controls; although well thought out, they're not as flexible as other controls and may confuse first-time users.
K – Model discontinued at the time the article was originally published in *Consumer Reports.* The test information has been retained here, however, for its use as a guide to buying.
L – According to mfr., a later model is **LA9800XP**, similar except for control panel.

Features

Top-of-the-line washers are all loaded with features, many of them similar. Still, the design and execution of a particular set of features might give you reason for choosing one washer over another.

Controls. These can usually be found on a console at the back of the machine and are clearly marked.

The controls generally consist of conventional rotary dials. Most of the washers prevent you from moving the dial forward while the machine is running, saving you from needlessly wearing out components.

An electronic touch panel may prove intimidating to first-time users. And resetting or advancing from one cycle to another may require turning the power off and on again and then pressing the desired function, a small inconvenience.

Dispensers. With any washing machine, you can put the detergent in the tub before you put in the laundry. A few machines, however, have a dispenser that releases the detergent when the water circulates—handy if you're using the soak cycle.

Higher priced machines generally can add fabric softener automatically.

Safety. Tangling with a washer's spinning tub and oscillating agitator is a losing proposition. Most washers minimize the risk by stopping the spinning within a few seconds of the lid being opened. Some reduce the risk by locking the door or lid during spin.

Special features. Some models have a prewash feature that provides an extra period of agitation. Most also have a soak cycle, which provides the soaking and spin-out that might be necessary for heavily soiled clothes.

Some less-expensive washers don't have the soak feature. But you can accomplish the same result manually: you turn off the washer after a few minutes of agitation in the regular cycle, let the wash soak about half an hour, and then set the machine to spin before the regular wash cycle.

Extra rinses are available on a number of models. They are useful for washing heavily soiled items and they can also be important for people who are allergic to residues of laundry products.

Recommendations

If conserving water or energy is a primary concern, choose a front-loader. Most people, however, would probably prefer a top-loader. Top-loaders hold more laundry and they can also adjust to small loads. The best of them are even efficient in using water.

Another consideration should be a brand's reliability. Consumers Union's

Annual Questionnaire surveyed subscribers to *Consumer Reports* to see how often brands they own have been repaired. The latest data indicate that **Maytag** models have been the most trouble-free. Better than average in Frequency of Repair are washers from Hotpoint, General Electric, Sears, and Whirlpool.

Washing-machine features

Brand and model	Price [1]	Normal/normal	Normal/slow	Slow/slow	Slow/normal	Hot/cold	Warm/cold	Cold/cold	Hot/warm	Warm/warm	Fill options	Timed soak	Extra rinse	Bleach dispenser (wash)	Softener dispenser (rinse)	Other
		Speed options [2] (agitation/spin)				Temperature options (wash/rinse)						Cycle options [3]				
Top loaders																
ADMIRAL W20B6	$570 [4]	Automatic				✓	✓	✓	✓	✓	3	—	✓	✓	✓	—
● ADMIRAL W20B8	620 [4]	Automatic				✓	✓	✓	✓	✓	Cont.	—	✓	✓	✓	—
FRIGIDAIRE WCDM	425	Automatic				✓	✓	✓	—	—	Cont.	—	✓	✓	[8]	A
● FRIGIDAIRE WIM	470	✓	✓	—	✓	—	✓	—	✓	✓	[7]	—	✓	✓	[8]	—
● FRIGIDAIRE WCIM	525	✓	✓	✓	✓	—	✓	—	✓	✓	[7]	✓	✓	✓	✓	—
GENERAL ELECTRIC WWA8320B	450	✓	—	✓	✓	✓	✓	✓	—	—	4	—	✓	✓	✓	B
GENERAL ELECTRIC WWA8350B	475 [5]	✓	✓	—	✓	✓	✓	✓	—	—	Cont.	—	✓	✓	✓	B
● GENERAL ELECTRIC WWA8480B	520 [4]	Automatic				✓	✓	✓	✓	✓	Cont.	✓	✓	✓	✓	—
GIBSON WA28D5WP	405	Automatic				✓	✓	—	—	—	Cont.	✓	✓	✓	✓	—
● GIBSON WA28D7WP	435	✓	✓	✓	✓	✓	✓	—	✓	✓	Cont.	—	✓	✓	—	—
HOTPOINT WLW3700B	434	✓	—	✓	✓	✓	✓	—	—	—	Cont.	—	✓	✓	✓	C

Model	Price														
HOTPOINT WLW4700B	449	✓	✓	—	✓	✓	✓	✓	✓	✓	—	Cont.	—	✓	✓	C
•HOTPOINT WLW5700B	499	Automatic		✓	✓	✓	✓	✓	✓	✓	✓	Cont.	—	✓	✓	C
KELVINATOR AW600A	490	Automatic		✓	✓	✓	✓	—	✓	✓	✓	Cont.	—	✓	[8]	—
•KELVINATOR AW800A	540	✓	✓	✓	✓	✓	✓	✓	✓	✓	✓	Cont.	✓	✓	✓	—
MAYTAG A612	569	✓	—	✓	✓	✓	—	—	✓	✓	—	Cont. [7]	—	✓	✓	—
•MAYTAG A712	614	✓	✓	✓	✓	✓	✓	✓	✓	✓	✓	Cont. [7]	—	✓	✓	—
NORGE LWA7120S	570 [4]	Automatic		✓	✓	✓	✓	✓	✓	✓	—	Cont.	—	✓	✓	—
•NORGE LWA9120S	620 [4]	Automatic		✓	✓	✓	✓	✓	✓	✓	✓	Cont.	✓	✓	✓	—
SEARS 23721	450	Automatic		✓	✓	✓	✓	✓	✓	✓	✓	3	✓	✓	✓	A
SEARS 23801	480	Automatic		✓	✓	✓	✓	✓	✓	✓	✓	Cont.	✓	✓	✓	—
•SEARS 23921	570	Automatic		✓	✓	✓	✓	✓	✓	✓	✓	Cont. [7]	✓	✓	✓	—
SPEED QUEEN HA6001	529	✓	✓	—	✓	✓	✓	✓	✓	✓	—	Cont.	—	✓	✓	—
•SPEED QUEEN HA7001	569	✓	✓	✓	✓	✓	✓	✓	✓	✓	✓	Cont.	—	✓	✓	—
WARDS 6542	[4]	✓	✓	✓	✓	✓	✓	✓	✓	✓	—	Cont.	—	✓	✓	—
WARDS 6641	[4]	Automatic		✓	✓	✓	✓	✓	✓	✓	✓	Cont.	—	✓	✓	—
•WARDS 6841	[4]	Automatic		✓	✓	✓	✓	✓	✓	✓	✓	Cont. [7]	✓	✓	✓	—
WHIRLPOOL LA5800XM	489 [6]	Automatic		✓	✓	✓	✓	✓	✓	—	—	Cont.	✓	✓	—	—
WHIRLPOOL LA7800XM	519 [6]	Automatic		✓	✓	✓	✓	✓	✓	✓	✓	Cont.	✓	✓	✓	—
•WHIRLPOOL LA9800XM	599 [6]	Automatic		✓	✓	✓	✓	✓	✓	✓	✓	Cont.	✓	✓	✓	D
WHITE-WESTINGHOUSE LA600E	459	✓	—	✓	✓	✓	✓	✓	✓	—	✓	Cont.	—	✓	[8]	—
WHITE-WESTINGHOUSE LA700E	499	✓	✓	—	✓	✓	✓	✓	—	—	✓	Cont.	—	✓	✓	E
•WHITE-WESTINGHOUSE LA800E	559	✓	✓	✓	✓	✓	✓	✓	✓	✓	✓	Cont.	✓	✓	✓	C,E

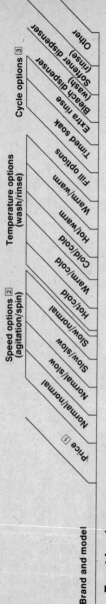

Front loaders

Brand and model	Price [1]	Speed options [2] (agitation/spin)				Temperature options (wash/rinse)							Cycle options [3]			
		Normal/normal	Normal/slow	Slow/slow	Slow/normal	Hot/cold	Warm/cold	Cold/cold	Hot/warm	Warm/warm	Fill options	Timed soak	Extra rinse	Bleach dispenser (wash)	Softener dispenser (rinse)	Other
●WARDS 5514	680	—	—	—	—	—	✓	✓	—	✓	Cont.	—	—	—	—	A
WHITE-WESTINGHOUSE LT600E	589	—	—	—	—	—	✓	✓	—	✓	3	—	—	—	—	A
WHITE-WESTINGHOUSE LT700E	659	—	—	—	—	—	✓	✓	—	✓	Cont.	—	✓	—	✓	A,E
●WHITE-WESTINGHOUSE LT800E	689	—	—	—	—	—	✓	✓	✓	✓	Cont.	✓	✓	✓	✓	A,E

[1] *Mfr. approximate retail; see Ratings for indication of actual prices.*

[2] *Manually selectable speeds; automatic models may have some or all speed combinations.*

[3] *All models have regular, permanent press, and delicate cycles.*

[4] *Model discontinued at the time the article was originally published in Consumer Reports. The information has been retained here, however, for its use as a guide to buying.*

[5] *Mfr. says model replaced by* **WWA8350G,** *similar except for color changes on control panel.*

[6] *Mfr. says model replaced by* **-XP** *line, similar to* **-XM** *line except for design of control panel.*

[7] *Can advance automatically to regular wash cycle.*

[8] *Optional.*

KEY TO OTHER FEATURES

A – Top not porcelain-coated.
B – Has built-in small-load wash basket.
C – Has separate agitator for small loads or gentle action.
D – Has electronic touch-pad controls.
E – Has built-in weighing scale.

Water filters

Chemically pure drinking water probably doesn't exist anywhere. Water forms solutions with thousands of substances, and what does not dissolve in water often remains suspended in it.

Impurities in water aren't necessarily bad. The "clean, delicious" taste of bottled spring waters often comes from the beneficial minerals they contain, such as calcium and magnesium.

Americans are becoming increasingly concerned with the quality of their water, as the booming sales of bottled water clearly show. But paying a lot for water from bottles isn't the only way to get good-tasting water. Many people have turned to activated-carbon filters to upgrade their drinking water.

The carbon material is formed by exposing a carbon-containing material (usually charcoal) to high temperatures and steam in the absence of oxygen. The resulting material is honeycombed with minuscule channels that branch and twist within. The channels greatly increase the surface area and thereby account for activated carbon's impressive filtering power. As water passes through this microscopic labyrinth, contaminants stick to the walls of the channels. "Adsorption" is the technical term.

Activated-carbon water filters are widely used to clean up water in bottling plants; on a more modest scale, carbon filters are used in the home to filter tap water. Some are sink-mounted and attach to the faucet outlet, others go under the sink and are connected to the cold water line there, and a few are independent of the house plumbing. Most manufacturers of home-use filters claim only that their products improve the taste and odor of water. A few claim that their products will remove toxic chemicals.

With the increase in filter sales, the U.S. Environmental Protection Agency (EPA) in the 1970s raised a number of questions about them: What exactly do these filters take out of water? How well do they do it? Which types work best? How often should the activated-carbon cartridge be replaced? Can bacteria multiply to hazardous levels on the filters? To get answers, the EPA contracted a private laboratory to test devices using activated-carbon filters of various types. The EPA has published the findings.

Limitations

There are some important water problems that activated-carbon filters will not affect. They won't help against:

Hard water. This contains large amounts of dissolved minerals, mainly magnesium and calcium. A water-softening device is needed to remove those minerals.

Other dissolved minerals. Activated-carbon filters will remove dust particles, but they have only a small effect on dissolved metals such as iron, lead, man-ganese, or copper. They won't remove chlorides, fluorides, or nitrates at all. A "reverse osmosis" or an "ion exchange" device, available at plumbing-supply stores, can be used to remove most of those minerals.

Hydrogen sulfide. This chemical gives water the taste and odor of rotten eggs. Chlorination usually takes care of the problem; carbon filters can remove only small amounts of hydrogen sulfide.

Filtering

Those limitations notwithstanding, activated-carbon filters can significantly improve water quality. First of all, the filters can quite effectively remove many objectionable tastes and odors. They can help clear sediments (turbidity) from tap water. Finally, many home filtering units can effectively remove organic chemicals, a class of contaminants discussed in the box on pages 346–47.

Three important filter characteristics are:

Organic chemical removal. Chloroform is a common organic contaminant in treated drinking water. It can also be harmful: it's known to cause tumors in animals and is suspected to be a carcinogen for humans. Because activated-carbon filters don't retain chloroform as well as they do more complex organics such as pesticides, it provides a tough test of a filter's ability to remove organic chemicals. To prevent chloroform absorbed by the filter from being returned to the water later on, avoid models that used powdered activated charcoal in a pad. Granular activated carbon is much more effective.

Chlorine removal. This indicates a filter's ability to remove a chemical that contributes heavily to water's taste and odor. To protect public health, chlorine is added at water-treatment plants—usually in sufficient quantity to maintain a residual amount at the tap. But even moderate levels of chlorine can contribute to characteristic and often disagreeable tastes.

Useful flow rate. As a general rule, filtration ability depends on the flow rate of the water. Gushing water isn't filtered as well as water flowing at slower speeds. Some filters are effective only when water trickles through; others can handle much faster flow rates. A filter's "useful flow rate" is the flow rate it can tolerate while still removing contaminants adequately.

There's a potential health problem associated with these filters: high levels of bacteria can accumulate on the carbon between uses. For that reason it's useful to know the extent of bacterial growth after varying periods of nonuse, and how long a filter should be flushed to reduce the bacteria level.

Under-sink filters

The under-sink units are the larger and generally more expensive type. But their size is also one of their main advantages. Most are effective against chloroform for extended periods. They all do well at removing chlorine.

For effective filtration that keeps the filtered contaminants within the filtering device and doesn't return them to the water later on, good cartridge design is essential.

The most effective units use granular carbon in a cartridge with hard plastic sides. This design forces the water to travel the length of the filter, ensuring longer contact time with the carbon. Some filter cartridges with mesh or wound-string sides allow water to pass through the sides of the filter, permitting much less contact with the carbon. But a wound-string type can incorporate the desirable features of long-time carbon contact.

To take advantage of the high useful flow rate of an under-sink filter and extend the life of the filter cartridge, connect the unit so it operates from a separate faucet. That permits running cold water through the filter only when you want it.

Sink-mounting

Sink-mounted filters are generally smaller, easier to install, and cheaper than an under-the-sink variety; but they also perform more modestly. Typically, they have a lower useful flow rate than an under-sink unit and the filter lifetime is also shorter—understandable in view of the smaller size. The **Hurley,** which contains much more carbon than the other sink-mounted models, is a notable exception.

The **Hurley** is unique, too, in that filtered water comes out of a spigot on the filtering unit and not out of the tap. That requires the filter to be positioned right at the edge of the sink.

The **Hurley** has another distinctive feature: it can be cleaned by "backwashing," a reverse-flow procedure in which hot water is run through the filter spigot and out the bottom of the filter. Backwashing flushes out some contaminants and can help to reduce bacterial levels for up to several days. To maintain the effect, however, you have to backwash regularly. And there is one possible hitch: The **Hurley** requires 145°F water for backwashing. If your water isn't that hot, backwashing might not do much good.

What about taste?

Does a filter really improve the taste of water?

If you ask people to compare two versions of heavily chlorinated water—one version unfiltered, the other filtered—the panelists are almost certain to prefer the filtered water overwhelmingly.

Evidently, an activated-carbon filter does help improve the taste of water; and, what's more, it doesn't have to be very effective to make the improvement noticeable.

Bacteria

A filter traps many contaminants, but it isn't effective against bacteria. In fact, a filter can allow the bacterial levels in water to multiply many times over. Wet activated carbon, richly infused with trapped organic matter, provides an ideal breeding ground. High bacterial levels are most likely when a filter has a saturated cartridge and when some time has passed since the filter's last use.

The tests run for the EPA showed that periods of stagnation—up to five days of nonuse—will increase bacterial counts in filtered water. (A filter can go unused even in a busy household—when the family takes a weekend outing or a vacation, for instance.) Consumers Union's tests indicate further that high bacterial levels are more likely if the filter has treated large quantities of water.

Bacteria get into the filter in the first place because disinfection at the water-treatment plant doesn't kill everything. A certain number of bacteria survive and end up in tap water.

The digestive system usually adapts to the low levels of "harmless" bacteria found in drinking water. Is there a health risk from washing large numbers of bacteria out of the filter and into a drinking glass? No one knows for sure.

But common sense suggests that drinking large doses of unidentified bacteria is best avoided.

Can disease-causing bacteria multiply on the filters? Studies have produced mixed results. EPA researchers have failed in attempts to grow one type of pathogenic bacteria on filters. But Canada's Health Protection Branch, a government agency, has identified nearly a dozen potentially harmful bacterial species that can grow on the filters.

One thing is certain: if bacteria don't get into the filter in the first place, they can't multiply inside. For this reason, filters should only be used to treat water that is microbiologically safe. (If you have questions about the safety of your water, you should ask the water company for its latest test report; if you use well water, you'll need to have it tested yourself.)

Some filters contain silver and claim to discourage growth of bacteria within the filter. But don't be impressed by any filter's "bacteriostatic" claim. The EPA has concluded that silver compounds in water filters show "no significant bacteriostatic effect" on drinking water.

When to reload

A light doesn't flash when it's time to change a filter cartridge. The manufacturers don't provide much guidance, either. They generally say it's time for a change if bad taste or odors return, or if water flow is severely reduced because sediment has clogged the filter. But it's not that simple.

For example, a filter may be able to control the large molecules that add tastes and odors long after it has lost the capacity to remove the small organic molecules such as chloroform.

In some cases, the filter may lose its chemical-removing ability long before water flow becomes sluggish. A filter can also be loaded with bacteria and not show a reduced flow rate.

Some manufacturers suggest filter lifetimes that range from one month to one year. A statement from the manufacturer of the number of gallons treated is probably a more useful guide to follow.

As a very general guide to converting gallons of use into days or weeks, you can figure that a filter treats about 4 gallons of water a day—a typical amount used for drinking and cooking. On that basis, a filter with an effective lifetime of, say, 1,200 gallons should last about 300 days.

But any such estimate can only be a rough guide—actual lifetimes may vary greatly depending on local water quality.

Recommendations

An activated-carbon water filter will remove chlorine and other chemicals that impart taste and odor to water; some filters are also effective at removing organic chemicals, including chloroform and others that have little taste. But no carbon filter we've seen can handle bacteria. Indeed, the wet carbon in these filters makes an excellent breeding ground for bacteria that can find their way into your drinking water. Because of the potential for bacteria you should buy a filter only if you're willing to change the cartridge regularly.

If you simply want to clear your water of rust and dirt, then a sediment filter may suit you just fine. Such a filter contains no carbon.

If your complaint about tap water is mainly aesthetic—an objectionable taste or odor—then a sink-mounted carbon filter would probably suit you.

An under-sink unit is for more serious problems—water that tastes awful unless it's treated, or water that contains small amounts of harmful organic chemicals. These units must be permanently mounted into the cold-water line; this means some effort if you do it yourself and may be costly if you hire a plumber. Then there are large, whole-house units with over 30 pounds of carbon that are used when a serious pollutant problem has been identified. Such units have a provision for regular backwashing.

Using a filter

How you use your filter is almost as important as which brand you buy. If you follow these suggestions you can maximize your filter's performance.

1. Flush out the filter before the first use of the day. Open the faucet wide and let the water run for at least 15 seconds for an under-sink model, at least 5 seconds for a sink-mounted filter. When you install a new cartridge, flush for several minutes to remove fine carbon particles.

2. Change filters regularly. A heavily used filter is more likely to contain high bacterial levels and to discharge organic chemicals previously trapped. An exhausted filter is worse than useless.

3. Don't filter hot water. A filter on the hot-water faucet won't remove contaminants very well. And the hot water may liberate chemicals previously trapped in the filter.

4. Use the slowest flow rate you can tolerate. The longer the water is in contact with the filter, the more impurities the carbon can attract and the cleaner your water will be.

5. After installing a new cartridge, circle on your calendar the date for the next replacement. Stick to your schedule.

Ratings of water filters

(As published in a November 1983 report.)

Listed by types; within types, listed in order of estimated number of gallons of water effectively filtered to reduce chloroform. Bracketed models are judged similar in ability to remove chloroform and are listed in order of increasing price. Estimated filter lifetime is for guidance; actual lifetime may vary greatly according to local water quality. All come with installation instructions and mounting hardware. Undersink models may require installation by licensed plumber. Except as noted, sink-mounted models fit on faucet end. Prices are suggested retail and do not include cost of installation; prices in parentheses are for replacement filter cartridges; *indicates that price is approximate; + indicates that shipping is extra.

Excellent	Very good	Good	Fair	Poor
●	◒	○	◓	●

Brand and models	Price	Dimensions (W x H) (in.)	Estimated filter lifetime [1] (gal)	Effectiveness	Useful flow rate	Chloroform removal / Chlorine removal [2]	Carbon type [3]	Comments
Under-sink models								
CULLIGAN SUPER GARD SG-2	$175* ($23*)	5x16	3000	●	◐	●	G	A,B,O
SEARS 3464	23*(7.50)	5x12	1200	◐	●	●	G	D,P,S
AMF-CUNO AP50T	67(12)	4½x14½	1200	◐	●	●	G	H
SEAGULL IV X-1F	290(40)	5x6½	1200	◐	○	●	Pb	A,B,C,E,F,K
FULFLO WC-12	51(8.85)	5x13	300	○	●	◐	G	D,H,J,R

- The following 3 models were judged much less effective than those preceding. In addition, they "unloaded," or returned, some chloroform to tap water when used for more than 300 gal. All may be satisfactory for removing some tastes and odors, however.

Brand and models	Price	Dimensions (W x H) (in.)	Estimated filter lifetime [1] (gal)	Effectiveness	Useful flow rate	Chloroform removal / Chlorine removal [2]	Carbon type [3]	Comments
AQUA GUARD AGT200	37+(6.25+)	4¼x12½	150	◐	●	●	P	H,R,T
KEYSTONE 3121	44(6.75)	5x13	150	◐	●	●	P	H,R
FILTERITE 1PC	55(9)	5½x12¼	150	◐	●	●	P	H,J
Sink-mounted models								
HURLEY TOWN & COUNTRY	190(47+)	6½x10½	3000	●	●	●	G	A,L,M,N,O
ECOLOGIZER WATER TREATMENT SYSTEM 5505	35+(10+)	4x6	250	○	◐	◐	G	A,H,I,L,U
POLLENEX PURE WATER "99"	35(4.95)	5½x5	150	◐	○	◐	G	A
AQUA GUARD AGT300	30+(3.50+)	4x4½	100	◐	○	○	G	H,T

- The following model was judged much less effective than those preceding because it never achieved chloroform-removal level of 50% in CU's tests.

| PEERLESS RP5506 | 20(4) | 3½x3 | — | ● | ◐ | ◐ | P | F,G,K,Q |

[1] Lifetime, as determined by CU, is number of gallons during which removal of chloroform is expected to be greater than 50 percent. Lifetimes of 1,000 gal. or more are combination of CU and EPA test results.

[2] Chlorine removal judged with unused cartridge for first few gal. only.

[3] G, granular; Pb, powder in block; P, powder in pad.

KEY TO COMMENTS

A – Water flow restricted by filter's design.
B – Requires $1/4$-in. tubing (supplied).
C – Supplied with outlet faucet fixture.
D – Supplied with wrench to open housing.
E – Registered as "purifier" with EPA.
F – Manufacturer claims that unit can filter out particles less than 1 micron dia. CU believes unit may clog prematurely in turbid water.
G – Silver incorporated into filter; bacteriostatic properties claimed but not verified by CU.
H – Transparent plastic housing; cartridge visible in use, an advantage.
I – Design incorporates cartridge change-date reminder, an advantage.
J – Lacks pressure relief button on top of filter housing, a disadvantage.
K – Would rate higher if used with water supplies that have no sediment problem.
L – Filter unit sits on counter and is connected by hose to diverter valve on faucet.
M – Backwashing to clean filter recommended by manufacturer at 145°F, an advantage if you have water that hot; backwashing can extend lifetime of filter.
N – No replaceable cartridge available; entire unit must be returned to manufacturer for refill ($50).
O – Filter contains much more carbon than others of this type. Estimated lifetime may exceed 3,000 gal. listed.
P – According to the company, filter housing tested has been replaced by a different model (not tested). Replaceable cartridge **34651** judged most effective of its type.
Q – Chloroform-removal level never reached 25 percent in CU's test.
R – **Sears** cartridge also fits this unit.
S – Cartridges sold only in packs of 2 for $15. Price given is for comparison purposes.
T – Available from Universal Water Systems Inc., 705 E. State St., Geneva, Ill. 60134.
U – Available from Rush Hampton Ind., P.O. Box 3000, Longwood, Fla. 32750.

Water flow-through under-sink filters

An activated-carbon filter can upgrade drinking water in a variety of ways: Sediment removal gives cleaner-looking water, while removal of chlorine and other chemicals can reduce unpleasant tastes and odors; removal of organic chemicals such as chloroform may be important for protecting health. As the arrows indicate, however, a filter won't remove *all* the impurities.

The best under-sink filter cartridges have hard plastic sides, as shown in the cutaway drawing below left. This design forces water to travel the length of the cartridge for good filtration. A cartridge (cutaway at right) shouldn't let water take a shortcut through the cartridge wall. In addition, the best under-sink filters use a large amount of carbon.

Organic chemicals in water: a major health concern

Many people are filtering their drinking water not just to improve its taste and odor but to try to rid it of chemical contaminants. So far, more than seven hundred organic chemicals have been identified in drinking water supplies, and some of them are suspected cancer-causing agents. But that certainly doesn't mean that everyone's water is suspect. The water source largely determines what types of chemical contaminants, if any, may be present.

Surface water

Surface water (rivers, lakes, and reservoirs) supplies about half the population with drinking water. It can be tainted with agricultural runoff, industrial effluents, and urban wastes. But the main problem comes from an unexpected source: chlorine disinfection.

Chlorine treatment of drinking water ranks as one of the great public-health advances of this century. Today, virtually all public water supplies in the United States are chlorinated to kill harmful microorganisms. As a result, once-common waterborne diseases such as cholera and typhoid have been eliminated. In recent years, however, research has found that chlorination of surface waters produces unexpected side effects.

Most surface waters contain some natural organic compounds from decaying vegetation. These compounds, and some man-made chemicals, often occur at such low levels that they are of little health concern. But chlorine can react with many of these chemicals, transforming them into substances that could threaten health.

One of the most common by-products of chlorination is chloroform, which belongs to a family of chemicals known as trihalomethanes (THMs). THMs produced by chlorination are the most common organic contaminants found in water supplies. Of the many organic compounds in drinking water that may pose health hazards, the THMs are among the most studied.

Chloroform produces cancer in laboratory animals and may increase the risk of some types of cancer in people whose water contains it. Several studies have linked increased cancer mortality with the use of chlorinated surface water, but the results are still inconclusive. The increased risk could be due to something other than THMs or chlorination.

Whatever risk does exist can be reduced by preventing or curtailing exposure to these compounds. Fortunately, an activated-carbon filter can significantly reduce THM levels, provided the filter cartridge is changed often enough.

If you're concerned about your water's safety, ask your water company for current water-quality data. Consider buying a water filter if the report shows that any organic contaminants are present.

Groundwater

Groundwater (water from wells) is the other main source of drinking water. Tests have shown that some underground reservoirs, known as aquifers, often contain toxic organic chemicals, some of which are possible carcinogens.

Contamination has forced the closing of hundreds of wells nationwide. Many health experts now regard groundwater pollution as the most serious problem affecting the nation's drinking water.

Organic chemicals in well water sometimes produce obvious "chemical" tastes and odors, but hazardous contamination is not always detectable to the senses. You should be aware of potential problems if you live near a toxic-waste landfill, or in an agricultural area where herbicides and pesticides are heavily used. Your local health department may know if there's a problem or may be able to test your water for you. But some health departments aren't equipped to do that, so a group of neighbors using the same aquifer may want to share the cost of testing by a private laboratory.

Cleaning up an aquifer is usually geologically impossible or prohibitively expensive. Laboratory studies suggest that activated-carbon filters can be quite helpful in removing organic chemicals at the point of use. At many sites, home water filters may be the most cost-effective way to treat contaminated groundwater.

Water heaters

A water heater uses more energy than almost any other appliance. Only a central heating or air-conditioning system demands more. According to the Department of Energy, a water heater consumes some 20 percent of the energy used in the home.

Like most people, you're probably not about to replace the water heater until it breaks down completely. But that doesn't mean you have to be stuck with high energy costs for hot water until you're ready to replace an old heater. The articles that follow discuss "energy saver" water heaters and other products and strategies that will help you lower your hot-water bills, including tips on ways to save energy and money with your present water heater.

Figuring costs

The total cost of heating water includes the cost of heating and the cost of storage (or standby) losses. There is also the initial equipment cost. These three major costs need to be considered separately.

Heating cost

It takes energy to raise the temperature of water. The more water you want to heat, and the hotter you want it, the more energy you need. There is a simple formula to express the interrelationship of these factors: 1 Btu will raise the temperature of 1 pound of water by 1°F. To raise the temperature of, say, 15 gallons of water (about 125 pounds) from 60° to 140° would require 10,000 Btu (125 x 80).

While that Btu requirement remains the same regardless of the source of the energy, the *cost* of the energy can vary quite a bit. At one extreme, there is the energy cost of using a solar water heater: As long as the sun is shining, the energy cost of heating 15 gallons of water is essentially zero. To heat the same 15 gallons, an electric water heater would require about 3 khw, or 23 cents, at the national average electricity rate of 7.75 cents per kwh. A gas water heater, which loses some of its energy up the flue, would have to burn 13,000 to 14,000 Btu—about 8.5 cents worth of gas—to deliver the 10,000 Btu required to heat the 15 gallons of water.

No matter what type of water heater you have, there are two simple ways to cut the operating cost: lower the temperature of the hot water you use, and use less hot water.

If water flows into an electric water heater's tank at about 60°F, heating 64.3 gallons of water (the daily use by the average American family, based on the Department of Energy's estimate of 450 gallons used per week) from 60° to 140° would require about 13 kwh, which would cost about 99 cents (with electricity billed at 7.75 cents per kwh). If you lower the heater's thermostat setting to 120°, however, you reduce the cost of heating that 64.3 gallons by about 25 percent, or 25 cents. (See the table for savings with other temperature settings.) And you reap a safety benefit: you reduce the risk of scalds.

In reality, though, the temperature setback probably won't yield a full 25 percent saving on your water-heating bill. It's usual to draw a mix of hot and cold water for bathing, showering, washing, and the like. Lowering the hot water temperature will simply mean that you use more hot water and less cold for your bath by opening your hot water faucet more and cutting back on the cold. But you will be able to save on the hot water going to a dishwasher or washing machine: these appliances don't "feel" temperature the way you do, they require a fixed amount of hot water for each use, and their performance may continue to be satisfactory even with water that's less hot.

A water heater's thermostat can't be set precisely, so turning down its thermostat is a matter of trial and error. To find the right setting, start by lowering the thermostat a little. If you still have enough hot water for your needs, and your dishwasher still does a satisfactory job, lower the setting further. When you finally reach the point where the water is no longer hot enough to wash dishes satisfactorily, raise the temperature setting a few degrees. If you don't have a dishwasher, use the amount of hot water available as a guide.

The second—and more obvious— half of reducing the cost of heating water is simply to use less. Taking shorter showers, using less bath water, substituting warm or cold clothes-washing cycles for hot, and postponing use of these appliances until you have full loads for them are all effective tactics.

A flow restraint inserted in a regular shower head sets a limit on the water you get from that head. Most low-flow shower heads should be able to cut water usage without making the flow unpleasantly feeble.

Storage cost

A conventional tank-type (storage) water heater is almost always filled with hot water. These tanks are invariably insulated, but there is still a continuous loss of heat to the surroundings. Unlike losses during heating, storage losses are largely unaffected by the amount of water you use and can run anywhere from $45 to $150 a year depending on how you keep the water, whether you use gas or electricity, and on the tank's efficiency. The most practical and efficient way to cut storage losses is to increase the effectiveness of the heater's insulation.

A well-insulated electric water heater will lose about $61 worth of heat in a year; a poorly insulated one perhaps twice that. A well-insulated gas model will lose about $37 worth of heat annually, as against $67 with a poorly insulated gas heater. (The reduction in storage losses is less marked with gas models because some of the heat loss is up the flue, and that heat loss can't be helped by insulation.)

Equipment cost

In addition to the two aspects of operating cost, there is another major factor to consider—the initial outlay for the equipment itself and for its installation. With conventional water heaters, there isn't too much to be said about this aspect of total cost—not because it represents a trivial investment, but because the differences in cost of installation among comparable models tend to be relatively small.

Conventional water heaters: gas and electric

A conventional water heater is basically just an insulated tank with a heater—either a gas burner or electric heating elements.

The most popular water-heater capacity is 40 gallons for gas and 50 or 52 gallons for electric. Those heaters take the form of an upright cylinder, 4 to 5 feet high and 1½ to 2 feet in diameter.

Water heaters fall into two broad categories, "standard" and "energy-saver." The major difference between these types is that an energy-saver is better insulated than a standard water heater.

Operating cost

One way water heaters are labeled is by their rate of energy input—Btu per hour or in watts. In theory, the higher the input, the faster a unit can heat cold water.

Electric water heaters commonly show up significantly higher in total operating cost than gas. But differences in operating cost among various electric models or among gas models are likely to be slight.

There are greater differences among heaters in purchase price and length of warranty. But you should be aware that a heater with a longer warranty may cost more to buy.

Heaters dubbed "energy-savers" live up to their name. A new standard water heater would cost about 18 percent more to run than an energy-saver model.

Hot-water quantity

Five gallons per minute is about the flow you'd get when running a washing machine or when more than one tap was being used for hot water at the same time.

At that flow rate, manufacturers' claims for tank capacity are optimistic. A "40-gallon" gas model may actually hold about 1 or 2 gallons less, and a "50-gallon" and "52-gallon" electric model may hold about 3 to 5 gallons less (such differences are within the limits of a voluntary industry standard).

As water is drawn from a tank, there is a fairly sharp stratification between the hot water still stored in the tank and the cold water entering the tank. (The cold water enters through a "dip tube," which directs the water toward the bottom of the tank.) That layering means you can get most of the stored water out of the tank as hot rather than warm water.

Most electric models have a heating element near the bottom of the tank, where the cold water enters, and a secondary element near the top, where the hot water is drawn off. After you use some hot water, the lower element comes on, but if you drain the tank of hot water, the lower element switches off and the top element comes on. That way, you don't have to wait for the entire tank to heat up before you can get hot water.

Heater features

Although cost of operation and delivery of hot water are most important, several other features are important, too.

Controls. Gas models have their thermostat and burner control on the outside of the water heater. It should be easy to make adjustments. Outside controls also make it easy to turn off the main gas burner and leave the pilot light on—something you might want to do before a vacation. Some models have a very low setting marked "vacation." But if the heater won't be exposed to freezing temperatures, you can save more by turning it off completely.

Controls on electric models are more difficult to adjust. They may be hidden behind one or two panels that have to be unscrewed. For safety's sake, turn off the electricity before removing the panels, because the controls are connected to the electrical supply.

Thermostats on some heaters are marked in degrees Fahrenheit. On others, the thermostats are marked only with words such as "hot," "normal," and "warm." Without degree markings, you need a thermometer to measure the temperature. (Even with degree markings, you should use a thermometer to check the setting.)

Heat traps. Normally, some hot water will circulate within the pipes even if you're not using any water. That's because hot water normally tends to rise and cold water tends to fall. Such circulation turns the first few feet of water pipes into a radiator, giving off heat from the water. That's where a "heat trap" can be useful. A heat trap is typically a loop of tubing that's mounted between the water heater and the water pipes to reduce the unwanted circulation.

Two heat traps (one on the hot-water side, another on the cold) can save quite a bit of heat—about $10 per year for

gas models and about $37 per year for electric models, based on average energy rates. When you buy a water heater that isn't equipped with heat traps, you should add a pair when you have the water heater installed.

Drain valve. You should drain off some hot water periodically to keep sediment from accumulating at the bottom of the tank; that will increase the life of the tank. In areas with hard water, draining is best done every month. In areas with soft water, every three or four months is sufficient.

The best type of drain is the brass sillcock valve, which looks like an outdoor faucet.

Some models have a sillcock valve with plastic threads for a hose connection. That type requires a little extra care when screwing on the hose to prevent cross-threading the fitting.

A common type of drain valve, and not desirable, is a threaded plastic pipe passing through the center of a large knob that controls the drain valve. Turning the knob to open the valve makes the hose twist—and possibly kink. And if you turn the knob too far the wrong way when trying to close the valve, the valve stem might come out and spill the entire contents of the tank.

Safety. Electric models are safe when properly wired. Gas heaters have a device to shut off the gas supply if the pilot flame goes out.

Water heaters have an energy cutoff that shuts off the heat if the water temperature gets too high because of a thermostat failure or other problem. Each also has a threaded hole for a pressure-temperature relief valve that can let some water drain off if the pressure or temperature inside the tank becomes too high.

Hydrogen, a highly flammable gas, is normally produced in water heaters—

usually in small quantities—as a result of the reaction between the heater and an anticorrosion device. But hydrogen can build up when hot water isn't drawn for several weeks—while you're away on vacation, for instance. After a vacation, it's a good idea to run some hot water from a faucet before using a dishwasher or washing machine; that will let the hydrogen escape. Don't smoke or light a match near the faucet while it's venting.

Installation. The safety of a gas or electric water heater depends on its installation. An experienced, licensed plumber or electrician should do the job.

Corrosion protection. The vast majority of water heaters have "glass" lining in the tank to protect the metal from rust. As further protection, heaters have at least one anode (a magnesium rod) suspended in the tank. Magnesium, a metal that corrodes readily, acts as a "sacrificial element"; corrosion tends to attack the anode and spare the steel tank.

Warranty. Heater warranties vary quite widely, with more expensive models having longer warranties.

Does it make sense to pay extra for a longer warranty? Water in some areas is unusually corrosive. If you live in such an area, it might pay to have a water heater with a long warranty. However, warranties don't cover transportation or installation charges should the tank fail. And if you moved from your home before the warranty expires, you would have paid the warranty premium on the tank for the next owner.

Recommendations

If you have natural gas supplied to your home, you should buy a gas water heater. Unless you live in the shadow of a hydroelectric plant and have an exceptionally low electricity rate, you're not likely to find a conventional electric water heater cheaper to run than a gas model. Whether you use gas or electricity, an energy-saver model is worth its price. A new standard gas water heater could cost slightly more than $30 a year to run at 1985 gas prices; a standard electric model would cost about $70 a year more than its energy-saving counterpart.

Cost aside, the electric models can present a drawback. They heat water more slowly if you drain the tank. That can be an important consideration if your family uses a lot of water.

If you don't have natural gas supplied to your house, are you fated to pay the stiff running costs of a conventional electric water heater? Not necessarily. Other parts of this report describe some unconventional alternatives.

Insulation kits for water heaters

You can cut your hot-water bills by wrapping an extra layer of insulation around the tank.

A water heater loses a considerable amount of heat while it's standing by. At average fuel rates, even a well-insulated energy-saver, storing 140°F water in a 70° room over a year's time, can accumulate about $61 worth of storage losses (for an electric water heater) or $40 (for a gas one); older water heaters or standard models, which are less well insulated, could lose nearly twice as much.

By adding insulation to the tank, you can reduce the amount of heat lost through the walls of the water heater, reducing the amount of energy required to keep the water in the tank hot.

The kits

A typical kit consists of a fiberglass blanket, about 1½ inches thick, that is wrapped around the heater and secured with tape. One kit, tested by CU's engineers, consisted of a sheet of flexible plastic foam, ⅜ inch thick, that was secured with straps and tape. Two kits also had an insulating lid, or top plate, to cover the top of an electric water heater.

CU engineers installed the kits on both gas and electric water heaters and then repeated the same test performed on the water heaters alone. The fiberglass kits reduced storage losses for the gas energy-saver water heater by about one-sixth, for a saving of up to about $6.94 per year at a natural gas rate of 61.7 cents per 100 cubic feet. On the electric water heater, you can expect to reduce storage losses for an energy-saver by about one-third, or up to $19 per year at an electricity rate of 7.75 cents per kwh.

Any kit will produce a greater saving if installed on an older water heater that has less insulation. When installing a kit, be sure to keep insulation away from the heater's controls and wiring. And keep insulation away from the top of a gas-fired water heater to prevent interfering with the flue and creating a safety hazard.

Fiberglass kits carry warnings concerning the handling of fiberglass. That's appropriate, for fiberglass is nasty stuff. When handling it, you should wear a dust mask, gloves, and a long-sleeved shirt. Wash your clothes separately so the fibers aren't transferred to other clothing.

In choosing an insulation kit, be sure it is big enough to fit around your water heater.

Instantaneous water heaters

A conventional storage-type water heater can meet heavy short-term demands for hot water without requiring enormous amounts of energy to heat the water at the time it's used. But to do that, the heater must keep a relatively large volume of hot water stored, which makes for relatively large standby losses.

One way to reduce or eliminate storage losses is to heat water only as it's needed. That approach is quite common in Europe, which has a long history of high energy costs, but it's relatively little known here.

Consumers Union engineers tested two such instantaneous water heaters, one gas and one electric. Neither came close to matching the performance of a conventional water heater, but both operated quite economically.

In concept, an instantaneous heater is a simple gadget. Basically, it's a tube that can be heated; cold water enters at one end and emerges hot at the other. When no more hot water is needed, the water and the heat supply are shut off. Result: no storage and no storage losses. Such heaters are usually located close to the place where hot water is needed, in the bathroom or kitchen.

The fundamental problem with instantaneous heaters involves safely supplying enough heat to the tube to get enough hot water. Consider, for instance, a conventional water heater, which uses heating elements rated at roughly 5,000 watts (about the maximum most utilities will allow for an electric water heater of any type). That wattage is equivalent to 17,000 Btu per hour. If an instantaneous heater supplied 17,000 Btu per hour to a low-flow shower head at a rate of 2 gallons per minute, the water would be heated only by a modest 17 degrees. With a water supply at 60°F—a not-unusual temperature—your shower would be a rather chilly 77°, even without any cold water added to the mix.

Gas-fired instantaneous heaters offer the possibility of adding energy at a higher rate, but at a cost in safety. A heater that would meet a dishwasher's typical requirements for hot-water temperature and flow rate would also produce boiling-hot tap water when the tap was opened only a bit. (High-capacity instantaneous heaters, however, do seem to work well in operations such as laundries, where flow rates tend to be constant.)

Instantaneous water heaters, then, don't appear to be the best way to cut storage losses in the home. They seem better suited to special purposes (for example, providing a dishwasher with 140°F water when the conventional water heater is set to 120°) than to serving as possible replacements for storage-type water heaters.

Tankless coil water heaters

Generally, people who heat their home with gas or electricity heat their water with a separate water heater that uses the same fuel. Many owners of oil furnaces, however, get their hot water directly from a "tankless coil" in their home heating plant—and as a result end up paying a lot more for the water than they should.

A tankless coil is a kind of instantaneous water heater built into an oil-fired furnace. (Occasionally, you might find a tankless coil on a gas furnace; usually, though, homes with gas heat use separate water heaters.) Cold tap water enters a coil of copper tubing immersed in hot water in the boiler. By the time the tap water leaves the coil, it has been heated enough to use in your shower or washing machine. For sheer simplicity, that arrangement is hard to beat. From the point of view of economy, however, a tankless coil is an entirely different matter.

Consider summertime operation. To deliver hot water on demand, the boiler must always be kept hot. (Typically, automatic controls fire the burner whenever the boiler water cools below a certain point.) In fact, a tankless coil system is usually kept quite hot. The instructions for the tankless coil–equipped test furnace in Consumers Union's laboratory said the boiler should be held at between 170° and 190°F. Standby losses would therefore be high: the test boiler lost an average of 105,000 Btu per day even when no hot water was drawn.

That 105,000 Btu represents several times the standby cost imposed by a standard gas-fired water heater.

Wintertime storage losses are lower because the furnace is on much of the time to heat the house. But, in some cases, use of a tankless coil in the winter months could bring an unnecessary increase in expenditures for oil. The coil can't store much water and it must be able to heat water as fast as it is drawn. To cover that base while still ensuring enough reserve capacity to heat the house even on a very cold day, service technicians often adjust the furnace to burn considerably more oil than it would need in most circumstances. That "overfiring" may save the technician some service callbacks. But an increased firing rate imposes greater off-time losses. In short, if you were to install a tankless coil, you could make your whole heating system less efficient.

Here's the very rough bottom line: If CU's coil-equipped test furnace was called on to deliver 64.3 gallons of water per day (the amount used daily by the average American family) that would come to an annual water-heating cost of $170 (134 gallons of oil at $1.24 per gallon). The annual standby losses would be an additional $150 to $200.

Accordingly, the total operating cost for heating water with a tankless coil would come to $320 to $370 per year, which would make the tankless coil one of the most expensive methods of heating water.

Recommendations

The section on conventional energy-saver gas and electric water heaters (see page 350) puts the figure for total operating cost into perspective. Some gas models can be bought for as little as $200 or so.

On the basis of Consumers Union estimates, all of the gas models could match the tankless coil's delivery for between $161 to $189 per year in operating cost. So if you are now using a tankless coil and have natural gas available to you, consider disconnecting the coil and having a gas water heater installed. The appliance should pay for itself fairly quickly.

You also have a no- or low-cost option for reducing costs with a tankless coil, once out of the heating season. Just switch off your oil burner before any span of time in which you expect not to need hot water. Results with CU's test boiler suggested that switching off a burner for, say, eleven hours daily would reduce storage losses by about 11 percent. That would save about 16 gallons of oil per year, or nearly $20 at average fuel rates. Most tankless coils will begin to provide hot water within a few minutes of turning on the furnace. (An automatic timer could be used to handle the switching job.)

Electric woks

In the right place, and at the right time, an electric frying pan can serve as an extra cooking element or as a warming dish. Its portability lets you cook snacks, or even meals where there's electricity but no kitchen range or hot plate—as in a family room or a college dormitory.

It isn't so easy to make a case for an electrified wok, an appliance that's more suited to Chinese cookery than to Western-style hash-slinging. Nevertheless, woks fry food too.

Stir-frying

The traditional Chinese wok is shaped like an inverted dome, with two metal handles attached near its rim. These days, such woks usually sit in a separate metal ring that supports the wok over a gas burner. (As a concession to modern electric stoves, woks with small, flat bottoms are also available.)

The traditional wok is ideally suited to stir-frying, which involves moving food quickly across an intensely hot, oiled surface.

As a technique, stir-frying owes its origin to an early Chinese concern with energy conservation. Chinese stoves had openings in their top into which the woks were placed. Thus, a small fire could be concentrated on the wok's rounded bottom.

An electric wok is a hemispherical aluminum appliance with an electric heating element embedded in its flattened bottom. A base or set of legs is attached, and perhaps a decorator color applied to the exterior and a nonstick coating to the interior.

Unlike frying pans, woks require no evenness of heating. On the contrary, wok cookery mandates a small but extremely hot cooking area.

Temperature recovery. A wok proves rather sluggish at heating cooking oil to a modest temperature, 335°, something you might want to do when deep-frying large items. Because of the wok's curvature, oil fills the utensils to a greater depth than in a frying pan. That, and the wok's relatively low wattage, makes for slower temperature recovery.

But the oil is deep enough to deep-fry relatively large items, such as pieces of chicken—something a frying pan can't do. Smaller pieces of food require much less oil to deep-fry in a wok than they would need in a frying pan.

Temperature overshoot. When you first turn on a wok, it will likely overshoot its set temperature by more than 150 degrees at the center of its cooking surface. That isn't necessarily a detriment in stir-frying, where food must be moved very fast. But it could be a problem with certain dishes, such as egg foo yung.

Wok cooking. To stir-fry in a wok, you cut food into bite-size chunks and whisk it through a small amount of hot oil in the bottom of the utensil. The concentrated heat seals in juices and cooks the food quickly and thoroughly.

Electric woks should all perform much the same as each other, and very much like a regular wok. Within three minutes, sliced chicken breasts should be tender and tasty, with a light-to-medium brown color. String beans should come out tender yet crisp, thoroughly cooked but still bright green. Onions, cooked for $1^1/4$ minutes, should also be quite good. But a regular frying pan, as well as an electric one, can stir-fry just as well and cook even faster than an electric wok can.

An electric wok is competent at cooking a heavy load (such as 8 spring rolls, miniature egg rolls) in oil at 375°. But a regular wok on a gas stove, or even a frying pan on an electric stove, can do just as well.

Hazards. Wok lids pose less of a burn hazard than most frying-pan lids. Woks' heat-resistant plastic knobs are larger, and the lids' curvature provides more clearance for fingers. Woks are also easier and safer to pour oil from; their sloping sides let you pour with little or no dripping.

Recommendations

Electric woks, though competent performers, are probably not a practical choice unless you cook often in the Chinese style. They are good for stir-frying relatively large quantities (fried rice, say) or for deep-frying. And they are ideal for steaming, another common method of Chinese cookery.

If you're tempted by wok cookery, consider starting out with a nonelectric wok. You can buy one for $15 or so.

Ratings of electric woks

(As published in a January 1982 report.)

Listed by types; within types, listed in order of estimated overall quality. Prices are suggested retail rounded to nearest dollar; + indicates shipping is extra.

Key: Excellent ● | Very good ◕ | Good ◐ | Fair ◔ | Poor ○

Brand and model	Price	Cooking area	Weight, pan/lid, without control	Evenness of heating	Speed of heating	Temperature recovery	Thermostat accuracy	Temperature overshoot	Ease of setting controls	Control legibility	Signal-light visibility	Advantages	Disadvantages	Comments
FARBERWARE 303	$70	—	2.7/1.4	◐	◐	◐	◐	○	○	◐	◐	A,B	a,b	A,B,C,D
WEAR-EVER A71500	58	—	2.4/1.0	◐	◐	●	●	◐	○	○	○	A	b	B
WEST BEND 5109	45	—	2.7/0.7	◐	◐	●	●	◐	◐	◐	◐	A,C	b	B,E
NORDIC WARE 85950	55	—	4.5/0.7	○	○	●	◐	◐	○	○	—	—	c	B
NORDIC WARE 85560	70	—	4.4/0.7	○	○	●	◐	◐	○	○	—	—	c	B,F

SPECIFICATIONS AND FEATURES

All: ● Have plug-in temperature control with signal light. *Except as noted, all have:* ● 2 buffet-style handles. ● Nonstick cooking surface. ● 30- to 35-in. cord, and ability to be immersed if control is removed.

All woks: ● Have domed lid and large knob; posed less of a burn hazard to a user removing the lid than most frying pans did. ● Were easier to pour oil from than any of the frying pans. ● Stand 7½ to 10½ inches tall when covered.

KEY TO ADVANTAGES

A – Exterior fairly easy to clean; handles fit snugly to pan, eliminating hard-to-reach areas.
B – Used less electricity than most.
C – Has cooking-temperature guide on handle or control.

KEY TO DISADVANTAGES

a – Control plugs in snugly; more difficult to remove from hot pan than most.
b – Lightness and height made unit somewhat less stable than other woks.
c – Exterior more difficult to clean than most; pan not immersible.

KEY TO COMMENTS

A – 55-in. cord.
B – Lid has no vent hole.
C – Stainless-steel interior; more difficult to clean than nonstick cooking surface of most but much less likely to be scratched.
D – Comes with rack and chopsticks.
E – Comes with useful 271-page Chinese cookbook.
F – Comes with rack, steaming tray, and wooden spatula.

Wood stoves

An efficient wood stove that promises to deliver as much heat as possible from expensive or hard-gleaned wood has obvious appeal. The stove looks better yet it also produces smaller amounts of noxious emissions and poses a minimum fire hazard. When wood-stove sales started to soar in the early 1970s, it soon became apparent that stoves were causing a disproportionate share of residential fires. Meanwhile, for the first time in a century or so, wood heat began to be popular in densely populated areas, some of which found themselves under a noxious pall of smoke.

The amount of unburned material in wood smoke can be quite large—for an older stove, it amounts to one-third or more of the energy contained in the wood. By contrast, unburned material from a gas or oil burner seldom amounts to even 1 percent of the fuel burned. Indeed, the U.S. Environmental Protection Agency has announced plans to regulate wood-stove emissions under the provisions of the Clean Air Act.

Stove designers and engineers have come up with a number of new solutions to the problems associated with burning wood. Many of these new stoves take a leaf from the automotive industry and incorporate a catalytic converter (stove people call it a catalytic combustor) to reduce emissions.

The new stoves hold out the promise of safer, cleaner, more efficient wood heat. And they represent the only types of wood stoves that will be legally available in many parts of the country, if present trends continue.

These advanced models are rather expensive—ranging from about $700 to $1,500. Those prices do not include installation, which can cost considerably more than the stove itself, depending on the amount of work needed to make the chimney and hearth safe.

Despite all the work done to improve wood stoves, the industry has yet to settle on a standard method of expressing a stove's heat output. That makes it difficult to choose a stove with a suitable output for the space it will occupy. Sometimes you can go by the manufacturer's figures for heat output; in other cases you have to guess, choosing a stove by its size or weight.

Stoves pollute

If wood burned completely, the gases traveling up the chimney would be an innocuous mixture of carbon dioxide, water vapor, oxygen, and nitrogen, with traces of sulfur dioxide and solid fly ash. Wood seldom burns completely, however. Strictly speaking, it doesn't burn at all. Rather, heat causes wood to decompose into charcoal, which ignites at about 500°F, and gaseous compounds (mostly hydrocarbons) that generally ignite at temperatures higher than 1,000°. Those gases, along with carbon particles and carbon monoxide, are the main pollutants from wood stoves.

A wood stove is a prodigious generator of carbon monoxide at any time, but especially when its air inlets are choked down to produce an overnight burn or a low heat output.

The hydrocarbon mixture consists mainly of pitch, creosote, and other tars. It includes a long list of organic chemicals, many of which are also found in cigarette smoke. As long as the chimney is hot enough, most of those hydrocarbons will reach the outdoors as gases. Then they condense into microscopic droplets, or particulates, which make up most of what you see as wood smoke.

The most modern stoves make separate provision for burning the gases, usually with a secondary burning chamber heated by the combustion of the charcoal in the wood. A secondary combustion chamber literally burns some of the smoke, yielding extra heat and reducing the stove's emissions.

Some stoves also add a catalytic combustor in the secondary burning chamber. The catalyst is typically a cylindrical ceramic honeycomb thinly plated with platinum or palladium. As smoke passes through it, chemical reactions lower the ignition points of the gases, allowing them to ignite at about 500°.

Measuring the pollutants

The particular mix and chemical composition of the compounds in wood smoke depend on several variables: the type of wood being burned, the size and shape of individual pieces, the amount of wood loaded into the stove and the way the logs are stacked, the amount of air allowed in through the dampers, and so on.

Despite its relatively clean burn, even a well-designed stove will deposit some creosote in the stove connectors and chimney. After installing a stove, it would be wise to check periodically for creosote buildup until cleaning is indicated, then set up a cleaning schedule. A chimney sweeping is likely to cost $50 to $100, and should be done at least once a year.

How much heat?

The amount of heat a stove can put out depends on the amount of wood in the firebox and on how the stove is adjusted.

The stoves tested by Consumers Union's engineers put out anywhere from about 7,000 to 42,000 British thermal units (Btu) per hour. By way of comparison, a 1,500-watt portable electric heater delivers about 5,100 Btu per hour at its highest setting. A stove delivering, say, 25,000 Btu per hour would therefore produce the same amount of heat as five electric heaters.

Any assessment of a stove's heat output is unrealistic unless it includes the contribution of the stovepipe. A five-foot stovepipe typically provides about 10 percent of the measured output. A longer pipe increases the amount of heat delivered to the room. Conversely, a stove connected directly into a masonry chimney delivers less heat. Be aware, however, that your stovepipe shouldn't be longer than about 10 feet. A longer pipe cools the flue gases too much and encourages the buildup of creosote deposits.

Even if a stove produces enough heat for more than one room, it's often hard to move that heat evenly from room to room. A blower in the stove or a portable fan can help. Even so, a wood stove can't match the uniform heating that a central system provides. A big stove is best suited for a large, unobstructed space—a vaulted living room in a ski chalet, say.

Overnight burning. If you adjust a fully loaded stove for its lowest burn rate, after eight hours there shouldn't be any problem restoring the fire to a healthy roar by raking the remaining wood together or adding another piece of wood.

Efficiency. Wood contains a known amount of potential heat; how much a stove can capture determines its efficiency. A high-efficiency wood stove runs at about 65 percent efficiency, on average.

Loading. Easy loading is an important consideration in choosing a stove. Factors to consider include the length of log a stove will take, usable internal volume, and the size of loading doors. In addition to the doors at the front, some stoves have auxiliary doors at the side or the top, an arrangement that gives extra flexibility for loading.

Doors and controls. On some stoves, handles and other controls are removable to minimize accidental burns. On others the handles are made of wood or a metal coil that stays reasonably cool when the stove is operating. Handle arrangements shouldn't lead to burned fingers.

Caution is the word when opening a stove's door. If you open it too quickly, air rushing in may make the stove belch smoke and, perhaps, flames. It's safer to open the air inlet (and the bypass damper, if the stove has one), then crack the door for a few seconds before swinging it open.

Some stoves have automatic, thermostatic air controls that generally work well. They minimize the number of adjustments you need to make to a hot stove.

Ash removal. This necessary evil is easier on models designed so that ashes fall through a grating into a drawer. With the others, you must shovel the ashes off the firebox floor or out from under a raised grate.

Window. A glass window is a pleasant feature that lets you view the fire as it burns. Even if the window clouds up, the grime can be scrubbed off—or burned off with a good hot fire. Several models also come with a fire screen that lets the stove mimic the open (and fuel-wasting) blaze of a fireplace.

Accessories. Some models come with a built-in blower; others make it available as an option. While a blower will improve heat distribution, its noise may detract from the rustic charm of the stove. Some blowers have automatic, thermostatically controlled switches that turn them on as needed. Manual blowers must be switched by hand; some have more than one fan speed. The most versatile blowers give you the option of automatic or manual operation.

Dimensions. A stove's physical size doesn't give an adequate idea of how much living space the unit will actually preempt. You must also figure in clearances from combustible surfaces.

A few stoves require as much as three feet of space all around. Others can be fitted into much less space and still meet fire codes.

Clearances can be reduced if heat shielding is applied to nearby combustible surfaces.

Wood and fuel savings

You can save on wood costs with a high-tech stove rather than a conventional wood stove. A well-built, airtight wood stove is about 50 percent efficient, overall—about half the heat contained in the wood goes up the chimney.

A higher-efficiency model will produce about a 30-percent saving in cordwood. If wood costs $125 a cord, and an older stove uses six cords of wood during a heating season, that's a saving of $250 a year.

But will a highly efficient stove save you anything significant on your central-heating bills?

If you merely shut off your furnace and take all your heat from the stove, the cost of wood heat is fairly easy to figure. Assume that you used 1,000 gallons of heating oil last winter, and that you paid $1.11 a gallon (the national average price). Your fuel bill would have been $1,110. A stove that's 65 percent efficient would provide the same amount of heat with about eight cords of hardwood. With wood at, say, $125 a cord, you have paid $1,000, for a saving of $110. The cheaper the wood, of course, the greater the saving. In our example, you would save $310 a year if wood cost $100 a cord.

Even if wood prices are very low, however, it's often impractical to think of a wood stove as a substitute for the furnace. A wood stove really can heat only one or two rooms, serving much the same function as a portable electric heater.

A stove will almost always cost less than an electric heater to heat, say, two rooms, if you consider only energy costs. In any event, either a wood stove or an electric heater can reduce the amount of heat you need from the furnace, thereby lowering its cost of operation. But it's hard to predict what your overall saving will be. That will depend on fuel prices, the number of rooms you want to heat, and how warm you like them.

Against the possible saving, you'd have to figure in the considerable cost of the stove itself, as well as installation costs that can run surprisingly high, plus the cost of regular chimney cleaning. You'd also be wise to check with your insurance company to see if a wood stove will affect your premium.

Recommendation

New-generation stoves burn cleaner than their predecessors and so contribute fewer pollutants to the air and less creosote to the chimney. They burn more efficiently and so consume less wood.

Still, the stoves are fundamentally auxiliary heaters. While a stove may cost much less to run than an electric heater (and adds infinitely more charm to a room), it costs considerably more to buy and install. A stove also requires much more of your time and attention than an electric heater.

You may not *need* a wood stove in order to cut your yearly heating costs. But you may decide that you *want* a stove, as much for its cheeriness as for its warmth.

A major consideration with a wood stove is the space it demands.

Whatever stove you buy, be extremely careful with its installation. Don't do the work yourself unless you're an accomplished do-it-yourselfer. Check with the fire department or the local building inspector to find out what minimum clearances are required in your area and what materials are recommended or required as heat shielding. Don't cut corners or skimp on materials. If you have a professional handle the installation, hire the company that has the best reputation and not necessarily the best price.

Ratings of high-efficiency wood stoves

(As published in an October 1985 report.)

Listed in order of estimated overall quality. Combustion efficiency was approx. 65% for all models. Except as noted, all have steel body, catalytic combustor, nonremovable handles. Prices are those CU paid, exclusive of installation.

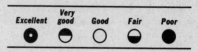

	Excellent	Very good	Good	Fair	Poor
	●	◒	○	◓	●

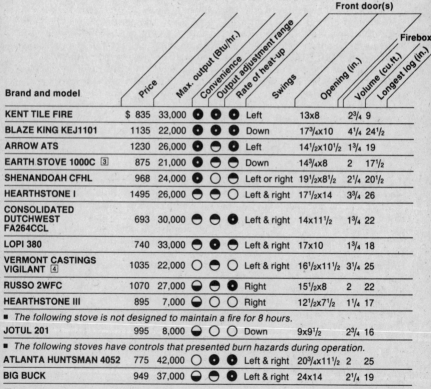

Brand and model	Price	Max. output (Btu/hr.)	Convenience	Output adjustment range	Rate of heat-up	Swings	Opening (in.)	Volume (cu.ft.)	Longest log (in.)
KENT TILE FIRE	$ 835	33,000	●	●	●	Left	13x8	$2^3/_4$	9
BLAZE KING KEJ1101	1135	22,000	●	●	●	Down	$17^3/_4$x10	$4^1/_4$	$24^1/_2$
ARROW ATS	1230	26,000	●	◒	●	Left	$14^1/_2$x$10^1/_2$	$1^3/_4$	19
EARTH STOVE 1000C [3]	875	21,000	●	◒	◒	Down	$14^3/_4$x8	2	$17^1/_2$
SHENANDOAH CFHL	968	24,000	●	○		Left or right	$19^1/_2$x$8^1/_2$	$2^1/_4$	$20^1/_2$
HEARTHSTONE I	1495	26,000	◒	◒	○	Left & right	$17^1/_2$x14	$3^3/_4$	26
CONSOLIDATED DUTCHWEST FA264CCL	693	30,000	◒	◒	●	Left & right	14x$11^1/_2$	$1^3/_4$	22
LOPI 380	740	33,000	◒	●		Left & right	17x10	$1^3/_4$	18
VERMONT CASTINGS VIGILANT [4]	1035	22,000	○	◒	○	Left & right	$16^1/_2$x$11^1/_2$	$3^1/_4$	25
RUSSO 2WFC	1070	27,000	◒	◒	●	Right	$15^1/_2$x8	2	22
HEARTHSTONE III	895	7,000	◒	○	○	Right	$12^1/_2$x$7^1/_2$	$1^1/_4$	17
■ *The following stove is not designed to maintain a fire for 8 hours.*									
JOTUL 201	995	8,000	◒	○	○	Down	9x$9^1/_2$	$2^3/_4$	16
■ *The following stoves have controls that presented burn hazards during operation.*									
ATLANTA HUNTSMAN 4052	775	42,000	○	●	●	Left & right	$20^3/_4$x$11^1/_2$	2	25
BIG BUCK	949	37,000	◒	◒	●	Left & right	24x14	$2^1/_4$	19

1 *As stated by mfr.*
2 *T = top; B = back; L = left; R = right.*

KEY TO COMMENTS

A – Does not use catalytic combustor (but stove burned about as cleanly as catalytic models in CU's tests).
B – Small fire may not activate catalyst, impairing stove's efficiency and cleanliness.
C – Cast-iron and soapstone body.
D – Cast-iron body.
E – Also has auxiliary loading door, hinged to swing toward rear, located on left side of **Hearthstone I** and **Consolidated Dutchwest**, on top of **Vermont Castings Vigilant.**
F – Firebrick lining
G – Comes with fire screen.
H – Has ash drawer.
I – Also burns coal. Coal grate provided for **Consolidated Dutchwest**; $100 extra for **Vermont Castings Vigilant.**.
J – Has window.

K – Window remains clean in operation.
L – Window clouds over in operation.
M – Some handles or controls can be removed.
N – Delivered heat quickly, unlike other large stoves.
O – Has glass-covered viewing port to check adjustment of fire.
P – Automatic blower control
Q – Manual blower control ($120 extra on **Lopi**).
R – Both manual and automatic blower controls.
S – According to mfr., this model was discontinued. Replacement is model **ATS-2**, which has a different combustor and air-supply configuration from model tested.
T – According to mfr., this model was discontinued. Replacement is functionally similar model **CFHL-86.**

| Weight (lb.) | Dimensions (WxDxH) (in.) | Min. safety clearances [1] | | | Flue | | | | Air controls | Comments |
		Back (in.)	Sides (in.)	Front (in.)	Flue (in.)	Location [2]	Height (in.)	Diameter (in.)		
250	21x27½x27½	11	24	16	14½	T	27½	6	Man.	A,J,K
410	30x30¼x32½	11	16	—	—	T	32½	8	Auto.	F,N,P
375	28x27¾x35	15	15	16	18	T	35	6	Man.	H,J,Q,S
325	28¼x27x35¼	13	16	16	—	T	33½	7	Auto.	F,Q
390	23½x25x31	36	36	18	—	T	31	8	Auto.	F,H,J,M,Q,T
670	37x28x34½	36	36	36	36	T	34½	8	Man.	C,E,G,J,L
380	25¾x19½x31	30½	36	16	—	T,B	31	7	Man.	D,E,G,H,I,J,M,P
425	24x27½x30	15	28	18	19	T	30	6	Man.	A,F
350	30x30½x30½	36	36	36	22	T,B	30½	8	Auto.	B,D,E,G,I,M
360	27x27x27	36	30	18	18	T,B	29½	8	Man.	F,J,K,P
355	24x30x24	30	24	36	31	T,B	24	6	Auto.	A,C,H,J,L
300	14x28x32½	24	38	18	18	T,B,L,R	36½	6	Man.	A,D,F,H,O
370	26½x25x33	20	33	16	24	T	33	8	Man.	F,I,J,L,R
540	42x33x31½	36	36	16	—	T	42½	10	Man.	F,G,J,R

[3] Performance data reflect modifications CU made.
[4] Tested with optional "Energy Extender" catalytic combustor.

Air pollution from wood stoves

Before 1973, wood stoves were largely relics of rural America. Then came the Arab oil embargo, and the wood stove became America's declaration of energy independence. Wood stoves have now become commonplace in metropolitan areas. Some 13 million stoves are now in use. But in communities from Bangor, Maine, to Eugene, Oregon, the rustic, romantic-looking plumes of wood smoke produce a heavy haze that threatens health and reduces visibility.

Burning wood produces much more pollution per unit of heat than does oil or natural gas. Wood yields copious amounts of carbon monoxide, hydrocarbons, and solid and liquid particulates, most of them small enough to reach the lungs when inhaled.

The particulates are the pollutants of greatest concern, for they contain many toxic organic chemicals. At high enough concentrations, exposure can cause coughing and chest discomfort, aggravate cardiovascular diseases, and may increase the risk of lung cancer.

Many of the toxic chemicals are in a class called polycyclic organic matter (POM). Benzo(a)pyrene, the best known POM, is a potent carcinogen in laboratory tests on animals. It is a suspected human carcinogen that is present in both wood and tobacco smoke.

Benzo(a)pyrene and other POMs are also emitted by coal-fired power plants, diesel-powered motor vehicles, and other sources. But wood stoves are now the nation's major source of POM pollution, accounting for nearly half the estimated emissions. The number of wood stoves used in metropolitan areas has increased in recent years, so POM emissions from stoves may affect many people. High levels of benzo(a)pyrene in urban air have been correlated with increased rates of lung cancer.

A study that measured outdoor particulate levels at seven sites in Oregon, Washington, and Idaho during the winter of 1980–81 attributed between 66 and 84 percent of the inhalable particles to wood-stove use. In several urban areas—including Missoula, Montana, Albuquerque, and Denver—wood-stove emissions have been blamed for violations of national air-quality standards for carbon monoxide and particulate matter during the winter months.

The impact on outdoor air quality has been particularly severe in Oregon, where more than half the households heat with wood to some extent. Air-monitoring studies in several Oregon communities have shown that wood-stove emissions contribute substantially to violations of air-quality standards.

In 1983, Oregon became the first state to enact an emissions-control law for wood stoves. The law requires that, by July 1986, all new wood stoves sold in the state must achieve a reduction in particulate emissions of 50 percent, compared with the average emissions of stoves sold in Oregon in the late 1970s and early 1980s. The emissions level for new stoves must be cut 75 percent by July 1988.

The federal government may soon develop nationwide standards for wood stoves. The U.S. Environmental Protection Agency has announced plans to regulate wood stoves under the Clean Air Act, possibly with an emissions-control program similar to Oregon's. The EPA hopes to issue its regulations by the beginning of 1987.

If operated properly, a stove should not increase a home's indoor pollution levels. But several studies comparing indoor and outdoor air quality for homes with wood stoves have found elevated levels of several pollutants indoors. The pollutants included carbon monoxide, particulate matter, and benzo(a)pyrene.

The studies show that indoor emissions usually occur when people open the stove door to add wood, which often produces back drafts. To avoid that, first open the damper to increase the draft up the flue and then open the door. Loose stovepipe joints and leaky, poorly gasketed stove doors can also allow significant emissions indoors. A chimney that's too short can lead to increased indoor pollution from "recycled" smoke. A short chimney doesn't allow smoke to drift up and away from the house; instead, some smoke comes indoors with the normal ventilating air.

Chimneys

Fires arising from solid-fuel heating hit an all-time high of 140,000 in 1983, according to the Consumer Product Safety Commission. Chimneys were a major culprit.

Flammable creosote is always deposited in a chimney, particularly by older airtight stoves that have been used for small, slow fires. If the temperature in the fire should soar to 1,500°F, which is quite possible when some stoves are started or reloaded, the creosote may well ignite.

A chimney fire is a frightening event. Air roars through the stove, and the stovepipe vibrates and glows red. Sparks or flames erupt from the chimney and threaten the roof. Fire may well find its way through cracks or chinks in a brick chimney and ignite beams or rafters. There's little you can do but shut down the stove, call the fire department, and leave the house—the blaze will continue, since enough air will usually leak through the stovepipe joints to feed the fire.

Even if the fire doesn't spread to the house, it may well cause serious damage to the chimney. The chimney should be inspected and repaired before it's used again.

A number of products compete as liners for damaged or deteriorated masonry chimneys: liquid cast-in-place products, poured down the chimney and pressed in place mechanically; flexible stainless-steel tubing that can snake around bends in the chimney; and rigid stainless-steel liners. A listing of major suppliers follows.

Underwriters Laboratories is the principal testing and certification organization for wood-burning products. UL has a standard for metal chimneys, and some lining systems are tested against that standard. (Compliance is voluntary on the part of the manufacturers; their products aren't required to meet any set of standards, although many do.)

If you don't already have a usable chimney, you can hire a local mason to build a new brick one lined with heat-resistant fire-clay tile. You're apt to find it quicker and less expensive to have one made of interlocking insulated metal sections. Such a chimney can be enclosed with house siding or brick facing.

Installing or repairing a chimney isn't cheap. In fact, the chimney can easily cost more than the stove. And trying to upgrade an old chimney can be far more complicated and time-consuming than starting from scratch.

Expect problems in reviving an old chimney, especially one built into the center of the house, where its construction can't be fully assessed.

To minimize the problems, try to get a firm, flat-rate contract specifying what is to be done—not one that allows a contractor to charge extra for unspecified "contingencies." Be sure that final payment for the work depends on the installation's passing inspection by the local building department. You should also arrange to be present for the work, to deal with questions and crises as they arise.

Among the major chimney liners and prefabricated chimneys are:

Poured liners

Ahrens (Ahrens Chimney Techniques, 2000 Industrial Avenue, Sioux Falls, South Dakota 57104)

Insulcrete (Insulcrete Relining Systems, P.O. Box 856, Delaware, Ohio 43105)

Perma Flu (Chimney Relining International, 105 West Merrimack Street, Manchester, New Hampshire 03108)

Solid Flue (American Chimney Lining Systems Inc., 9797 Clyde Parkway, Byron Center, Michigan 49315)

Supaflu (National Supaflu Systems Inc., Route 30A, Central Bridge, New York 12035)

Thermocrete (Thermocrete Chimney Lining Inc., P.O. Box 119, Mountain Road, Stowe, Vermont 05672)

Flexible stainless-steel liners

Columbia-A-Flex (Columbia-A-Flex, P.O. Box 48, Ghent, New York 12075)

Mirror Stainless Flex (Mirror Stove Pipe Co., Drawer A, Bloomfield, Connecticut 06002)

Super Flex (Sleepy Hollow Chimney Sweeps & Supply Ltd., 103 West Main Street, Babylon, New York 11702)

A-Flex (Z-Flex Inc., P.O. Box 4035, Manchester, New Hampshire 03108)

Rigid stainless-steel liners

Elmer's (Elmer's Pipe Inc., 214 Minot Avenue, Auburn, Maine 04210)

Mirror Stainless (Mirror Stove Pipe Co., Drawer A, Bloomfield, Connecticut 06002)

Saf-T Liner (Heat-Fab Inc., 38 Haywood Street, Greenfield, Massachusetts 01301).

Security (Security Chimneys Ltd., 2125 rue Montery, Lavel, Quebec, H7L 3T6 Canada)

Standex (Standex Energy Systems, P.O. Box 588, Medina, Ohio 44258)

Ventinox (American Boa Inc., P.O. Box 1743, Albany, N.Y. 12201)

Prefabricated metal chimneys

Dura-Vent (Simpson Dura-Vent, P.O. Box 1510, Vacaville, California 95688)

Metalbestos SS II (Delkirk Metalbestos, P.O. Box 372, Nampa, Indiana 83651)

Super Chimney 21 (GSW Jackes-Evans, 4427 Geraldine Avenue, St. Louis, Missouri 63115)

Super Flue-2100° (Standex Energy Systems, P.O. Box 588, Medina, Ohio 44258)

Temp/Guard (Metal-Fab Inc., P.O. Box 138, Dept. W, Wichita, Kansas 67201)

Index

Monte Florman was Technical Director of Consumers Union from 1971 to 1982, when he took early retirement. Now a Consumers Union consultant, Florman began his long CU career in 1948, two years after graduating from New York University with a degree in electrical engineering. Initially a test engineer in CU's Appliance Division, Florman headed the division from 1955 to 1965. He then served as CU's Associate Technical Director until 1971, when he was named Technical Director.

When You Order Three or More Books Postage and Handling Are Free

Use this coupon to order paperbound books, binders, and bound volumes
*Free postage and handling when you order 3 or more items(U.S. orders only).

QTY.		AMT.
57	**Love, Sex, and Aging.** CU's report on relationships over 50. 1984 . . . . $14.95	$___
62	**The Essential Guide to Prescription Drugs.** For safe drug use. 4th ed. 1985 . . . . $12.95	$___
73	**The Consumer Reports Books Guide to Appliances.** Covers 67 kinds of home appliances. 1986 . . . . $6.95	$___
68	**The Columbia University College of Physicians and Surgeons Complete Home Medical Guide.** Comprehensive family reference on illness, good health. 1985 . . . . $22.95	$___
65	**The Best Guide to Allergy.** Understanding allergic symptoms and how to live with them. Rev.1985 . . . . $8.95	$___
70	**Understanding Arthritis.** The most authoritative source available. 1984 . . . . $9.95	$___
24	**Freedom From Headaches.** For help with headache pain. 1981 . . . . $8.95	$___
67	**The Consumer Reports Books Guide to Housing Alternatives for Older Citizens.** Provides options, strategies, resources. 1985 . . . . $9.95	$___
63	**The Heart Attack Handbook.** A guide to treatment and recovery. 2nd ed. 1985 . . . . $6.95	$___
61	**Cookbook for Kids.** 40 fun-to-make recipes. 1984 . . . . $10.95	$___
60	**Where Does It Hurt?** Symptoms and causes explained. 1983 . . . . $10.95	$___
59	**The Pregnancy Book for Today's Woman.** Dependable guidance. 1983 . . . . $10.95	$___
58	**CR Guide to Electronics in the Home.** CU's Ratings and advice. 1984 . . . . $11.95	$___

QTY.		AMT.
56	**What Did You Learn In School Today?** Help your child do better. 1983 . . . . $8.95	$___
55	**Infants and Mothers.** Advice to parents on baby's first year. Rev. 1983 . . . . $12.95	$___
53	**Physical Fitness for Practically Everybody.** CU's exercise report. 1983 . . . . $12.95	$___
52	**Carpentry for Children.** For kids who like to build things. 1982 . . . . $11.95	$___
66	**Vitamins and Minerals: Help or Harm?** Advice on use—and hazards—of supplements. 1983 . . . . $11.95	$___
50	**Top Tips from Consumer Reports.** CU's handy book of helpful hints. 1982 . . . . $6.95	$___
48	**Putting Food By.** The best ways to can, freeze and preserve. 3rd ed. 1982 . . . . $12.95	$___
47	**More Kitchen Wisdom.** More advice for kitchen chores. 1982 . . . . $7.95	$___
45	**You and Your Aging Parent.** Advice on needs, resources. Rev.1982 . . . . $9.95	$___
23	**My Body, My Health.** Guide to gynecologic health. 1979, updated 1981 . . . . $11.95	$___
21	**Soup Wisdom.** How to make soup + 30 CU-tested recipes. 1980 . . . . $6.95	$___
15	**James Beard's Theory & Practice of Good Cooking** 1977 . . . . $11.95	$___
12	**Ulcers.** Who's likely to get them and why; treatments that work. 1978 . . . .'$8.95	$___
08	**Kitchen Wisdom.** A kitchen book to make things easier. 1977 . . . . $8.95	$___

Detach here.

____**Binders** For January through November issues of Consumer Reports
10000 $9 each, 2 for $16 (please indicate quantity).

Cloth-bound Volumes of 11 issues of Consumer Reports, January through November; $15.95 each.

___ '85,	___ '84,	___ '83,	___ '82,	___ '81,	___ '80,	___ '79,	___ '78,	___ '77
10015	10014	10013	10012	10011	10010	10009	10008	10007

Mail with payment to:
**Consumer Reports Books
540 Barnum Avenue
Bridgeport, CT 06608**

Name_____
(please print)

*Postage/handling: In U.S., add $2.15 to your entire order; in Canada and elsewhere, add $4.

Address _____ Apt.___

Books $_____

City_____

*Postage/handling $_____

Total enclosed $_____

State_____ Zip_____

Please allow 4 to 6 weeks for shipment   CU publications may not be used for any commercial purpose.   AP-3